Quantitative Viral Ecology

MONOGRAPHS IN POPULATION BIOLOGY

EDITED BY SIMON A. LEVIN AND HENRY S. HORN

A complete series list follows the index.

Quantitative Viral Ecology

Dynamics of Viruses and Their Microbial Hosts

JOSHUA S. WEITZ

PRINCETON UNIVERSITY PRESS

Princeton and Oxford

Copyright © 2015 by Princeton University Press
Published by Princeton University Press, 41 William Street,
Princeton, New Jersey 08540
In the United Kingdom: Princeton University Press,
6 Oxford Street, Woodstock, Oxfordshire OX20 1TW
press.princeton.edu

Library of Congress Cataloging-in-Publication Data

Weitz, Joshua, 1975–
    Quantitative viral ecology : dynamics of viruses and their microbial hosts / Joshua S. Weitz.
        pages cm. — (Monographs in population biology)
    Includes bibliographical references and index.
    ISBN 978-0-691-16154-9 (hardback)
    1. Viruses—Ecology.   I. Title.
    QR478.A1W45 2015
    579.2'17—dc23
                                    2015012967

British Library Cataloging-in-Publication Data is available

This book has been composed in Times Roman

Printed on acid-free paper. ∞

Typeset by S R Nova Pvt Ltd, Bangalore, India
Printed in the United States of America

10 9 8 7 6 5 4 3 2 1

for Maira

Contents

Acknowledgments

I am deeply grateful to the many colleagues and collaborators who have influenced the way I think about viruses and viral modeling. The text of this book reflects insights I gleaned in the process of many stimulating discussions and long-term collaborations. My research on viral ecology has been facilitated by the support of key institutions, in particular Princeton University and the Georgia Institute of Technology, as well as grants from multiple funding agencies, including the Burroughs Wellcome Fund, James S. McDonnell Foundation, Defense Advanced Research Projects Agency, the National Science Foundation Programs of Biological Oceanography, Dimensions of Biodiversity and Physics of Living Systems, Army Research Office, and the Simons Foundation. In addition to these, I owe a debt of gratitude to all the participants of the National Institute of Mathematical and Biological Synthesis working group on "Ocean Virus Dynamics," whose conversations have helped deepen my understanding of the ecological role of ocean viruses. I want to thank Steven Wilhelm, in particular, for his crucial efforts in making the working group happen as well as for the insightful comments and feedback he provided on an early draft of this monograph.

I am grateful to the students and postdocs in my group whose work, conversations, and debates have spurred many of the ideas contained in this book: Alexander Bucksch, Lauren Childs, Michael Cortez, Cesar Flores, Richard Joh, Luis Jover, Andrey Kislyuk, Tae Lee, Gabriel Mitchell, Yuriy Mileyko, Olga Symonova, Bradford Taylor, Hao Wang, and Charles Wigington. Of these, a particular note of thanks to Alexander Bucksch, Michael Cortez, Cesar Flores, Luis Jover, Bradford Taylor, and Charles Wigington, who reviewed chapters and provided many useful suggestions to help improve the text. I am also thankful for the feedback of my collaborators Bart Haegeman and Mark Young; colleagues Daniel Goldman, Michael Goodisman, and Andrew Zangwill; the editorial assistance of Verene Lancaster; and the editorial team at Princeton University Press; as well as for the comments and suggestions arising from a careful line-by-line read by both Deborah Cook and Barbara Liguori. Finally, I am grateful for the comments, suggestions, and perspectives on the entire text—words, equations, and ideas—by both Mike Barfield and Robert Holt.

This book would not have been possible without the generosity of Matthew Sullivan, who hosted my sabbatical visit to the University of Arizona in 2013–2014, where the bulk of this monograph was written. Matt has been a wonderful sounding board for ways in which theory and experiments can, together, bring new insights into the ecological role of viruses. Matt also took the time to provide feedback on multiple chapters, improving and sharpening the text throughout. Likewise, I'd like to thank the Sullivan group for providing feedback that helped influence how I communicate what theory can do for virus ecology.

A final note of thanks to each of the advisers who mentored me and welcomed me to their labs at various stages of my scientific career: Thomas Fu, Ronald Cohen, Haye Hinrichsen, Eytan Domany, Lyman Page, Robert Austin, Lydia Sohn, John Joannopoulos, Daniel Rothman, and Simon Levin. This book is one modest attempt to give back.

Preface

A PLANET OF VIRUSES

Viruses have a nasty reputation as killers, at worst, and annoyances, at best. This reputation is deserved, at least in part. The viruses that cause HIV/AIDS, smallpox, and Ebola virus disease infect human cells, damage cells and tissue, and lead—in many instances—to the death of the infected individual. Other viruses, such as rhinoviruses, also infect human cells but are eventually cleared by the immune system, with only mild to moderate effects—for example, the sniffling, sneezing, and fatigue referred to in ubiquitous commercials for common-cold remedies. These viruses are very much the minority on Earth in one obvious respect: most viruses do not infect humans; rather, they infect all manner of life, from animals, plants, insects, to microbes. For example, even if everyone on Earth had a case of the common cold, the number of cold viruses inside humans would likely be billions of times smaller than the number of viruses circulating in the global oceans! Viruses such as those that cause Ebola may also be the minority in another, more subtle, respect: what viruses "do" is not confined to killing cells or individuals. There are, in fact, "good" viruses, that is, those that help their hosts, as well as many viruses that affect their hosts in a way that may not fit into the simple categories good or bad.

This broad view of viruses is laid out in *Planet of Viruses*, a pithy, highly readable book by Carl Zimmer that illustrates through a series of vignettes the diversity of places and hosts that viruses affect (Zimmer 2012). We, the scientific community interested in environmental viruses, still know far less than we would like about the direct effects of viruses on the living organisms they encounter and, by extension, the indirect effects of viruses on the environment. Scientific studies of the interface between living organisms and the "environment" is what many scientists categorize as *ecology*. In that sense, the present book sets out to explore the ecology of viruses from a quantitative perspective. The central objective is to act as a guide for students and scientists who are interested in the many ways that viruses affect their (indeed, our) environment. There are many ways to facilitate such a goal. In thinking about this book, I turned to others who have set strong precedents in advancing the study of virus-host interactions.

One landmark work is Martin Nowak and Robert May's book, *Virus Dynamics: Mathematical Principles of Immunology and Virology* (Nowak and May 2000). This monograph set out to explain and interpret simple models of the interaction between the human immune system and viruses that compromise it, with a focus on HIV and hepatitis B. *Virus Dynamics* was inspired by collaborations between mathematical biologists—including the authors—and virologists who wanted to characterize and control the postinfection dynamics of viruses inside infected patients. This work was seminal both in its contributions to modeling—inferring mechanisms from data—and in demonstrating that modeling should be considered an essential component of what is considered first-rate virology. This view, although not universally adopted, is no longer heretical nor confined to the obscurity of technical journals. However, I contend that a similar book on environmental viruses would be challenging. HIV and hepatitis B virus have clearly identified negative effects on human health. In both cases, the infection of a human host begins with a single or a highly limited number of viral genotypes. Moreover, an enormous amount of structural, biochemical, and immunological information is available on the relationship between these viruses and the human host. When considering environmental viruses of microbes, we would be hard pressed to decide which environmental virus is archetypal or which virus or viruses should be the subject of a book. The answer is likely to be, many!

And so instead of developing the mathematics involved in modeling each virus, I have, instead, turned for inspiration to the volume *Bacteriophage Ecology*, edited by Stephen Abedon (2008). This book focuses on a type of virus—the bacteriophage—that exclusively infects bacteria. *Bacteriophage* means "bacteria eater"; the word *phage* has its etymological roots in the Greek word *phagos*, "to devour." Of course, bacteriophage (or phage, for short) do not devour their hosts. Bacteriophage, like all viruses, depend on host replication machinery to multiply, a process explained in greater detail in Chapter 1. Bacteriophage ecology might seem like an obscure topic in the academic hinterlands. To the contrary, studies of phage have had major influences on the foundations of molecular biology, were an early and now recurring theme in efforts to develop antibacterial therapeutics, and are now a central theme in emerging studies of the microbial world and its impressive genetic and functional complexity. *Bacteriophage Ecology* anticipates many of these areas, addressing, largely in empirical terms, many of the topics essential to a deeper understanding of the effects of viruses in the environment. However, *Bacteriophage Ecology* also focuses on the interaction of a single or small number of phage genotypes with their hosts, whereas, viruses are diverse at genetic, functional, and taxonomic levels. Even more challenging is the fact that many of the viruses that are important in shaping the environment are not yet culturable. Canonical methods must be adapted specifically to the diversity of the viral world and its relative recalcitrance to traditional culture-based approaches. Moreover, both canonical and new methods must be placed in context—linking models to data whenever possible. This is what I have tried to do here.

ORIGINS OF THIS MONOGRAPH

In early 2003, Hyman Hartman stopped by on one of his regular visits to the graduate student offices of Daniel Rothman's group at MIT. Daniel Rothman, a geophysicist by training and a geobiologist in practice, was my PhD adviser. Hyman is a biochemist with expertise in the origins of life who, for many years, has challenged colleagues with questions, sometimes insisting that they take an entirely new look at a problem. Hyman had heard I was close to finishing my PhD and was considering new directions for my postdoctoral research. I described my plans to study the relationship between the structure of plant vascular networks and their function. His paraphrased reaction: "Forget plants. You need to work on the problem of viruses in the ocean." In the course of our discussion, Hyman insisted that ocean viruses were abundant and diverse. And, importantly, that their abundance and diversity belie laboratory studies and theories suggesting that viruses of microbes could be easily eliminated in culture. I was intrigued, but not quite convinced.

A few months later, just weeks before I graduated, I discovered colleagues at the forefront of ocean virus research working just steps away, in the Parsons Laboratory at MIT. One of them, Matthew Sullivan—then a PhD candidate working under the supervision of Penny Chisholm—agreed to meet for coffee. We chatted about our mutual interests, including the role of theory in biology. Matt gave me a preview of some of the new discoveries coming out of the study of ocean virus ecology, enabled, in part, by advances in technology that could allow scientists to see and characterize a previously hidden world. We agreed to keep in touch, particularly if I ended up focusing on viral ecology. A few months later, in August 2003, Matt, Penny, and John Waterbury published a seminal paper characterizing a suite of novel phage isolates that could infect cyanobacteria that were ubiquitous in the surface ocean (Sullivan et al. 2003). I was behind, that was evident, but so was nearly everyone else. And, more important, this was an area entering a new age of discovery.

Whenever one enters a new field, it takes time to orient oneself. What are the central ideas and outstanding problems? Who are those that founded the field and who are currently pushing it forward? And, crucially, what can I do to make a difference? More than 10 years have elapsed since my conversation with Hyman. I did, in my own way, follow his advice—not to forget plants but to think about the problems of viruses in the ocean. The problem, as it turns out, was not singular. Why are there so many viruses, both in terms of number and in terms of diversity? Whom do viruses infect? How do viruses modify the fate of infected cells? How do viruses and microbes coevolve? How do viruses shape the structure of microbial communities? And, finally, how can theory help deepen our understanding of the ecological role of the viruses of microbes? This book reflects my attempt to shed light on each of these questions, and perhaps even a few more. I have tried to do so by synthesizing field and lab studies of virus-host interactions via the lens of theory and models. Its origins reside in conversations at the end of my PhD program, but its structure and, indeed, its subject require some elaboration. I did not formally begin to write this monograph until I entered my new sabbatical offices at the University of Arizona, in fall 2013, just steps away, once again, from Matt Sullivan.

IDEAS AND TOPICS COVERED

The term *quantitative viral ecology* can and will mean different things to different readers. Here, the focus is on the dynamics at multiple scales that ensue as a result of the interaction between viruses and their *microbial* hosts. There are many excellent books and monographs on the population biology of infectious diseases, of which viral diseases may form one part. In such texts, the hosts are usually human, animal, or plant, which implies different timescales and issues of concern. Here, I have not drawn a firm line in the sand as to what constitutes a microbe—in practice, this book focuses on the viruses of bacteria and, to a lesser extent, those of archaea and microeukaryotes. Nonetheless, it is my hope that the models and concepts herein may prove useful to those working at other scales, or perhaps where virus-microbe interactions take place on the surface of or inside a larger organism, as is the case in animal-associated microbiomes.

This monograph on quantitative viral ecology has seven core chapters, motivated by three major questions: (i) what are viruses of microbes and what do they do to their hosts? (ii) how do interactions of a single host-viral pair affect the number and traits of hosts and virus populations? (iii) what do viruses of microbes do in natural environments, when interactions take place between many viruses and many hosts? Throughout, I have tried to emphasize the ways in which theory and models can provide insights into all these questions and provide a cohesive framework to the study of the ecology of viruses of microorganisms, that is, quantitative viral ecology.

The book begins with an introduction to the fundamentals of virology, albeit from an ecologist's perspective. Chapters 1–2 are therefore highly complementary to the introduction to viral biology found in molecular biology and virology textbooks. Both of these chapters make evident one of the central themes of this book: that viruses differ not only in which hosts they infect but also in the functions—qualitative and quantitative—that they carry out. These preliminary chapters enable the book to function as a stand-alone text for those readers new to the field, either from a biology or modeling perspective. In doing so, I have also tried to synthesize many analyses of viral diversity in the hope of providing new connections and directions for future research on the physical and chemical basis for viral traits.

With this foundation in place, Chapters 3–5 form the theoretical core of the book, results and insights of which are of value in their own right but will prove useful in efforts to characterize complex communities. These chapters introduce core models to address how virus-host interactions affect population dynamics (Chapter 3), evolutionary dynamics (Chapter 4), and coevolutionary dynamics (Chapter 5). These chapters focus on relatively low diversity communities, showing how a combination of physical and mathematical models can provide insights into how viruses shape microbial populations and potentially stimulate diversity. The data relevant here spans more than 70 years—from the earliest days of molecular biology to the origins of the modern study of microbial population biology and, finally, to ongoing studies of eco-evolutionary dynamics of viruses and hosts. In this theoretical core, I have tried to communicate how much there is to learn from studies and models of the interactions of populations of viruses and hosts, often grown from a single isolate.

However, the complexity in the natural world demands new methods and new perspectives. I move in this direction in the final chapters by building upon the results of the theoretical core. In doing so, I chose to focus on the ecological role of viruses in a particular type of environment: marine surface waters. This choice reflects, in part, my involvement in the analysis of ocean virus dynamics. More important, it reflects the fact that many of the seminal findings of viral effects on ecological processes stem from the study of marine viruses. Moreover, the methodology for examining viruses of marine microbes is, for now, better developed than methods for examining viruses of microbes in alternative contexts, for example, in soils, hot springs, or microbiomes.

Chapter 6 begins this final part, by introducing quantitative methods to infer the abundance and diversity of viruses as well as the network of interactions with potential hosts. These methods are then applied to datasets from the oceans, largely from the marine surface. As I will explain, there are likely more than 10^{29} viruses in the marine surface waters—what they do to the global Earth system is a topic of both foundational and practical concern.

Finally, Chapter 7 extends the core models to examine the basis for viral effects in marine surface waters. This complexity includes new types of interactions, like the regeneration of organic matter by viral-induced lysis of hosts, and new trophic complexity, like the effect of predators on the very hosts that viruses infect. These two chapters lead naturally to the conclusion, in Chapter 8, in which I present a perspective on the types of developments, both technical and theoretical, that are forthcoming and the influence they are likely to have in shaping our view of the diversity and functions of environmental viruses. While ocean viruses may provide a model for discovery, they are unlikely to be the only source of surprises and discoveries on the diverse role of viruses in shaping the environment—whether in the oceans or inside us.

As a whole, this monograph includes analysis of models, synthesis of laboratory manipulations of cultured host-virus systems, as well as inferences of viral effects when culturing is not possible. All these necessitate the use of many different kinds of theoretical and modeling approaches. In this sense, I have tried to embrace an approach to be more like the fox who, according to Archilochus, knows many things, and less like a hedgehog who knows one big thing. The benefit of this approach is that there are very few prerequisites to understand the theoretical concepts discussed herein. A reader with physics, mathematics, computer science, or engineering training need not have any background in virology to understand the material. The text does expect familiarity with probability and statistics and some recollection of differential equations—increasingly a part of the standard training in the biological sciences. To the extent that methods begin to stretch beyond a core competency in modeling and analysis, I have included appendixes linked to each chapter that explain methods when they first appear and are used in the text.

A final note before beginning—the working title I had in my mind while writing the book began slightly differently:

an invitation to
Quantitative Viral Ecology

The lowercase preamble captures my intention throughout this book, that is, to draw in readers who may want to engage in this field in a meaningful way, and to accelerate the pace at which theory can be connected to empirical studies of virus-host interactions. This working title also captures an important feature of my perspective on the field, which is that many, if not most, of the important problems are not yet identified, to say nothing of being solved! The study of the viruses of microbes is at its infancy when viewed in an ecological and evolutionary perspective. I view this book as an opening salvo—the beginning of something that will require a mix of theorists and empiricists.

PART I

VIROLOGY: AN ECOLOGICAL
PERSPECTIVE

CHAPTER ONE

What Is a Virus?

1.1 WHAT IS A VIRUS?

Efforts to define a virus inevitably raise the question of exceptions. Nonetheless, a definition or two can help guide us in identifying what is common to all viruses.

> Merriam-Webster's Online Dictionary: an extremely small living thing that causes a disease and that spreads from one person or animal to another.
> Introduction to Modern Virology: submicroscopic, parasitic particles of genetic material contained in a protein coat (Dimmock et al. 2007).

These two definitions are useful, as they reflect the difference in perception as well as current understanding of what a virus is. In that respect, the roots of the term *virus* are also revealing:

> Oxford English Dictionary: late Middle English (denoting the venom of a snake): from Latin, literally "slimy liquid, poison."

Irrespective of source, it would seem that viruses have a bad reputation. Informal surveys tend to yield similar results. For example, when I ask undergraduates to name a virus, some of the most common answers are HIV, influenza, Ebola, chickenpox, herpes, rabies—not a friendly one in the bunch. The answers represent a typical conflation of the disease with the virus. Nonetheless, this conflation is not entirely inappropriate, as viruses do often negatively affect their hosts, whether by causing disease in humans, plants, or animals or killing their microbial hosts.

In fact, one version of the history of viruses begins more or less as follows (Dimmock et al. 2007)—with smallpox. Smallpox is one of the most vicious of diseases, with historical estimates of mortality rates on the order of 30%. Smallpox is caused by a virus, so-called variola, from the Latin *varius* or *varus* meaning "stained" or "mark on the skin," respectively (Riedel 2005).

In 1796, Edward Jenner, a surgeon and scientist, made a bold hypothesis based on the common lore that dairymaids did not suffer from smallpox, perhaps because they had been exposed to an apparently similar disease that affected cows, that is, cowpox. Jenner hypothesized that exposure to cowpox led to protection against smallpox. To test this hypothesis, he transferred material from a fresh cowpox lesion of a dairymaid to an 8-year-old boy who had no prior signs of having been exposed to either cowpox or smallpox. The transfer was likely done with a lancet, directly into the arm of the young boy, who then had a mild reaction—similar to the side effects of modern vaccines—but quickly recovered. Then, Jenner did something remarkable, ghastly, but ultimately providential: two months later he returned and inoculated the same boy with material taken from a new smallpox lesion! Remarkably, the boy did not get sick. This event was widely credited, after Jenner's death, as being the first example of a successful vaccination—as it turns out, a vaccination against what later became known as the smallpox virus.

Viruses as agents of disease and death seem to be the common theme, both in the popular and historical understanding. This bad reputation is similar to that ascribed to bacteria, that is, until recently. Bacteria, which were once considered exclusively "bad" because they cause such diseases as cholera, meningitis, gonorrhea, and chlamydia have had their image redeemed, at least in part. That yogurt companies can market the benefits of products enriched with additional naturally occurring *Lactococcus* cells, that fecal transplants are being considered as a means to stimulate normal digestive tract function, and that the American Society of Microbiology now regularly convenes a meeting on beneficial microbes suggests a reformation in both the scientific and popular opinion of bacteria.

Now, imagine for a moment a yogurt enriched with viruses. This does not seem like a good sales pitch. Or imagine instead, an ocean of viruses. Do you want to go swimming? In fact, a swimmer entering coastal waters for a dip could fill up a single liter bottle and find more than 10 billion, if not 100 billion, virus-like particles. This swimmer is unlikely to get sick, at least not from the viruses. The reasons include the strength of the human immune system and the type of viruses that are found in seawater. Ocean viruses are predominantly viruses of microorganisms and do not have direct effects on human cells. What they do to associated microbes remains an important but ongoing question. Indeed, an alternative history of viruses begins with the viruses of bacteria and constitutes the basis for a far more nuanced view of the range of effects that viruses may have than what is now considered the norm.

This history begins in the late nineteenth/early twentieth century, when microbiologists—also known as "microbe hunters"—such as Louis Pasteur and Robert Koch were trying to identify the causative agents of disease and to find cures for them (de Kruif 2002). Two microbiologists of the next generation of microbe hunters, Frederick Twort, a British microbiologist, and Felix d'Herelle, a French physician, independently observed a curious phenomenon of clearing in solutions and on plates otherwise replete with bacteria (Twort 1915; d'Herelle 1917). Both Twort and d'Herelle passed the material through a series of filters and chemical preparations that should have eliminated any bacterial or predatory organisms like protists. The filtered material derived from the remains of killed bacteria continued to kill newly grown cultures of cells. Twort thought it was an enzyme that killed bacteria, whereas d'Herelle speculated that a small organism was responsible. He called the small, unseen organism a *bacteriophage* or "bacteria eater," from the Greek word *phagos* meaning "to devour." The notion that viruses could kill bacteria suggested the possibility of phage therapy— the application of viruses to treat human diseases caused by bacterial pathogens. Phage therapy was championed by d'Herelle and became a focus of scientific investigation and a subject of public discourse. Indeed, Dr. Arrowsmith, the protagonist of Sinclair Lewis's *Arrowsmith*, published in 1925, discovers a phage capable of killing the microbe that causes bubonic plague. It would seem that viruses, at least those that infect bacteria, could be forces of good.

Despite these advances, neither Twort nor d'Herelle had seen a virus. This happened later, after the electron microscope was developed and applied to the study of bacteria and viruses in the late 1930s. Mice-associated pox viruses (Von Borries et al. 1938) and the tobacco mosaic virus (Kausche et al. 1939) were two of the first viruses analyzed with an electron microscope. Similar visualizations of bacteriophage followed in 1940 (see discussion in Ackermann 2011). In the debate over whether viruses of bacteria were enzymes or distinct particles, the latter camp ruled the day, helped by the direct observations of viruses. In summary, d'Herelle had been prescient in one significant respect: bacteriophage are a kind of virus—the kind that infects bacteria. But not all the early predictions were realized. Phage therapy is not nearly as commonplace as is the application of antibiotics to treat illness. The reasons for this are treated wonderfully in a book on d'Herelle and the origins of modern molecular biology (Summers 1999). Nonetheless, the promise of phage therapy and phage-enabled therapeutics remains (Sulakvelidze et al. 2001; Merril et al. 2003; Fischetti et al. 2006; Abedon et al. 2011). To the extent that phage therapy can work, it does so because of virus-host dynamics. Similarly, it fails because virus-host dynamics also include evolution (Levin and Bull 2004). Mathematical models that underpin phage therapy models will be introduced and analyzed in Chapters 3–5.

Uncharacteristically for biology, mathematical models were very much part of the formative studies of phage that were designed and executed by luminaries such as Emory Ellis, Max Delbrück, and André Lwoff. Of these, Delbrück was a physicist, and papers from the early days of phage biology (certainly those with his name attached) reveal quantitative thinking that helped build intuition regarding the dynamics that could be seen only at scales far larger than those at which the actual events were unfolding. These early studies provided the foundation for subsequent diversification of the study of phage: the basic concepts of what happens subsequent to infection, experimental protocols for inferring quantitative rates from time-series data, and methods for interpreting and disentangling alternative possibilities underlying the as-yet-unseen actions taking place at micro- and nanoscales (Delbrück 1946; Lwoff 1953). Two recent books revisit these early days, including one written by Summers (1999), mentioned previously, and another by Cairns et al. (2007), *Origins of Molecular Biology*. Both place phage and phage biologists where they deserve to be: at the center of the historical development of molecular biology. These early studies also provided another output: raw material. Phage biologists isolated many of the phage and bacterial strains that have since been disseminated globally for use in many branches of biology (Abedon 2000; Daegelen et al. 2009).

Finally, there is a third story of viruses to tell, one that began only in recent years. It reveals how fraught with difficulties efforts are to define what a virus is and how important it is to think carefully about this seemingly semantic question. In 1992, a French research team identified a previously unknown parasitic organism that infected amoeba. The organism had been observed at least a decade earlier but had not been characterized (for more discussion of this history refer to Wessner (2010)). Each particle was nearly 0.5 μm in size, with a large genome approximately 10^6 bp (base pairs) in length, encapsulated in a membrane vesicle that was itself encapsulated by a protein shell and was further surrounded by fibers. Morphologically the organism had much in common with bacteria, or should have had. As it turned out, this organism is a virus—a *giant* virus. The virus was called a "mimivirus," because of features that suggested it was a *mi*micking *mi*crobe virus (la Scola et al. 2003). Previous research on giant viruses, for example, that by James Van Etten and colleagues, had characterized large *Chlorella* viruses with genomes exceeding 300,000 bp (Van Etten et al. 1991). What made these giant mimiviruses even more remarkable, beyond their size, was that they seemed to constitute a hybrid, chimeric, or seemingly new form of life. Once they infected amoeba cells, these viruses did not depend exclusively on host machinery to produce their component parts. Instead, they carried with them nearly all the genes to do so themselves. We are now, it seems, at a moment where discoveries call into question the long-standard definition of viruses—these things that live, die, and multiply, just like other organisms.

What, then, is a virus? There are those who say viruses are not alive and others who argue that they are. In the present context, I would prefer to focus attention on ecologically relevant questions, for example, what do viruses do to the hosts and host populations they infect? This question has implications for how entire microbial communities change and function, in part because viruses infect microbial (and metazoan) hosts from the three kingdoms of life (see Figure 1.1). To understand the effect of viruses on microbes and microbial communities it is important to first ask, what are the physical, chemical, and biological dimensions across which viruses differ? These dimensions of viral biodiversity are crucial to understanding viral life history traits and, ultimately, the effects that viruses have on shaping the microbes and the environments in which they persist.

1.2 DIMENSIONS OF VIRAL BIODIVERSITY

1.2.1 PHYSICAL

Tobacco mosaic virus, one of the first viruses viewed under an electron microscope, is a rodlike virus approximately 300 nm in length and 20 nm across. In contrast, phage λ, a subject of formative studies of gene regulation (Ptashne 2004), has a capsid approximately 50 nm in diameter with a tail fiber extending approximately 150 nm. Although viruses are "small," their range of sizes is larger than is widely recognized, spanning at least one, if not two, orders of magnitude, with significant morphological variation when considering viruses that infect all kingdoms of life (Figure 1.1). That size varies by two orders of magnitude is due to the recent discovery of "giant" viruses that can reach $0.5\,\mu$m in size that infect amoeba, ciliates, and perhaps other eukaryotes (Van Etten et al. 2010).

The physical disparity in size might seem curious, or simply a curiosity (akin to the "Rodents of Unusual Size" from *The Princess Bride*). However, differences in the physical size of virus particles are linked to many aspects of viral life-history traits. The study of the relationship between size and function is one of the oldest in science. Leonardo da Vinci is considered the first to develop an argument for allometric scaling in biology (see discussion in Brown and West (2000)). Da Vinci hypothesized that the sum of cross-sectional areas of tree limbs should be equal before and after branching and,

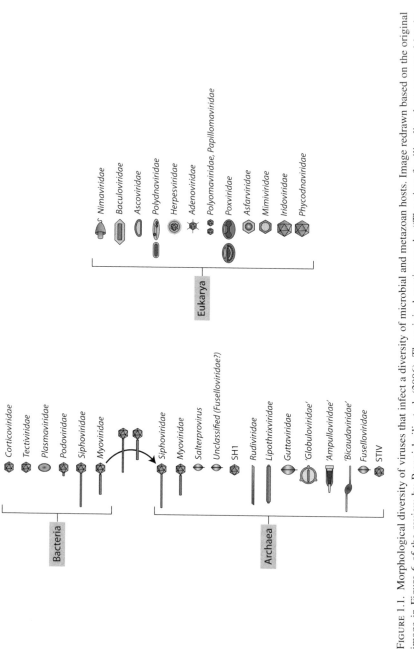

FIGURE 1.1. Morphological diversity of viruses that infect a diversity of microbial and metazoan hosts. Image redrawn based on the original image in Figure 6 of the review by Prangishvili et al. (2006). The original caption reads: "The virus families listed are approved by the International Committee on Taxonomy of viruses, and the schematic representation of virions (not drawn to scale) are presented as in Fauquet et al. (2005). Proposed families are shown in inverted commas. SH1, *Haloarcula hispanica* virus 1; STIV, *Sulfolobus turreted* icosahedral virus."

further, that limb lengths scale with limb diameter. This isometric change in limb component size is not quite accurate, given that limb widths increase in size faster than do limb lengths (McMahon 1973). The modern history of the study of *allometry*, the change of organismal structure and function with body size, has its origins in the late 1800s. In 1883, the German physiologist Max Rübner claimed that the metabolic rate of dogs could be estimated accurately based on knowledge of their size alone (see discussion in Kleiber (1947)). The reasoning assumed that an organism's metabolic rate was mediated via exchange with the surroundings. Organismal surfaces were hypothesized to scale with body size to the 2/3 power, scaling in some sense like spheres.[1] If exchange area scaled to the 2/3 power, then so, too, would metabolic rate. This simple hypothesis is not that far off, though how far off such a prediction is depends on whether mice or elephants are being considered. The analysis of the scaling of metabolic rate is a matter of long-standing scrutiny and debate (Kleiber 1961; McMahon 1973; Peters 1983; Schmidt-Nielsen 1984; Brown and West 2000). Indeed, linking organismal body size to organismal function, such as rates of locomotion, predation, and even death, is the basis for the study of macroecology (Brown 1995).

What is the analogous link between size and function in the case of virus–microbe interactions? Here, there are two sizes to consider: the size of the virus and the size of its host. This chapter largely focuses on virus size, which has two key components: the size of the virus particle and the length of the virus genome. These two sizes are interrelated. Viral genomes are packed under pressure inside a protein capsid. In the case of dsDNA nonlipid-containing phage, the genome is highly organized; for example, there is evidence that DNA can be coiled (Purohit et al. 2005) or even folded toroidally (Petrov and Harvey 2007). The total volume of the genome can be approximated as the sum of the volumes of the nucleotides. The available volume inside the capsid is $4\pi r^3/3$, where r is the internal radius of the capsid. How does the realized volume of the genome change as the available volume increases?

Figure 1.2 shows the measured empirical relationship between the number of base pairs and capsid internal radius, r. The relationship is linear on a log-log plot, with an exponent of 3 and a prefactor of 2. The prefactor is not universal, and requires use of nanometers as units for length. In other words,

$$n_{bp} = 2r^3 \tag{1.1}$$

---

[1] An old joke about physicists and spherical cows comes to mind; here it seems to apply to physiologists and spherical dogs. Nonetheless, such spheres are not a bad starting point.

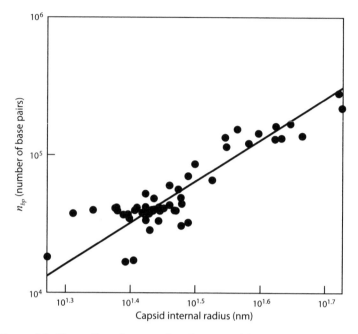

FIGURE 1.2. The scaling of genome length measured in number of base pairs versus the internal radius of the capsid. Quantitative relationship between genome length and capsid size for nonlipid-containing dsDNA bacteriophage for 54 phage types (see Jover et al. (2014) for full accession information). The solid line corresponds to the best-fitting cubic; that is, $n_{bp} = cr^3$, where $c \approx 2.0 \pm 0.2$. Reprinted from Jover et al. (2014).

for icosahedral capsids. Hence, a virus with an internal radius of 25 nm is predicted to have approximately 31,250 bp, and a virus with an internal radius of 50 nm is predicted to have approximately 250,000 bp. This relationship between viral size and genome length has been tested against nonlipid-containing dsDNA phage (Jover et al. 2014). The relationship is a useful baseline for quantifying other life history traits, as explained in the next section. The empirical relationship can also be used to derive the *fill* of a capsid, that is, the fraction of volume within a capsid taken up by the phage genome:

$$n_{bp} = \text{fill}\,\frac{v_{ic}}{v_{bp}} = \text{fill}\,\frac{4\pi}{3v_{bp}}(r_c - h)^3, \tag{1.2}$$

where v_{ic} is the volume inside the capsid, r_c is the outer radius of the capsid, h is the thickness of the capsid, and v_{bp} is the volume of a base pair. Noting that $r = r_c - h$, and given the constants $h \approx 2.5$ nm and $v_{bp} \approx 1.07$ nm^3, one can infer that fill $= 0.51 \pm 0.04$ (Jover et al. 2014). In summary, scaling analysis reveals that phage genomes take up approximately 50% of the available volume inside the capsid, irrespective of whether the capsid is small or large. Similar scaling arguments can be used to estimate the number of proteins making up the capsid, n_{pr}. The core notion is that the capsid can be represented as a spherical shell with volume v_c and uniform thickness h. The expected number of proteins in the capsid is

$$n_{pr} = \frac{v_c}{v_{pr}} = \frac{4\pi}{3v_{pr}} \left[r_c^3 - (r_c - h)^3 \right] \tag{1.3}$$

$$= \frac{4\pi}{3v_{pr}} (3r_c^2 h - 3h^2 r_c + h^3). \tag{1.4}$$

The number of base pairs increases, to leading order, as r_c^3, whereas the number of proteins increases, to leading order, as r_c^2. Such information is also key to estimating the elemental content of virus particles.

Viruses of microbes vary in other ways as well. As should be apparent from Figure 1.1, viruses differ not just in size but also in shape. This is true whether considering the viruses of bacteria, of archaea, or of microeukaryotes. The study of environmental phage isolates often begins with the question, is the isolate a myo, a sipho, or a podo? This lingo stands for myoviruses, siphoviruses, and podoviruses. The question means, is the virus from the family Myoviridae, Siphoviridae, or Podoviridae, respectively. All three are tailed viruses but are distinguished most readily by their tails. Myoviruses have long, contractile tails; siphoviruses have long, noncontractile, flexible tails; and podoviruses have short tails (Figure 1.3). Perhaps the best known representatives of these three families are T4, phage λ, and T7, which have tail lengths of approximately 140 nm (Kostyuchenko et al. 2005), 150 nm (Katsura and Hendrix 1984), and 20 nm (Krüger and Schroeder 1981), respectively. By comparison, environmental assays of tail lengths of marine viruses can be used to infer that lengths are typically 150 nm for myoviruses, 210 nm for siphoviruses, and 15 nm for podoviruses (Brum et al. 2013)—though many viruses are non-tailed. The tails of viruses have functional roles, particularly in defining host specificity and infection initiation (e.g., see the detailed analysis of entry in the case of T7 in Hu et al. (2013).

1.2.2 CHEMICAL

Virus particles can comprise a head and, sometimes, a tail. The head is a protein capsid surrounding genetic material, either RNA or DNA, whereas the tail is made up of proteins. The protein capsid or "coat" may include various

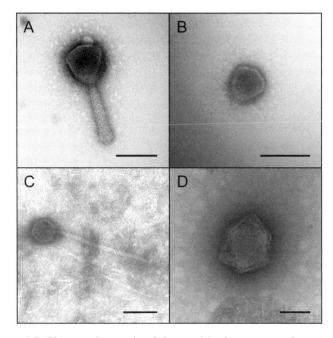

FIGURE 1.3. Electron micrographs of virus particles from ocean-surface samples: (a) myovirus; (b) podovirus; (c) siphovirus; (d) nontailed virus. The samples were negatively stained. Scale bar denotes $100 \, \mu$m. Images are reproduced with permission of the copyright holder, Jennifer Brum.

decorations. Structural virologists have developed elegant systems for describing the types of folds and configurations observed in capsid assembly (Jiang et al. 2003). The structure of virus particles may itself be informative with respect to the evolutionary history and origins of viruses (Bamford et al. 2005). Yet, the protein stoichiometry of viral capsids may be less important than the elemental stoichiometry of virus particles, from the perspective of ecology and ecosystem functioning. Why is this the case? In natural environments with scarce resources, there are limited opportunities to ingest and digest a particle that is rich in both phosphorus (P) and nitrogen (N). Hence, a virus may be a tasty snack to certain nanoflagellates (Gonzalez and Suttle 1993), because viruses contain a relatively high proportion of nitrogen and phosphorus per unit mass, compared with alternative food sources (like bacteria or other phytoplankton). How often this snack is ecologically relevant depends on the availability of other potential prey items and the ability of the consumer to utilize a virus as food. The snack also involves a certain risk: taking up a virus may lead to infection.

How much carbon (C), nitrogen, and phosphorus is in a virus particle? The capsid is made up of proteins and has an external radius r_c. The external radius is the distance from the center to the outer boundary of the capsid. The number of base pairs inside the capsid, n_{bp}, scales as $(r_c - h)^3$, where h is the thickness of the capsid (Eq. 1.4). This scaling presumes that the genome occupies a fixed fraction of the available volume. In contrast, the expected number of proteins in the capsid, n_{pr}, scales as r_c^2. This means that the relative number of proteins to base pairs decreases with increasing viral size. Biochemically, this relationship has an important consequence, as nucleotides and proteins have distinct molecular compositions.

The average molecular formula of a base pair (i.e., a pair of nucleotides), expressed in terms of C:N:P is 19.5:7.5:2. That is, there are 19.5 molecules of carbon for 7.5 molecules of nitrogen for 2 molecules of phosphorus in every base pair. The "average" here assumes an equal probability of having an A:T base pair as a G:C base pair, whose molecular compositions are distinct. Deviations are expected for any given genome sequence. Nonetheless, assuming 50% GC content provides an important baseline for assessing the elemental composition of viruses. In contrast, the amino acids that constitute proteins have no phosphorus, but they do contain carbon and nitrogen—again, the particular ratio depends on amino acid composition. Analysis of primary sequence information for more than 2000 viral proteins reveals that they have, on average, 31 molecules of carbon and 8.7 molecules of nitrogen/nm^3 of protein (Jover et al. 2014); that is, a C:N ratio of 3.6:1—slightly more carbon rich and nitrogen poor than DNA.

Jover et al. (2014) demonstrated how to combine the scaling of molecular composition at the level of nucleotides and proteins with the elemental composition of such molecules to arrive at a predictive model for the elemental composition of virus heads. The combination involves the addition of the elemental composition of the two components of a virus head, its genome and its capsid:

$$C_{\text{virus head}} = C_{\text{genome}} + C_{\text{capsid}}, \tag{1.5}$$

$$N_{\text{virus head}} = N_{\text{genome}} + N_{\text{capsid}}, \tag{1.6}$$

$$P_{\text{virus head}} = P_{\text{genome}}, \tag{1.7}$$

After the size-scaling relationship and the chemical composition of molecules are combined, the specific predictions are

$$C_{\text{virus head}} = 39(r_c - 2.5)^3 + 130(7.5r_c^2 + 18.75r_c + 15.63), \tag{1.8}$$

$$N_{\text{virus head}} = 15(r_c - 2.5)^3 + 36(7.5r_c^2 + 18.75r_c + 15.63), \tag{1.9}$$

$$P_{\text{virus head}} = 4(r_c - 2.5)^3, \tag{1.10}$$

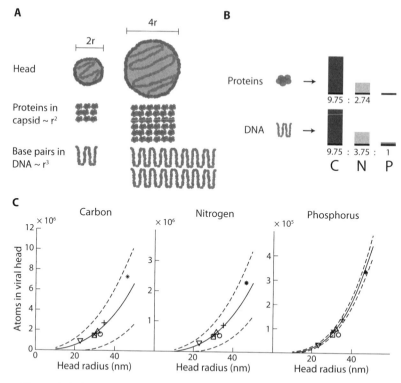

FIGURE 1.4. C, N, and P content of the viral head as a function of its external radius (solid lines). The data correspond to experimentally obtained contents of C, N, and P for different viral heads: * T4, + N4, × Syn5, △ λ, ○ HK97, □ T7, ▽ φ29. Full information on data sources can be found in Jover et al. (2014). Reprinted from Jover et al. (2014).

where r_c is in units of nanometers. Predictions for the number of atoms of carbon, nitrogen, and phosphorus were then compared against the total elemental content as enumerated for seven viruses: T4, N4, Syn5, λ, HK97, T7, and φ29. The criterion used in selecting these viruses was that their genome sequence and complete capsid structure was available. The latter requirement proved more restrictive. Many viral genomes have been sequenced, but very few entire 3D models of dsDNA viruses of microorganisms are available in which the corresponding amino acid sequences of each protein are known. Model predictions, and variation due to uncertainty in model parameters, are shown in Figure 1.4. As should be evident, the model captures the size dependence of elemental content in actual virus heads, yet the formalism can easily be extended to virus tails. Because tail proteins do not contain phosphorus, the total P content is bound in the head only, whereas the C and N inside a virus

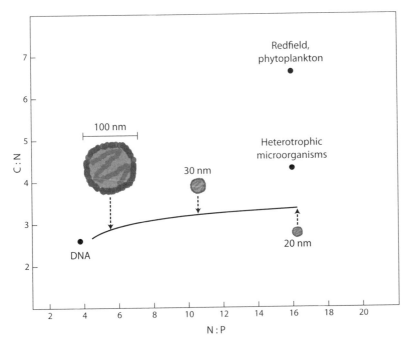

FIGURE 1.5. Size dependence of elemental stoichiometry for viral heads. The *x*-axis is the N:P ratio, and the *y*-axis is the C:N ratio. The three reference points plotted correspond to DNA (given a G:C ratio of 0.5), the Redfield ratio of 106:16 (for C:N) and 16:1 (for N:P), and an approximate consensus for heterotrophic bacteria. Reprinted from Jover et al. (2014).

head underestimates the total C and N inside a virus particle by approximately 10%–15% for this group of myoviruses and siphoviruses (Jover et al. 2014).

Analyzing the elemental content of viral heads leads to an important conclusion: the elemental stoichiometry of virus particles differs from that of microbial hosts. The term *elemental stoichiometry* refers to the ratio of atoms of distinct elements inside an organism (Sterner and Elser 2002). Viruses, owing to their high concentration of DNA, are predicted to have C:N ratios that remain relatively invariant. This relative invariance reflects the similar C:N ratios of proteins and nucleotides—the building blocks of most virus particles with the exception of those that contain lipids and carbohydrates. In contrast, the N:P ratio of virus particles is predicted to range from approximately 16 to 4. Together, these ratios imply that there may be as few as 15 atoms of carbon for every atom of phosphorous in a virus particle. This composition is remarkably nutrient rich, given that there are usually greater than 50, if not greater than 100, molecules of carbon for every molecule of phosphorus in a microbial host cell (Figure 1.5). The differential scaling of carbon, nitrogen, and phosphorus in a virus compared to that in its host has a number of implications for marine biogeochemical dynamics—issues that will be revisited in Chapters 6 and 7.

Infection of host cells provides another avenue for exploring chemical diversity associated with viruses. Virus infection of host cells redirects host cell metabolism toward production of viruses. Yet, in some virus-host systems, the structural elements of virus particles include novel macromolecules that are produced only in the course of virus infections. A prominent ecological example is the interaction of viruses with the algal phytoplankton host *Emiliana huxleyi*—the most abundant coccolithophore in the global oceans, and a key population in driving carbon cycling (Iglesias-Rodríguez et al. 2002). When infected by the algal virus EhV86, host cells undergo a dramatic change in chemical profile. Of note, the profile of lipids changes from host-associated lipids to virally encoded glycosphingolipids (Bidle and Vardi 2011). The molecular weight of virally encoded lipids reaches levels of up to 200 fg per cell (similar to the total mass of an *E. coli* bacterium). Hence, the chemical diversity associated with viruses can represent novel biomarkers for identifying infection as well as mediating chemical arms races in natural systems. Underlying such chemical diversity is the biological diversity encoded in viral and host genomes.

1.2.3 BIOLOGICAL

It has been said—repeatedly—that viruses infect all manner of life. This makes for a good initial hypothesis. But the statement itself is vague and fundamentally unanswerable. Alternatively, one could ask, are all living organisms *potentially* infectable by a virus? Or are all organisms currently infected by a virus? Or perhaps, will all organisms, at some point, become infected by a virus? One could also ask, which *types* of organisms can viruses infect? Are there viruses that infect individual microorganisms from the three domains of life: Archaea, Bacteria, and Eukaryota? The answer to this last question is yes for Archaea (Prangishvili et al. 2006; Prangishvili 2013), yes for Bacteria (Calendar 2005), and yes for Eukaryota, including microbial eukaryotes (Van Etten et al. 1991, 2010). Because there is significant diversity within these three kingdoms of life, it is unknown, how many types of viruses infect any given host sampled from the environment. A follow-up question of relevance is, do viruses infect each type of archaea, bacteria, or eukaryote at finer taxonomic scales, for example, family, genus, or even species?

This question—currently and likely permanently unanswerable—serves as a useful *Gedankenexperiment*, "thought experiment." The reason why the problem is intractable has something to do with the nature of viruses and the nature of the diversity of their microbial hosts. The gold standard for viral identification is *isolation*, that is, the separation of a particular virus from the rest of the community. To isolate a virus, one must first isolate a host upon which the virus can replicate. This precondition leads to many potential biases in what is known about the ecology of viruses and their microbial hosts, because current information is largely based on the few examples of viral isolates, which requires isolating bacteria, archaea, and/or microeukaryotes. Such isolation is difficult (Rappé and Giovannoni 2003). There are many more microbial individuals in the environment than can be isolated as yet in the laboratory. This discrepancy is known as the "great plate count anomaly" (Staley and Konopka 1985). The anomaly refers to the large difference between estimates of microbial abundances using culture-based techniques and culture-independent techniques. Estimates of the number of bacteria in seawater from culture assays typically yield 10^4/ml, when grown on seawater-based sterile media. A culture-independent approach to stain the sample so that particles that have DNA and are approximately 0.5–2 μm in cross section typically yields estimates of microbial abundances of 10^6/ml. As a consequence, culture-based estimates often undercount the true abundances of microbes by two orders of magnitude (Rappé and Giovannoni 2003). The reasons for this paradox are many, yet can be summarized as follows: the conditions necessary to cultivate microbial populations are not yet known (Leadbetter 2003). These conditions may even include a requirement to live with, or at least among, other organisms—suggesting fundamental limits to current culture-based efforts at isolation.

The diversity of viruses extends to, and indeed is encoded in, viral genomes. The genome structure of viruses differ from that of hosts—in similarity to one another, size, and compactness. Comparisons of viral genomes are made more difficult given that they do not possess universal marker genes, like those that encode 16S rRNA within microbes and metazoans (Pace 1997). Instead, the differences among viruses can be compared at coarse and fine scales. Coarse-scale comparisons are facilitated by comparing the genomes or imputed proteomes of viruses and using these distances as the basis for constructing trees (Rohwer and Edwards 2002). Fine-scale comparisons among viruses focus on microevolutionary changes in viral genomes, such as may occur during adaptation of influenza viruses (Koelle et al. 2006) to bacteriophage λ (Meyer et al. 2012). In either case, variations in viral-encoded genes help determine which hosts a virus can infect and what happens to those hosts after they are infected.

Viruses face pressure to reproduce rapidly while evading host defenses and are thought to have evolved a highly compact genome. One measure of this

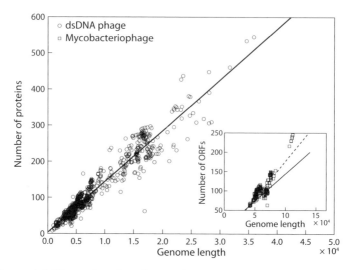

FIGURE 1.6. The number of putative proteins as a function of genome size. (Main) Relationship as estimated from 1124 sequenced dsDNA bacteriophage genomes. The solid line is the line of best fit (slope of 1.41×10^{-3}, corresponding to one open reading frame (ORF) for every 709 nucleotides, assuming no gene overlaps). For this fit, $R^2 = 0.93$. Genome sequences and annotations were downloaded from the National Center for Biotechnology Information (NCBI) in March 2014. (Inset) Relationship as estimated for mycobacteriophage. Each square in the inset represents one of 253 sequenced phage genomes designated as a siphovirus for which an ORF annotation had been completed. The dashed line is the line of best fit (slope of 1.94×10^{-3}, corresponding to one ORF for every 515 nucleotides, assuming no gene overlaps). For this fit, $R^2 = 0.74$.

compactness is evident in Figure 1.6, in which the number of proteins is compared against genome length for dsDNA phage (see Kristensen et al. (2013)). The data are well fit by a line whose slope implies that viral genes are approximately 709 nucleotides in length. The linear fit explains 93% of the variation. Another way to interpret this result is that given a broad sample of viruses of differing sizes, the number of genes can be estimated by dividing the genome length by 709. In fact, viral genes can overlap, suggesting viral genomes are even more compact than a strict division would suggest. Figure 1.6 inset shows further evidence that viral genomes are compact. The analysis focuses on 253 siphoviruses from the well-studied mycobacteriophage clade that have been isolated as part of the HHMI Phage Hunters effort (Hatfull et al. 2008; Hatfull 2012). Increases in viral genome length correspond to a proportionate increase on average in the number of genes, unlike human genomes, which are largely noncoding. Given viral genomes that vary from a few kilobases to hundreds of thousands of bases, what are those genes and what do they do?

Estimates of the total number of phage-encoded genes presents one view of the scope of the problem. More than a decade ago, Forest Rohwer suggested that there are likely more than 2 billion genes encoded among phage on Earth, the vast majority of which have not been discovered (Rohwer 2003). In 2013, Matthew Sullivan and colleagues performed a similar analysis based on clustering of viral proteins, revising estimates to a range of 0.6 million to 6 million genes encoded in the global phage virome (Ignacio-Espinoza et al. 2013). This discrepancy between estimates of viral gene diversity is due, in part, to the use of different thresholds by which two genes are considered part of the same or different groups. Moreover, both estimates rely on inference methods meant to extrapolate the diversity of a community from that of the sample. Then, and now, the global phage virome was and is deeply undersampled, which poses a problem to estimation based on extrapolation. For now, it is sufficient to point out that the total number of phage genes, and indeed of viral genes, is large—certainly larger than a few million. Definitive estimates of the phage gene pool size require careful consideration of the problems inherent in such deep extrapolations. This point will be developed further in Chapter 6.

Irrespective of the total size of the viral gene pool, understanding the function of putative viral genes is also difficult. Such understanding depends, in some sense, on being an expert first, that is, identifying functions for those open reading frames that code for putative proteins. The diversity of viral genes is immense compared with the diversity corresponding to a relatively few well-studied model systems. As a consequence, databases of documented viral functions remain sparsely annotated. For those unfamiliar with what makes up a virus, it is worthwhile to revisit the "usual suspects" of functions that are commonly found in viral genomes, and for which a function can be hypothesized. It is not yet practical to analyze the genomes of known viruses one at a time, but thankfully, Eugene Koonin's group has been thinking about related problems for a long time (Koonin et al. 2002; Koonin and Wolf 2009). By analyzing thousands of bacteriophage genomes, David Kristensen, Eugene Koonin, and collaborators have proposed a system for categorizing putative viral proteins into *phage orthologous groups*, or POGs (Kristensen et al. 2011, 2013). POGs denote viral genes that are thought to have similar functional roles and common evolutionary origins. POGs should be thought of as analogous to *clusters of orthologous groups*, or COGs. The concept of COGs has proved instrumental in categorizing genes found in diverse organismal types in terms of similar function (Tatusov et al. 2000). When these clustering approaches are applied, the "top" categories of POGs include some genes coding for functions that should be familiar to virologists. These functions

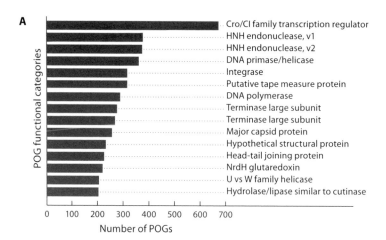

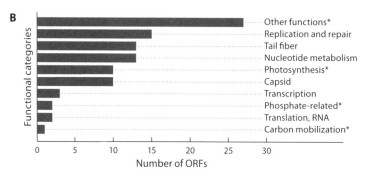

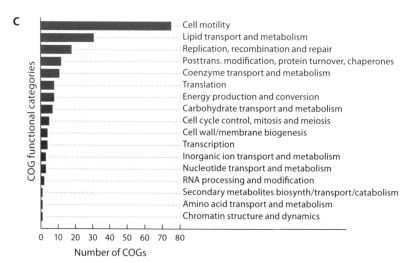

include structural proteins, enzymes that help viruses integrate into their host genomes, enzymes that help viruses replicate, and regulatory proteins that shape the fate of the infected cell. The relative frequency of appearance of the top 20 POGs compiled by Kristensen et al (Kristensen et al. 2013) can be seen in Figure 1.7. This list represents a starting point, particularly for those outside the field of viral biology, for recognizing some of the most common types of genes inside viral genomes.

What is less often appreciated is the extent to which viruses encode genes that encode for a diversity of functions. Consider the myovirus P-SSM2, which infects ubiquitous cyanobacteria of the genus *Prochlorococcus* (Sullivan et al. 2005). This cyanophage genome, and others closely related to it, includes a number of "unusual" genes, such as *psbA*, *psbD* and *pstS* (Figure 1.7). *psbA* and *psbD* are phage-encoded photosystem II genes. These genes are expressed during infection and augment the production of photosynthetic machinery, which is then redirected toward the viral pathway. Importantly, these genes were not "stolen" just recently from host cells—phylogenetic comparison of gene sequences suggests evolution inside both phage and host genomes, as well as mixing between phage and hosts (Sullivan et al. 2006). Similarly, *pstS* is a phosphate-inducible gene that has been hypothesized to increase the uptake of phosphorus by infected cells, a relevant factor in low-nutrient environments. In summary: viruses encode genes that don't pertain "just" to structural components, integration, and escape. Viruses also encode genes that modify metabolic pathways during the infection cycle.

Another example of unusual genes encoded in viral genomes derives from the mimivirus, a so-called giant virus that infects amoeba (la Scola et al. 2003). The mimivirus genome is nearly 1.2 Mbp in length with an estimated 1184 genes (Raoult et al. 2004). The mimivirus genome includes many genes that are highly unusual for viruses, though seemingly common for microbes. Functional categories of these genes include associations with cell motility,

---

FIGURE 1.7. Viral gene diversity. (A) Functions of the top 20 phage orthologous groups (POGs), ranked in terms of the number of viral genome isolates in which they appear. Adapted from Figure 2 of Kristensen et al. (2013). The most highly represented POGs correspond to the usual suspects of phage genome composition. Nonetheless, they do not reflect all the variety in viral gene function, as illustrated by the presence of photosystem and nutrient-inducible genes found in cyanophage (B) and many cellular-analogue genes found in mimiviruses (C). (B) Functional categorization of cyanophage, including common and "unique features" marked with *-s. Adapted from Table 5 of Sullivan et al. (2005)). (C) Functional categorization of mimivirus based on clusters of orthologous genes (COGs). Adapted from Table 3 of Raoult et al. (2004).

energy production, lipid transport, lipid metabolism, cell wall/membrane biogenesis, and even chromatin structure (Figure 1.7). The steps of a mimivirus infection unfold quite unlike those following phage infection of a bacteria, in which host cell machinery is responsible for the bulk of transcription and translation. In contrast, during a mimivirus infection the site of transcript production resides in a localized structure called the "virus factory" (Suzan-Monti et al. 2006)—similar to that in virus infection cycles of other amoebae. The mimivirus and other giant viruses harbor genes to generate new virus particles that emerge from the viral factory. The infection cycle includes many of the steps that would otherwise be associated with a typical cell cycle, with many steps still the topic of active research.

These two examples serve to illustrate a larger point: the diversity of functions in viral genomes annotated in sequence databases reflects the ability of researchers to isolate viruses. Isolated viruses are not a random sample of the environment. Viruses are highly diverse, in that they can persist in hot springs (Held and Whitaker 2009; Snyder and Young 2011), soils (Williamson et al. 2007), lakes (Heldal and Bratbak 1991), oceans (Breitbart 2012), and inside microbiomes (Minot et al. 2013). Adaptation to hosts in such varied environments is associated with concomitant genetic diversity. These discoveries are ongoing. Indeed, in 2014 a French team of researchers reported the discovery of giant viruses related to the mimivirus family. These viruses were frozen in the Siberian tundra for more than 30,000 years, were revived, and still retained the ability to infect extant amoeba cells (Legendre et al. 2014). There is evidently more to discover on all three components of viral diversity outlined in this chapter. To understand the ecological role of viruses requires moving outward, from virus particles to virus–host interactions. That is the subject of Chapter 2.

1.3   SUMMARY

- The study of virology is relatively recent. The conclusive discovery that viruses were a causative agent of infection in plants, animals, and microbes was not made until the advent of electron microscopy in the 1930s.
- Viruses vary in size from genomes of a few thousand to more than a million nucleotides.
- Viral capsids vary in linear dimensions from approximately 20 nm to more than 400 nm in diameter.
- The elemental composition of viruses can be predicted based on simple scaling arguments.

- Virus particles are relatively nutrient rich compared with their hosts.
- Viral genomes are compact, with the number of putative genes scaling linearly with genome size.
- The functional diversity of viruses includes many canonical viral genes, such as those that code for capsid proteins and transcriptional regulators.
- The functional diversity of viruses includes many noncanonical genes, such as those that code for proteins that are part of photosystem pathways or cell-wall pathways.

CHAPTER TWO

Viral Life History Traits

2.1   LIFE HISTORY TRAITS IN ECOLOGY

The primary aim of this chapter is to introduce a number of key commonalities and differences among viral life history traits. Viruses come in a myriad of shapes and sizes, differing in what they look like in electron micrographs and in the biological functions their genomes encode. This diversity, in turn, has consequences for virus-host interactions.

The term *life history trait* was introduced in the early twentieth century by founding population biologists to describe those traits of organisms that have a direct effect on the net number of offspring of an individual. The interested reader may consult the brief synopsis in Morris (2009), the formative review of mid-twentieth-century advances in life history traits and tactics (Stearns 1976), or recent books on life history traits and evolution (Charnov 1993; Roff 2002). In population ecology, the suite of life history traits might include the age at first reproduction, the number of offspring produced at maturity, and the probability of surviving from a juvenile stage to maturity. Qualitative differences in traits may reflect adaptation to distinct ecological conditions, such as the production by salmon of nearly all their offspring at the end of their life versus the production by humans of offspring across a wide age range of reproductive maturity. Quantitative differences between traits can also evolve given a shared set of life history stages; for example, the average clutch size of birds varies between species and across different environments (Lack 1947; Cole 1954). The overall growth rate of a population and its age- or stage-structure then depends on the qualitative nature and quantitative distribution of such life history traits (see worked examples in Hastings (1997)).

Viruses may not necessarily have all the characteristics of classic populations with multiple life history stages, nonetheless, they do multiply, decay (a form of death), and have at least two distinct life history stages: inside and outside of hosts. As I will show, identifying those life history traits and life history stages will be useful in considering those life history trait differences between viruses that are quantitative versus those that are qualitative.

It is the term *quantitative* that motivates a key aim of this chapter: to provide a baseline set of parameters for virus-host interaction models that span evolutionary, ecological, and ecosystem scales. The scope of these models is often quite general. Such models also have the potential to help facilitate understanding the effects that viruses have on their hosts and their environment in *specific* systems. That specificity requires not only models but numbers. Surprisingly, such numbers are often hard to identify, whether in the empirical or in the modeling literature focusing on environmental viruses.

Indeed, the diversity of viruses hampers such quantification. Any choice of a particular set of examples will necessarily focus on some systems, at the expense of others. This may leave some readers unsatisfied. I have tried to balance the need for specificity and generality by illustrating the quantitative diversity of viral life history traits through a series of distinct examples of virus infection of host organisms. The biological, chemical, and physical bases for these traits are then explored, revealing principles by which the traits of viruses may be predicted from some minimal information about the virus and the host of interest, such as, their respective sizes. To some, estimating viral life history traits at the cellular scale provokes challenges of intrinsic interest, whether in molecular or evolutionary biology. To others, these traits are a means to an end, that is, helping to inform the consequences of virus-mediated traits in terms of dynamics and patterns that emerge at "higher" ecosystem scales.

2.2   VIRAL LIFE CYCLE

Viruses of microorganisms are morphologically and genetically diverse. They also vary in their life history. Nonetheless, the life history traits described in this section are motivated by phage-bacteria interactions but are relevant to most virus-host interactions, even if there are features of such interactions that are not covered. For example, giant viruses have a qualitatively different series of life stages from those of viruses of bacteria or of archaea (Van Etten et al. 2010). Similarly, the infection cycle of archaeal viruses have many unique features, including novel modes of entry and exit (Prangishvili 2013). Even if the molecular details differ, there are at least two distinct stages in the life history of a virus: inside and outside of hosts. Here, the description of the life cycle focuses on events that happen when the life cycle is successful, beginning with the extracellular stage. With apologies to Tolstoy: "Happy viruses are all alike; every unhappy virus is unhappy in its own way."

A virus particle cannot persist indefinitely. It must eventually find a host in order to multiply. Common to all known viruses is their lack of self-propulsion.

Viral movement is governed by diffusion and, possibly, advective flows, at least until the virus comes into "contact" with a cell. *Contact* is a euphemism for the biochemical interactions that occur between viral surface proteins and the outer surface of a cell. The contact is often reversible. The site of adsorption is termed a *receptor*. Cellular receptors include proteins, carbohydrates, and lipids—which may themselves be specifically located as part of some larger cellular structure, such as, a flagellum or pilus (Lindberg 1973; Labrie et al. 2010). In some instances, the initial contact point does not represent the final point at which the virus genome will be injected. For example, phage λ has been observed to perform a constrained random walk on the surface of the target *E. coli* cell, preferentially following a path linked by virus-host–specific LamB receptors (Rothenberg et al. 2011).

Following successful adsorption, the virus genetic material, either DNA or RNA, enters the host. For phage, the genetic material is injected as a result of biophysical and biochemical processes. Viral heads are pressurized; for example, the internal pressure exerted by the phage λ genome on its capsid is estimated to vary between 20 and 60 atm (Ivanovska et al. 2007). Internal pressures inside capsids are higher than that of the bacterial cytoplasm, estimated to vary from a few atmospheres to approximately 10 atm (Mitchell et al. 2010). The pressure in a virus head represents the end results of packaging, with the aid of a "packaging force" (Inamdar et al. 2006). After contact with a receptor, conformational changes in the virus particle can lead to what has been described as an "explosive injection" or "popping a champagne cork" (Grayson et al. 2007). Complete virus injection is not always accomplished by biophysical means alone. For example, spooling proteins, often encoded by the virus, are then transcribed and translated and participate in "pulling" the remainder of the viral genome into the host cytoplasm (Molineux and Debabrata 2013).

Once part or all of the virus genome is inside the cell, the following events take place. First, viral genes are transcribed and translated by the host cell. Note that viral gene expression does not necessarily require integration with the host chromosome. In the early stages of virus infection, many viruses catalyze the hydrolysis of the host DNA, making it unlikely that the host cell can continue to divide given that DNA replication requires unbroken strands (Adams et al. 1959). In this way, cellular function is redirected to produce viruses rather than to sustain and produce more host cells. Some viral genes have a regulatory function, modulating not only the timing of events postinfection and redirecting host gene expression but also fundamentally altering the potential fate of the host cell. The two primary fates are lysis or lysogeny (Figure 2.1). *Lysis* is the release of intact virus particles with the

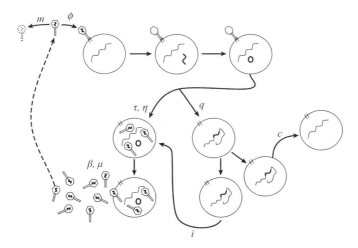

FIGURE 2.1. Schematic of the two primary pathways for viral replication inside bacterial hosts, and associated traits. Outside of host cells, virus particles are modeled as decaying at a rate m and adsorbing to host cells with a per-bacterium rate of ϕ. In infection events that lead to lysis, the time between infection and complete assembly/release of virions is represented as τ, the latent period, or as a rate constant η, corresponding to the probability of lysis per unit time. The number of viruses released is β, with a per-virion probability of mutation of μ. In infection events that lead to lysogeny, the probability of entering lysogeny is q, the probability per generation of spontaneous induction leading to lysis is i, and that of curing is c.

simultaneous death of the host cell. *Lysogeny* is the integration of the viral genome with that of the host, so that the virus sequence is inherited with each cell division—with the possibility of reentering the lytic pathway at a later stage. Other options are also possible; for examples, cells infected by bacteriophage M13 release virus progeny one at a time. These viruses bud off from the host cell, without necessarily destroying the integrity of the cell in the process (Trun and Trempy 2009).

These features of the virus infection cycle are represented in Figure 2.1. Crucially, Figure 2.1 also represents the organizing schematic for the rest of the chapter. Each of the arrows in Figure 2.1 denotes a possible event in the life history of a virus. Each of the variables, m, ϕ, and so on, denotes a quantitative rate or level associated with each process. These variables represent the life history traits that, in turn, affect the outcome of virus-host interactions and that may themselves evolve over time. A key goal here is to provide intuition for the range of quantitative values associated with these traits. As the reader may expect, the value of a viral life history trait is not strictly encoded by the virus; it is both host and context dependent. The coupling between environmental conditions and the fate of an infected cell is a key theme that will recur throughout this book. Finally, technical details associated with the estimation of life history traits can be found in Appendixes A.1, A.2, and A.3. The reader is strongly encouraged to examine these technical details to better understand the connection between observations, inferences, and methods.

2.3   TRAITS ASSOCIATED WITH LYSIS

The lytic pathway of a virus involves the following primary steps (Figure 2.1): viral adsorption to cell surfaces, injection of genetic material, replication of genetic material and production of virion components, packing of viral genetic material into newly produced viral heads, and release of intact virus particles back into the environment. Here, the focus is on explaining the basis for variation in the following traits associated with the lytic pathway: (i) latent period; (ii) burst size; and (iii) mutation rate.

Latent period (τ, η): The latent period is the time between injection of a viral genome into the cell and the lysis of the host cell. The processes that occur during the latent period include (i) translocation of viral genetic material from the periplasm into the cytoplasm; (ii) replication of genetic material and production of virus particle components; (iii) packaging of viral genomes into viral heads; (iv) disruption of the cell surface and release of viral progeny. It is worthwhile to consider a few key features of the latent period, τ. The appearance of intact virus particles, that is, packaged with viral genomes, represents the moment at which cellular lysis would lead to the potential continuation of a viral infection. This period, τ_e, is termed the *eclipse period*; it is less than the latent period. The total latent period varies significantly among virus-host systems. For example, when phages T4 and λ, infect *E. coli*, they can exhibit latent periods as fast as 20 minutes and 50 minutes, respectively (De Paepe and Taddei 2006). In contrast, algal viruses that infect *Chlorella* and *Phaeocysti pouchetii* take 3–4 hours and 12–18 hours, respectively (Jacobsen et al. 1996). Such long latent periods are not restricted only to algal viruses. The latent period of cyanoviruses infecting ubiquitous marine cyanobacteria can be on the same scale, for example, 9–17 hours for cyanophage S-BBS1 infecting *Synechococcus BBCl* (Suttle and Chan 1993), and 8 hours for podovirus P-SSP7 infecting *Prochlorococcus* MED4 (Lindell et al. 2007). The nearly two orders of magnitude variation in latent period raises the question, what determines the time between adsorption and lysis for viruses of microbes?

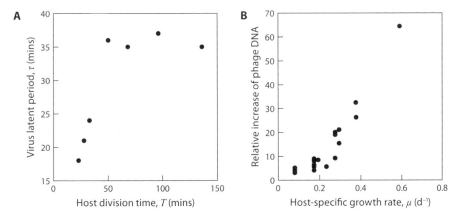

FIGURE 2.2. Correlation between host growth rate and phage latent period. (A) Latent period of bacteriophage T4 increases with cell division period of *E. coli* B/r (Hadas et al. 1997). (B) The buildup of intracellular phage DNA of the podovirus P-SSP7 as a function of the growth rate of cyanobacteria MED4, as measured via the increase of *DNApol* copies 8 hours after infection relative to the concentration at 1 hour (Lindell et al. 2007).

A complete answer to this question requires subdividing each step of the viral infection and lysis process. A working assumption is that measured latent periods are often strongly correlated to division times of the host organism. Such correlations are evident when considering variation in latent periods in the same host–phage system grown in different conditions, and when considering variation in latent periods across distinct host-phage systems. For example, bacteriophage T4 exhibits a twofold increase in latent period when infecting *E. coli* B/r under varying growth conditions that coincide with an approximately sixfold increase in the division time (Hadas et al. 1997). Similar results are found in environmentally relevant marine systems; for example, phage infecting *Synechococcus* had longer latent periods when grown on phosphorus-depleted hosts (Wilson et al. 1996), just as phage had longer latent periods when infecting bacteria of *Vibrio* sp. SWAT3 grown in growth medium diluted 1000-fold (Baudoux et al. 2012). These findings have mechanistic support; for example, for P-SSP7 infecting MED4, intracellular copies of viral DNA increased nearly collinearly as host division rates increased past a critical division rate threshold (Lindell et al. 2007). That is to say, a hallmark of viral production is that it is often positively correlated with host growth rate (Figure 2.2).

In retrospect, the correlation between latent period and host division time may not seem surprising. If synthesis of viral genes is controlled by bacterial metabolism, then it would seem logical to hypothesize that decreases or increases in core bacterial physiological processes would have an effect on intracellular viral production. However, the latent period involves many steps, not all of which are necessarily linked to transcription or translation. It is therefore worthwhile to ask, how should one component of virus production— viral genome synthesis—scale with host generation time? If the host genome is of length L and the division period is T, then the average replication rate per nucleotide is L/T. For many *E. coli* strains growing in "mid-log" phase, $L \approx 4.5 \times 10^6$ bp, and $T \approx 0.5$ hr. The average replication rate is therefore 2500 nt per sec—far faster than the upper limits of replication of a single strand, 600 nt per sec (Xie et al. 2008). However, DNA replication occurs bidirectionally and, under optimal growth conditions, with multiple forks, exceeding the per-strand "speed limit." In contrast, the cyanobacteria MED4 ($L \approx 1.6 \times 10^6$ bp) divides approximately every 24 hours, such that the average replication rate is 6.7×10^4/hr, or 18 nt/sec—more than 100 times slower than *E. coli*! In fact, DNA replication for marine autotrophs growing in the upper ocean is restricted to part of the diurnal cycle, approximately a third of the day (Zinser et al. 2009), leading to a revised average replication rate of approximately 54 nt/per sec. Nonetheless, this rate is still 50 times slower than that of *E. coli*.

How can DNA replication rates of host DNA inform baseline estimates of latent periods associated with viral infections? Referring to the preceding examples, consider the estimates of *E. coli* replication of 9×10^6 bp/hr and *Prochlorococcus MED4* replication of 2×10^5 bp/hr as a baseline. If these replication rates are unchanged during a viral infection, then it would take a period of $\tau = T\,(\beta l/L)$ to replicate β viruses each of length l. Given viruses with a burst size of 50 and a genome of 50 kbp, the predicted time to replicate DNA would be 17 minutes and 12.4 hours for viruses infecting *E. coli* and MED4, respectively. These estimates are qualitatively consistent with actual latent periods. Rescaling the latent period based on host division period appears to lead to reasonable first estimates, particularly when virus-host interactions are compared across systems that vary significantly in host division period. However, caution is necessary before this rescaling argument is applied indiscriminately. Phage replication may be limited by factors other than DNA replication: phage infections may drive DNA replication at rates different from those in uninfected cells; and, finally, there is significant variation between upper and lower limits, which is a function of both the host and virus. For example, the podovirus Syn5 releases between 20 and 30 viral progeny less than one hour after infecting the relatively slow-growing cyanobacteria host *Synechococcus* sp. WH8109 (Raytcheva et al. 2011). Indeed, the topic of what limits phage latent period, its relationship to host physiology, and its evolution is a subject of ongoing interest that will be revisited in subsequent chapters.

The latent period of viruses plays a crucial role in models of viral ecology. However, owing to stochastic viral gene expression and variation in cell physiology, it is not expected that every virus will lyse its host exactly τ after adsorption. A recent analysis demonstrated how gene expression noise leads to latent period variation in phage λ (Singh and Dennehy 2014). Viral lysis times may vary, even if their average is τ. Ideally, one would characterize viral infections in terms of an age a (Smith and Thieme 2012), such that the probability of lysis is $p(a)$, where $\int_0^\infty da\, p(a) = 1$, and $\tau = \int_0^\infty da\, p(a) a$. Modeling age-dependent infections may not necessarily be required to describe the type of dynamical data on virus-host interactions currently available. Models that incorporate a delay between infection and lysis generally use one of the following limiting alternatives. First, it can be assumed that the variation in lysis times is minimal, such that each successful infection releases viruses exactly τ after adsorption. This is the limit of a unimodal, narrow distribution of lysis times. Second, it can be assumed that lysis occurs at an exponential rate $\eta = 1/\tau$, such that the average time to lysis is τ. These two models likely under- and overestimate, respectively, the extent of variation in lysis time. The consequence of such assumptions requires integrating models of virus latent periods with dynamics at ecological and evolutionary scales.

Burst size (β): The number of viruses released upon lysis is termed the *burst size*. Studies of infection of *E. coli* by enterophage, including λ, T4, and T7 often give the impression that burst sizes are constrained to approximately 50–250 virions. These estimates are consistent with foundational studies, including those by Ellis and Delbrück who measured a burst size of 60 for the first "anti-*Escherichia coli* phage" to be isolated and studied (Ellis and Delbrück 1939). Similar ranges are reported in environmental studies. In one of the classic papers on marine viruses, Jed Fuhrman describes the process of infections "typically producing 20–50 progeny viruses" (Fuhrman 1999). This principle is supported by subsequent work. The average burst sizes of viruses infecting aquatic heterotrophs was estimated to vary from about 20 in oligotrophic marine systems to nearly 40 in eutrophic freshwater systems (Parada et al. 2006). Any notion of what is typical is highly influenced by the choice of study system, as is rapidly becoming a theme in this monograph. The true variation in burst sizes is much greater. Just as in the analysis of latent periods, there are two kinds of diversity to highlight: variation in burst sizes given the same or similar virus-host systems and variation in burst sizes in comparisons across systems.

As an example of intrasystem diversity, burst sizes of viruses that infect *E. coli* vary over two orders of magnitude, from estimates of 50 for phage PRD1 infecting strain MG1655 RP4, to 3570 for phage R17 infecting *E. coli* MG1655 with the F-plasmid (De Paepe and Taddei 2006). Smaller degrees of variation have been observed using a smaller sample of viruses for a given reference host. For example, three phage that infect *Mycobacterium smegmatis*, B01, B02a, and B02h, had burst sizes of 61, 11, and 20, respectively (Kraiss et al. 1973). In addition, variation in burst size for the same virus-host system is possible owing to stochastic and physiological differences. Burst sizes of bacteriophage T4 are reported to be 150 (De Paepe and Taddei 2006) but were found to vary from 9 to 260 when the host, *E. coli* B/r, was grown under different media conditions. Growth medium with succinate led to the smallest burst sizes, while medium containing glucose and penicillin-G led to the largest burst sizes (Rabinovitch et al. 1999). Conditions that coincided with the smallest burst sizes corresponded to the longest latent period, which suggests a general metabolic stress in the host. Conditions that coincided with the largest burst sizes corresponded to some of the shortest latent periods. Similar results were observed in other systems. Cyanophage burst sizes varied from 4.4 to 21.5 per host cell when infecting *Synechococcus* sp. WH7803. The range corresponded to growth conditions in which phosphorus was depleted versus replete in the medium, respectively (Wilson et al. 1996).

In contrast to these studies focused on a single virus-host system, Chris Brown and colleagues compiled genomic and trait information associated with interactions among a suite of viruses and their prokaryotic or eukaryotic phytoplankton hosts (Brown et al. 2006). Burst sizes varied over three orders of magnitude, from a low of 41 for virus S-PM2 infecting *Synechococcus* sp. WH7803, to a high of 21,000 for virus HaRNAV infecting *Heterosigma akashiwo*, a eukaryotic autotroph. Brown and colleagues found that burst size was correlated with host genome size. Figure 2.3 shows the correlation between burst size and the ratio of the lengths of the host genome to the viral genome. The rationale for this choice was that viral genome size correlates strongly with capsid size, as discussed in Chapter 1. There also is long-standing evidence that viral size and burst sizes are negatively correlated (Weinbauer and Peduzzi 1994). This is the basis for the present hypothesis that burst size and viral genome length should be negatively correlated. For each virus-host system, the measured burst size was compared against the ratio of host genome length to viral genome length. This ratio accounts for variation in virus burst size that might be driven by variation in host size, using genome length as a proxy. Yet, burst size is unlikely to be a result of biophysical and physiological constraints alone. Indeed, the "optimal" trait likely depends on ecological and

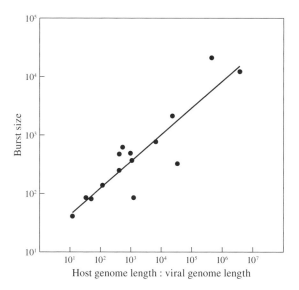

FIGURE 2.3. Burst size in viruses can vary over three orders of magnitude. Burst size (y-axis) is plotted against the ratio of the host genome length to the viral genome length (x-axis). The filled circles denote the burst sizes of marine viruses as compiled by Brown et al. (2006)). The line represents the best-fit power law, $\beta \propto (L/l)^{0.45}$

evolutionary pressures. What does, in fact, limit the burst size of viruses? As a corollary, what range of burst sizes is suitable for modeling virus-host interactions at ecological and evolutionary scales?

To frame the issues at stake and to provide additional intuition on baseline values of burst size, consider an oft-repeated explanation for the upper limit of burst size: burst size is limited by cell volume. This explanation is not implausible. A firm limit on the number of virus particles produced should result from the limitation of physical space. Moreover, electron micrographs of infected cells can give the impression that host cells are "filled completely" (Weinbauer and Peduzzi 1994). Be prepared to be disappointed or, perhaps, to have your skepticism affirmed. Multiple lines of evidence suggest that burst sizes are not limited by cell volumes, even in instances where some micrographs would suggest that they are. A cell is a three-dimensional object, and yet electron micrograph techniques involve dehydration steps that usually flatten cells so that all viruses can be seen and counted. Even if there are enough viruses to "fill" a 2D projection of a cell, this is not sufficient to conclude that they necessarily fill the host cell. Instead, evaluating whether viruses fill their hosts prior to bursting can be assessed by comparing the expected volume of viruses in a burst. The total volume of virus capsids provides an initial estimate: $V_{\text{burst}} \approx \beta \left(4\pi r_c^3/3\right)$, where r_c denotes the radius of the capsid, to the volume of a host cell, V_{cell}. What are the predicted size fractions taken up by viruses?

To begin, consider infections of *E. coli*, a target rodlike cell 1 μm in diameter and 2 μm in length that is approximately 1.5 μm^3 in volume (Milo et al. 2010). The physical dimensions and burst sizes of viruses that infect *E. coli* are well documented. Virus heads range from below 30 nm to over 100 nm in diameter (De Paepe and Taddei 2006). Burst sizes range from below 50 to more than 3500 (De Paepe and Taddei 2006). The expected volume fraction taken up by the head components of virus particles can be estimated by combining information on burst size and virus capsid diameter in those instances where information on both is available. The expected volume fraction is 2% (on average) for 11/13 viruses with approximately spherical capsids whose dimensions are reported in De Paepe and Taddei (2006). The volume fraction are as follows: $\phi X 174$: 0.21%, MS2: 0.27%, PRD1, 0.48%, P4: 0.95%, λ: 1.0%, Mu: 1.1%, P2: 1.2%, T3: 1.5%, T7: 2.0%, R17: 2.5%, T5: 2.8%, $\phi 80$: 4.8%, and P1: 8.6%. These percentages increase marginally given the inclusion of tail fibers. So, these viruses tend not to fill their host cell; to the contrary, they seem to occupy a rather small fraction of the host cell volume. Secondary evidence indicates that viruses can evolve increased burst sizes. Phage λ mutants with delayed lysis were shown to yield more than 500 viruses per burst (Reader and Siminovitch 1971), a phenomena later attributed to delays in the production of holin enzymes, a precursor to eventual lysis (Chang et al. 1995). A full analysis of relative cell volume fraction taken up by virus particles of λ mutants would require further information on the relationship between burst size and infected cell size upon lysis.

It is possible to apply the same type of approach to environmental infections. Weinbauer and colleagues evaluated the relationship of viral burst sizes and virus particle sizes to cell volume in the case of aquatic virus infections of heterotrophic microbes (Weinbauer and Peduzzi 1994). The total volume of viruses in bursts from host cells of different sizes can be estimated by multiplying measured burst sizes by capsid volumes. This volume can be compared against the cell volume of target hosts, the reanalysis of which can be seen in Figure 2.4. Estimates of the fraction of the total cell volume taken up by viruses are all less than 2%, irrespective of cell volume or of virus size. This environmental analysis is consistent with the that of *E. coli* and its viruses.

This analysis should help dissuade readers from believing that viruses of microorganisms invariably "fill up" their hosts—the evidence suggests they do not. In all fairness, direct comparisons of total virus capsid volumes with total cell volume makes an implicit assumption that cells are not already crowded.

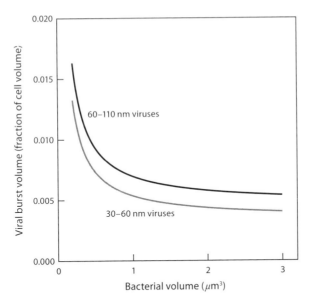

FIGURE 2.4. Estimates of burst sizes suggest that the total volume of virus particles does not take up a substantial fraction of the infected cell volume. Solid lines denote the predicted fraction of cell volume in a viral burst as a function of infected cell volume. The predictions are derived from the best fit between burst size and cell volume for two size classes of viruses (Weinbauer and Peduzzi 1994), $\beta_{30-60nm} = 41.42 + 70.70x$ and $\beta_{60-110nm} = 7.29 + 14.32x$, where x is cell volume in μm^3. *Key point: The range of fractional volume occupied is 0.5% to 2%.*

David Goodsell's lovely images of the structure of cell surfaces and cyto-plasms (Goodsell 2010) should make it evident that packing viruses requires that they fit in among many cellular components, even as some of those components may be degraded during lysis. Moreover, recent work on the physical nature of the cellular cytoplasm suggests that particles larger than 30 nm, like viruses, do not experience a fluidlike environment (Parry et al. 2014). That is, larger particles experience crowding, caging, and other processes related to metabolically driven transport. It may be that there is only space to fit in a much smaller number of virus particles than there would be in a similar volume of water. This limit and its mechanism are worth further exploration. This analysis also raises an important possibility: that virus burst sizes are limited by factors other than physical space. These factors include the physiology of the cell and its response to its local environment, that is, due to ecology. Classic studies of growth suggest that the rate and productivity of organismal growth often depends on some limiting nutrient or resource. In ecology, one such concept is Justus von Liebig's "law of the minimum" (Legovíc and Cruzado 1997).

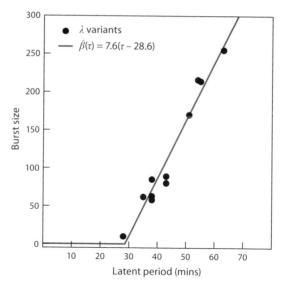

FIGURE 2.5. Measured relationship between burst size and latent period for phage λ variants. Experimental data from Wang (2006). The statistical fit assumes that $\beta = 0$ for $\tau \leq \tau_e$ and then increases linearly with a "rise" of 7.6 virions per minute thereafter until release at time τ. Here, $\tau_e \approx 0.5$ hr.

Another line of reasoning is that burst sizes do not reach their upper limits because to do so would be selectively maladaptive in the environment. Virus burst sizes may have evolved away from physiological maximums!

To begin to think in this direction, consider that within a given virus-host system, viral variants with longer latent periods also have concomitant increases in burst sizes (Wang 2006). The conventional interpretation of this relationship is based on the steps taken after adsorption but before lysis. Initially, no infectious virus particles are produced prior to eclipse period τ_e. Infectious virus particles are produced at a nearly constant rate thereafter, such that $\beta = \rho(\tau - \tau_e)$, where ρ is the "rise" (Figure 2.5). The rise denotes the rate of new, intact virus particle production. Eventually, there will be diminishing returns in terms of viral production with extended latent periods. Nonetheless, these data imply that latent period and burst size are evolvable traits. Subsequent chapters will return to this question and provide further evidence and stimulate further thinking on the combination of physiological, ecological, and evolutionary factors that might limit the burst sizes of viruses of microorganisms.

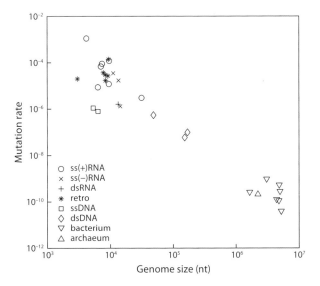

FIGURE 2.6. Mutation rates as a function of virus genome length for different virus types. Mutation rate measured in terms of the substitutions per nucleotide per replication cycle for viruses, and per cell division for bacteria and archaea. Symbols denote rates estimated for viruses with distinct genome types. Full caption and source details available in Sanjuán et al. (2010). The present analysis is based on data provided by Rafael Sanjuán.

Mutation rate (μ): Some viruses released via the lytic pathway will be different from the original virus. The variation may be genetic, as the DNA or RNA packed into the virus head may differ between virus particles. The variation may also be structural. The assembly of viral capsids may have defects or other imperfections that lead to differences in adsorption or decay rates even if the genome packaged inside is identical. Restricting attention to the basis for genetic variation, consider that viruses replicating inside cells depend on host metabolism to synthesize their genome and structural components. Yet, viruses may also encode their own faster, more error-prone polymerases or otherwise reduce the extent to which errors in virus replication are corrected. The effective mutation rates of viruses have been found to be significantly higher than those of their host organisms—whether bacteria, archaea or eukaryotes (Duffy et al. 2008). Mutation rates of viruses vary significantly from $\approx 10^{-3}$/nt per replication to $\approx 10^{-8}$ (Duffy et al. 2008; Sanjuán et al. 2010) (Figure 2.6). As is evident, mutation rates are higher for RNA-based viruses and lower for DNA-based viruses, yet the rates for both are higher than those of their hosts. Mutation rate is also inversely correlated with genome length. One way to begin to understand this inverse correlation is to ask, what is the probability, p_0, that there will be no mutations in a genome of length l given a mutation rate of μ in one replication? There will be no mutations anywhere in the genome if a single mutation does not occur in each of the l base pairs; that is, $p_0 = (1 - \mu) \cdot (1 - \mu) \ldots (1 - \mu) = (1 - \mu)^l$. Consequently, the probability that there will be *at least one* mutation is $p_{\geq 1} = 1 - p_0 = 1 - (1 - \mu)^l$. For values of $\mu l \ll 1$, then $p_{\geq 1} \approx \mu l$. This implies the existence of a critical mutation rate, μ_c, beyond which most viruses are likely to have at least one mutation somewhere in their genome per replication. The critical mutation rate is $\mu_c = 1/l$ and increases as genomes get shorter.

This logic is a linchpin of the "error threshold" hypothesis for the evolution of viral mutation rates. The error threshold was first introduced by Manfred Eigen to predict the potential for persistent replication of a set of sequences that could change from one type to another (Eigen 1971). If one sequence is the master sequence and the other some "mutant," then the sequence dynamics begin to resemble the dynamics of RNA viruses for which mutations are frequent and often lead to inferior offspring. The error threshold represents the critical mutation rate beyond which the master sequence is unlikely to persist. The complete theory includes not just mutation but selection.

In brief, consider two sequences x and x' with populations N and N', such that N grows at a rate 1 and N' at a rate $1 - s$, where s is the fitness cost of the mutation. Let N denote the wild-type population and N' the mutant population. If mutations occur at a rate μ, then the dynamics of the two populations can be written as

$$\frac{dN}{dt} = \overbrace{N}^{\text{growth of WT}} - \overbrace{\mu N}^{\text{mutations}},$$

(2.1)

$$\frac{dN'}{dt} = \overbrace{\mu N}^{\text{mutations}} + \overbrace{(1 - s)N'}^{\text{growth of mutants}}.$$

These dynamics are unbounded, but from a comparison of their relative abundances it is evident that $N(t) > N'(t)$ in the limit that $t \to \infty$, so long as $s > \mu$. In words, a master sequence can persist in a population as long as the relative fitness cost for mutations exceeds the mutation probability. The error hypothesis has many proponents and critics (Eigen et al. 1989; Wilke 2005). The fact remains that a strong negative relationship exists between mutation rate and genome length. The realized mutation rate approaches that of $1/l$ for RNA viruses and is usually an order of magnitude lower for dsDNA viruses.

2.4   TRAITS ASSOCIATED WITH LYSOGENY

Lysogeny is a truly fascinating topic. The study of alternative fates of infected cells was a key part of formative discussions on the nature of heredity, the basis for modern-day genetics, and molecular biology (Cairns et al. 2007). Studies of lysogeny prior to 1953 are covered in a highly readable review by Lwoff (Lwoff 1953). Giuseppe Bertani's review, completed 50 years after that of Lwoff, is also worth consulting (Bertani 2004). A *lysogen*, that is, a bacterial cell with a virus integrated into its genome, can be functionally different from the uninfected host cell. While it infects the bacterium, the virus benefits from changes in cell physiology that increase the fitness of the host. Over the long term, the integrated virus must also get out—resulting in the release of progeny viruses and the death of the host—if it is to avoid an evolutionary dead end.

The genetic switch of phage λ is perhaps the most well studied of all genetic switches within viruses of microbes (Ptashne 2004). Here, it will serve the purpose of illustrating some of the basic phenomena, both molecular and ecological, associated with lysogeny. As is becoming a theme in this book, there is not a single mechanism by which lysogeny occurs. The breadth of applicability of the phage λ–host system remains unknown, given that lysogeny in environmental phage is severely understudied in natural environments (McDaniel et al. 2008). Further, viruses may also persist inside the host cell without integrating into its chromosome—termed *pseudolysogeny*, which remains somewhat of a mystery. An old joke at a virus conference:

Q: What do you call a virus-host interaction that exhibits
     pseudolysogeny?

A: A bad experiment.

Phage λ has a viral genome approximately 48 kbp in length with 73 open reading frames. After phage injection, a series of "early" genes are transcribed by host RNA polymerases. These early genes include transcriptional regulators that promote one of two alternative pathways: lysis or lysogeny. Depending on the relative balance of transcription factors, lysogeny ensues, following the expression of an integrase gene that facilitates the site-specific integration of phage λ into the host genome. Once integrated, the virus is known as a *prophage*. In the prophage state, viral gene expression is largely down regulated, with the notable exception of the *cI* repressor gene. The expressed protein, CI, cooperatively forms dimers that bind to a promoter region of the prophage. These dimers then bind together to form tetramers that block the binding of RNA polymerases that would otherwise initiate pathways in the phage genome leading to reactivation of the lysis pathway. Occasionally, owing to stochastic events or to changes in host cell physiology, a prophage may be induced or become inactive. Prophage induction involves excision of the viral genome from the host, reactivation of the lytic pathway, and the eventual release of a viral burst. Curing involves the deactivation of the virus, because the prophage can no longer induce or is degraded upon induction.

The preceding description of lysogeny is necessarily abbreviated, as it involves insights accumulated over 60 years of study (Ptashne 2004), as well as ongoing active research (Court et al. 2007; St. Pierre and Endy 2008; Maynard et al. 2010; Golding 2011). Life history traits associated with lysogeny include the probability of initiating lysogeny, q, the induction rate, i, and the curing rate, c.

Probability of entering lysogeny (q): The probability of entering lysogeny, $0 \leq q \leq 1$, varies with viral strains. *Virulent* viruses are incapable of integrating their genome with that of the host because they lack integrases. In contrast, *temperate* viruses can form lysogens, and this classification normally refers to bacteriophage only. In contrast, viruses of eukaryotes nearly always integrate their genome with that of the host to initiate transcription—this is not a form of lysogeny. However, for even the most well-studied temperate phage, the success of entering lysogeny also varies with host, physiological conditions, and even the concentration of viruses. What, then, is the evidence for a particular value(s) of q for phage?

The (modern) quantitative study of lysogeny has its roots in a series of papers led by Philippe Kourilsky in the early 1970s (Kourilsky 1973, 1975; Kourilsky and Knapp 1975). Kourilsky examined the variation in the probability of lysogeny among phage λ variants that infect *E. coli*. In the first of these papers, Kourilsky varied the ratio of viruses to hosts and measured the subsequent change in the probability of lysogeny. The ratio of viruses to hosts was quantified in terms of the "average phage infectivity," what is now termed the *multiplicity of infection*, or MOI $= V_0/N_0$. The MOI (denoted here as M) can be used to estimate the typical number of viruses infecting a given host cell in standard phage assays or population studies. If all viruses adsorbed to host cells, then this ratio would represent the average number of viruses adsorbed to any given cell. However, in practice, the actual number of viruses adsorbed to target hosts is less than the initial pool of viruses. Kourilsky observed that the probability that a cell would be lysogenized increased as more viruses were added relative to the number of hosts available (Figure 2.7, left). The rate

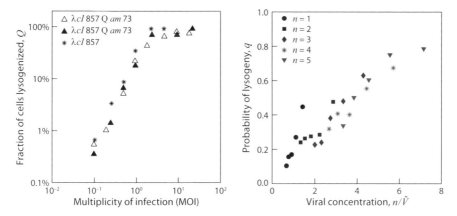

FIGURE 2.7. Fraction of lysogens increases as a function of the multiplicity of infection. (Left) Population-scale data from Kourilsky (1973), in which the percentage of cells lysogenized is plotted as a varying function of the ratio of virus to cell titer (or population-level multiplicity of infection). (Right) Single-cell-scale data from Zeng et al. (2010), in which the probability of lysogeny, q, increases with the number of infecting virus particles (here, $n = 1, 2, 3, 4,$ or 5) as well as with virus concentration $n/\tilde{V}$, where $\tilde{V}_{cell}$ is a proxy for cell volume, measured in terms of length (μm). Data from Kourilsky digitized from Figure 3 of Kourilsky (1973). Data from Zeng et al. (2010) courtesy of Lanying Zeng.

of increase varied among strains, but the overall conclusion was robust. The fraction of lysogeny increased in all cases from less than 1% to nearly 100% when MOIs increased from 0.1 to 10.

It is important to distinguish between the cellular-level probability of lysogeny, q, and the population-level probability that a cell will be lysogenized, Q. In particular, the increase in lysogeny, $Q(M)$ (Figure 2.7, left), belies the microstate decision that occurs within each cell (Figure 2.7, right). Each individual host has a discrete number of coinfecting viruses, and each inserts its genome into the cellular cytoplasm, which is equivalent to having multiple infections by distinct virions. The actual decision process at the cellular scale must be reconciled with population-level observations. To see how, consider that the probability that a given cell will have exactly n infecting particles can be approximated by the Poisson distribution: $P(n|M) = M^n e^{-M}/n!$. Thus, the observation of $Q(M)$ averages the outcomes occurring in cells infected by different numbers of viruses. Denoting $q(n)$ as the probability of lysogeny of a single host cell with n infecting viruses gives

$$Q(M) = \sum_{n=1}^{\infty} q(n)P(n|M) \tag{2.2}$$

at the population scale. The overall incidence of lysogeny at the population scale is the sum of the probabilities of all possible events that could lead to lysogeny, infections of either $n = 1, 2$, or more viruses. How is this information useful? One insight from this approach is that the possible microstate decisions, $q(n)$, are constrained by the population-scale measurement, $Q(M)$. Similarly, constraints on a predicted/observed cellular-scale response $q(n)$ enable predictions of population-scale outcomes, $Q(M)$.

Consider the following example of phage λ infecting *E. coli*. The fraction of original host cells that become lysogens is defined as

$$Q(M) = 1 - \overbrace{P(0|M)}^{\text{\% of uninfected cells}} - \overbrace{P_L(M)}^{\text{\% of lysed cells}} . \tag{2.3}$$

The probability that an individual cell is not infected is $P(0|M) = e^{-M}$. In his original study, Kourilsky hypothesized that "lysogenization of exponentially growing cells by λ or λQ^- would require infection of individual cells by at least two phages" (Kourilsky 1973). In the current notation that constraint is equivalent to the hypothesis that $q(n) = 1$ if $n \geq 2$, and $q(n) = 0$ if $n = 1$. Thus, the fraction of lysed cells should correspond to only those cells that were infected by a single phage, which occurs with probability $P_L(M) = Me^{-M}$. Substitution of $P_L(M)$ and $P(0|M)$ into Eq. 2.3 yields

$$Q(M) = 1 - e^{-M} - Me^{-M} \tag{2.4}$$

$$\approx M^2 \quad \text{when } M \ll 1. \tag{2.5}$$

The incidence of lysogeny should increase faster than linearly as the ratio of viruses to hosts increases given low MOI. Specifically, a log-log plot of the fraction of lysogens versus MOI at the population scale should have a slope of 2 (Figure 2.7, left). Kourilsky reported population-level increases of lysogeny with MOI consistent with this prediction (Kourilsky 1973). The data at the time were consistent with the mechanism of a sharp threshold between lysis and lysogeny between single and doubly (or more) infected cells. Other phage variants had different slopes, and this population-scale analysis does not rule out a smoother dependency of $q(n)$.

Single-cell experiments conducted by multiple labs over the past decade have refined these early conclusions. Francois St. Pierre and Drew Endy showed that viruses that infect smaller cells tend to enter lysogeny rather than to lyse the cell, even when infecting singly (St. Pierre and Endy 2008). A series of parallel studies led by Ido Golding showed that the probability of lysogeny increased, when measured at the single-cell level, as a function of the intracellular virus concentration (Zeng et al. 2010) (Figure 2.7, right). Although many features of this switch are still being examined (Golding 2011; Høyland-Kroghsbo et al. 2013; Svenningsen and Semsey 2014; Bednarz et al. 2014), these two single-cell studies can be viewed as demonstrating that the likelihood of lysogeny increases with the intracellular concentration of phage. This result is akin to one of Kourilsky's possible explanations for his original observation: "The number of phage DNA copies exerts some function" (Kourilsky 1973). It would seem that a single copy of a gene regulatory circuit behaves differently than do two copies. Yet, if one thinks of the concentration of copies as a relevant quantity for measuring binding rates that might lead to transcription/translation, then a working hypothesis is that gene regulatory circuits with nonlinear feedback can couple in ways that lead to qualitative changes in behavior (Arkin et al. 1998; Weitz et al. 2008; Zeng et al. 2010; Joh and Weitz 2011). The two major conclusions of this analysis are that (i) the probability of lysogeny varies within and among strains, and (ii) the governing factors underlying variation in q include the degree of coinfection. Given that the degree of coinfection depends on the ratio of viruses to hosts, which itself is a function of population dynamics, this dependency provides a further point of feedback and coupling between cellular-scale and population-scale dynamics.

The question of *why* lysogeny would increase with the multiplicity of infection remains a long-standing mystery. It would seem that coinfection could promote higher virulence, because of within-host competition among competitors (Nowak and May 1994; May and Nowak 1995), yet the data for phage λ suggest the opposite, as if the viruses are acting in a prudent fashion. Such prudence may be evolutionarily feasible if coinfection occurs among conspecific phage. In that case, coinfection would indicate that most hosts are already infected, so releasing new progeny is not beneficial, particularly if extracellular mortality of virus particles is high. This topic remains of ongoing interest from a cellular biology perspective, as highlighted here, but also from an evolutionary perspective (Gudelj et al. 2010; Maslov and Sneppen 2014).

Induction and curing probability (i, c): The induction probability, i, denotes the probability per division that a viral genome integrated in a host genome will emerge and enter the lytic pathway. *Induction* refers to an irreversible process concluding with the release of virus progeny and the death of the host cell. Two dogmas have emerged in studies of the induction rate. The first is that spontaneous induction is rather rare. Estimates are in the range of less than 10^{-6} for phage λ, phage Mu, and phage P1 (Rosner 1972; Rokney et al. 2008). The second dogma, paradoxically, is that induction can approach a per-generation probability of nearly 1 when cells are "stressed." Stressors include cell starvation, exposure to the mutagenizing agent mitomycin C, and exposure to UV light. What are some of the mechanisms proposed to explain how induction can be either very rare (in the absence of stress), or very common (in the presence of stress)?

First, the basis for the incredible stability of lysogens has been most thoroughly examined, again, in phage λ (Oppenheim et al. 2005; Rokney et al. 2008). Once integrated, the CI repressor protein is expressed at relatively high levels (for a virus), such that approximately 200 CI dimers per cell are expected (Ptashne 2004). A single molecule in a cell the size of *E. coli* has an effective molecular concentration of 1 nM, and CI dimers bind to a promoter region preferentially over alternative molecules ($K_m \approx 10$ nM). Given 200 dimers, a 200 nM concentration of CI dimers will far exceed the half-saturation threshold, such that the promoter region is likely to be blocked. Owing to stochastic events, one would expect a regulatory protein associated with lysis to bind, occasionally, to an unoccupied promoter region, thereby turning on the lytic pathway. The exceptional stability of phage λ lysogens is ascribed to the formation of long-range octamers, wherein CI dimers that couple to form tetramers at one promoter site bind to a distal region of the prophage genome (Révet et al. 1999; Dodd et al. 2001; Aurell and Sneppen 2002; Aurell et al. 2002; Santillán and Mackey 2004; Morelli et al. 2009). Further development of model virus-host systems exhibiting lysogeny will be necessary to identify other mechanisms.

Second, although spontaneous induction may be rare, certain stressors can lead to dramatic increases in induction rates. The rationale is that viruses have evolved mechanisms to detect when the cell they have infected is in danger of dying or becoming nonfunctional. And, in response, they have evolved mechanisms to induce a viral burst before it is too late (Stewart and Levin 1984). The molecular pathway for such a response has been examined in phage λ, for which certain stressors can raise the probability of induction to nearly 1. When in the lysogenic state, the prophage expresses the *cI* repressor gene. The product CI proteins are relatively stable. When a cell experiences DNA damage, it activates a DNA repair pathway, like the SOS response in *E. coli*. This repair pathway includes expression of the RecA protein. In uninfected cells, the RecA protein binds to the site of a transcriptional regulator of the repair pathway, enabling the host to repair itself (Ptashne 2004). However, in lysogens formed by infection of *E. coli* by phage λ, the CI dimer acts like a decoy; it is targeted and cleaved by RecA. Viral repression of the signal to keep itself in the host activates the lytic pathway, leading to induction. Similar mechanisms, including CI homologs, are found in many other temperate phage, e.g., in *Lactobacillus phage* A2 (Ladero et al. 1998), in the enterophage P22 (Sauer et al. 1982), and in *Streptococcus thermophilus* bacteriophage ϕSfi21 (Bruttin et al. 1997). Although molecular mechanisms underlying stress-induced induction of environmental phage are not necessarily known, similar stimulation has been shown to have a similar effect. The exposure of environmental bacteria to UV radiation and agents like mitomycin C leads to induction and release of viruses (Paul and Weinbauer 2010).

Finally, not all lysogens invariably induce to kill their host cells, releasing viral progeny. To the contrary, lysogeny can be an evolutionary dead end—an event that happens when the bacterium "cures" itself of the prophage. *Curing* means that a previously inducible host is no longer inducible. Curing rates are thought to be low; for example, they are estimated to be $< 10^{-6}$ per generation for phage λ (Campbell 1976; Echols 1975) and approximately 10^{-5} for P1 (Rosner 1972). Although multiple mechanisms might lead to curing, consider that with every cell division, there is a probability μ_0 of mutation per nucleotide. Assume that the mutation rates of all nucleotides of a host chromosome, including those of the prophage, are unchanged by infection. The mutation probability μ_0 of the host is usually much lower than that of the mutation probability during viral replication (see Figure 2.6). The probability that no nucleotides of a phage genome will be mutated in a single reproduction event is $(1 - \mu_0)^l$, where l is the number of nucleotides of the viral genome. When $\mu_0 \ll 1/l$, then the probability that no mutations occurred in a single generation can be approximated as $1 - \mu_0 l$. The probability that at least one mutation occurred in a single generation is approximately $\mu_0 l$. These mutations do not necessarily lead to deactivation of phage viability, but they do suggest that viruses with longer genomes will deactivate more readily, as only the viral genes that are expressed are under selection during lysogeny. In contrast, the rest of the genome undergoes neutral drift until induction. In phage λ at least, the physical process of induction requires two primary enzymes, an integrase and an exonuclease (produced from a single transcript encoding the *int* and *xis* genes). The total length of the primary sequence is approximately 1300 bp. The probability that one or more nucleotides of either gene will mutate in a single generation is approximately 1300μ. Most nonneutral mutations are deleterious. Assuming $\mu_0 \approx 10^{-9}$ (Sanjuán et al. 2010), one can hypothesize that mutation accumulation in key induction-related genes will make it impossible for a prophage to induce, with a probability of approximately 1 in every 10^6 lysogens. This value is consistent with the low levels of curing described earlier. It might seem rare, but such curing can be a mechanism for viral degradation even inside microbial genomes. These cured lysogens might then actively wardoff future infections without danger of dying themselves, an issue that warrants further investigation.

2.5   EXTRACELLULAR TRAITS

Viruses are assembled within host cells and released into the environment. This extracellular stage is likely the only subsequent life history stage that most viruses will experience. If 100 viruses are released from a host cell, then the long-term survivorship of those 100 viruses must be approximately 1/100 at steady state; otherwise, the virus population would increase or decrease successively over time. The vast majority of virus particles do not go on to successfully contact, infect, and lyse another host cell. It is the rare virus that adsorbs to a target cell and successfully infects and then lyses it. In that sense, virus particles are akin to seeds of a giant oak tree: many acorns fall, but most are "lost" to seed-consumers and other natural causes. What, then, is the fate of 99% of viruses? There are many possibilities. The virus particle may bind with debris from a previously lysed or ghost cell with intact receptors (Fischetti et al. 2006) or to particulate organic matter that need not have a receptor. The virus particle may also be consumed by filter-feeding organisms, such as microeukaryotes (Gonzalez and Suttle 1993). Or the virus particle may bind to another virus particle, forming an aggregate unlikely to successfully bind to a cellular receptor (Langlet et al. 2007). All these examples include some form of two-body interaction; that is, the virus particle interacts with another particle. Spontaneous decay is another possible mechanism, by which, virus particles lose their infectivity, owing either to inherent instability or to other interactions (e.g., with light or with a molecule) that destabilize the virus. The following sections provide a basis for estimating the rates of two alternative outcomes for a virus particle: decay or adsorption.

Virus particle decay (m): A virus particle exists in a *metastable state*, which means that the assembled virus particle is at a local, but not global, free-energy minimum. Because virus particles do not have an active metabolism, in the absence of repair mechanisms, viruses can spontaneously deactivate. Therefore, the "simplest" model of virus decay is one of density-independent decay caused by thermal fluctuations that drive an intact virus over the energy barrier separating its metastable state from a free-energy minimum. The result is a broken, disassembled, and noninfectious particle. Another relatively simple model of virus decay is one in which interaction with a photon or small molecule changes the nanoscale structure of the capsid or the packed genome.

Damage to the capsid may decrease the probability of adsorption to host cells, while damage to the genome may decrease adsorption and/or the likelihood of propagation inside the target host. Either way, such damage decreases the number of infectious viruses in a population.

Evidence in support of spontaneous decay of viruses comes from the study by Marianne De Paepe and Francois Taddei of enterophage infecting *E. coli*, in which the authors investigated the stability properties of more than a dozen phage strains suspended in cell-free media (De Paepe and Taddei 2006). They measured the number of infective phage particles over time in terms of plaque-forming units, or PFUs (Appendix A.2) and found that the concentration of PFUs decayed exponentially. The exponential decay suggests a "one-step" thermal-induced transition from an infectious to a noninfectious state occurring at a rate m, or, in chemical reaction notation, $V \overset{m}{\rightarrow} 0$. Further support for the mechanism is found by examining the temperature dependence of the decay rate. From a thermodynamics perspective, the probability of decay per unit time should scale as $m(T) \propto e^{-E_a/(k_b T)}$, where E_a is the energy barrier, and k_B is the Boltzmann constant. This is indeed what De Paepe and Taddei found, as shown in Figure 2.8 for examples from R17 and PRD1 (De Paepe and Taddei 2006). These two phage represent those with the fastest and slowest decay rates, respectively, of the phages studied. Extending the analysis by plotting the logarithm of the measured decay rate m against $1/T$ should result in a straight-line fit on semilog axes. The inset of Figure 2.8 shows that loss of infectiousness is temperature dependent, as predicted for a mechanism induced by thermal fluctuation, at least for these two viruses evaluated at three temperatures. The energy barrier over which thermal fluctuations operate is estimated from the slope of the relationship between $\log m$ and $1/T$. It is 1.11 and 1.49 eV for R17 and PRD1, respectively. The energy associated with thermal fluctuations at room temperature is 0.02 eV, which gives a sense of the magnitude of these energy barriers. These particular phage are predicted to be highly stable with respect to thermal fluctuations and increasingly so when the temperature decreases. This feature enables the successful storage of many viruses in dark conditions either on the benchtop or under cold conditions.

The energy barrier of approximately 1 eV also suggests that light absorption can damage virus particles. The energy of a photon varies with its frequency; for example, the energy of a photon associated with yellow light is 2 eV, while the energies of photons with frequencies in the UV spectrum range between 4 and 10 eV (from far-to near-UV light). Thus, the interaction of virus particles with photons can result in a net absorption of energy sufficient to damage the virus particle. Interestingly, the majority of studies on energy

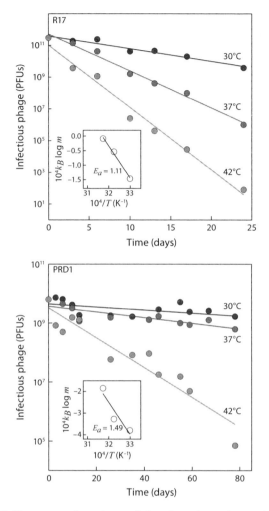

FIGURE 2.8. Temperature dependence of virus decay is consistent with a single-step decay process for R17 (top) and PRD1 (bottom). Both insets denote the relationship between the measured decay constant, m, and temperature. The plot of $10^4 k_B \log m$ versus $10^4/T$ is expected to have a slope equal to the activation energy, E_a (in units of eV, as denoted in the text). Decay data courtesy of Marianne de Paepe. The analysis is adapted from methods presented in De Paepe and Taddei (2006).

and virus capsids have focused on assembly (Hagan 2014). In contrast, there is very little research on the "weak points" of an intact virus particle, such as energy barriers separating an intact from a broken virion, and the absorption of light by viruses. UV exposure is a standard method of deactivating pathogens, including viruses, first introduced over 100 years ago (Bolton and Cotton 2008). Furthermore, environmental virologists working with ocean viruses have long understood that one of the primary mechanisms for the loss of virus particles in the water column is light-induced deactivation (Suttle and Chen 1992). The rate of decay suggests that virus lifetimes may range from a few hours to days in the marine surface, rather than weeks to years in conditions meant to preserve them in the lab. Such rates will prove essential to population-scale models of virus-host interactions.

As a final caveat—there always are caveats in biology—it is worth pointing out that damage to a virus genome may be repairable if the virus is still able to adsorb to and infect a host cell. Light-induced "reactivation" of otherwise "inactivated" phage was first discovered more than 60 years ago (Dulbecco 1949, 1950). The reactivation takes place inside cells by photo-induced repair mechanisms. A "defective" virus may be repaired if and only if it can inject its genome into a host cell. In contrast, genomic defects or capsid disassembly is not actively repairable outside a host cell. Fifty years after this initial discovery a group of aquatic virus researchers demonstrated the environmental relevance of photoreactivation: it is hypothesized to stabilize the abundances of virus populations exposed to UV light in marine surface waters (Weinbauer et al. 1997). The decay rates used in most models are operational, site-specific constants, representing the aggregation of the many factors leading to a net decrease in infectivity. The spontaneous decay of viruses may be negligible in the limit that other clearance rates are high.

Adsorption rate (ϕ): Viral adsorption is assumed to be the result of two processes: (i) random movement of a virus particle in the environment far from a target cell, followed by (ii) adsorption to specific receptors on the surface of a host cell. Thus, adsorption rate on a per-virus basis depends on both host and viral type as well as on host density. The adsorption rate constant, ϕ, is quantified in units of inverse host concentration: ml/(cells $\times$ hr). The product of ϕ with the host density, N (cells/ml), yields the rate (hr^{-1}), at which free virus particles adsorb to host cells.

The diffusion-limited adsorption of small molecules onto a cell surface is a classic problem in chemoreception theory (Berg and Purcell 1977). Chemoreception theory assumes that a cell surface has a number of patches where chemicals adsorb irreversibly. Then, the adsorption rate of chemicals onto a cell can be derived given the fraction of the cell surface that is covered with adsorbing patches and additional assumptions regarding the size of cells and of the target molecules. The same methods can be applied to virus-host interactions. The total flux of virus particles onto a surface should be a product of the diffusion rate of the particles, D, their concentration, c_∞, and the radius of the cell, a, modified by a cell-specific, dimensionless constant reflecting the concentration and specificity of receptors on the surface, C. Dimensional analysis[1] can be used to argue that the form for the total flux of particles onto (and then into) the cell should be

$$J = Cc_\infty Da, \qquad (2.6)$$

which has units of number of particles per unit time. Here the particles are viruses, such that $c_\infty \equiv V$, where V is the virus density. Then, the adsorption rate measured on a density basis should be, J/V or

$$\phi = CD_{\text{virus}}a_{\text{host}}, \qquad (2.7)$$

where the maximum value of the prefactor is $C \approx 4\pi$ for a spherical cell whose entire surface is covered in perfectly absorbing receptors. The actual value of C can vary substantially depending on the virus and host pair (Berg and Purcell 1977). Nonetheless, this upper limit is useful, as it shows that both the viral and host size determine the diffusion limit of viral adsorption.

How, then, does the diffusion rate of a virus depend on its size? This is a well-studied problem, the roots of which date to early calculations by Albert Einstein on Brownian motion (Einstein 1956). The application of these methods to the adsorption of viruses to hosts in natural environments is due to both Berg and Purcell (1977) and Murray and Jackson (1992). Following the convention of both groups, diffusion is approximated via the Stokes-Einstein relation: $D = k_b T/(6\pi\eta r)$, where r is the hydrodynamic radius of the viral capsid. The value of r should be thought of, to a first approximation, as the viral capsid radius, assuming icosahedral viral capsids. The true hydrodynamic radius will depend on the detailed shape of the entire virus particle including

---

[1] Dimensional analysis is often used in the analysis of scaling phenomena in which a small number of governing variables can be used to explain a property of a system. It is of particular use for identifying appropriate combinations of variables with different fundamental units (e.g., length, time, temperature) such that their combination yields an output with the appropriate units. For example, the frequency of oscillation of a pendulum with mass M_{pend} positioned a distance L_{pend} away from a fixed point would seem to depend on the mass. Frequency has units of time^{-1}. The relevant factors are M_{pend} (units of mass), L_{pend} (units of length), and g (standard gravity, 9.8 m/s^2, i.e., units of length per time squared). Dimensional analysis suggests that the frequency of a pendulum should be $\omega \propto \sqrt{g/L_{\text{pend}}}$—the only combination of M_{pend}, L_{pend}, and g which has units of time^{-1}. Note that the mass is not a factor. As a consequence, the period of a pendulum should be proportional to $T = 2\pi/\omega = 2\pi\sqrt{L_{\text{pend}}/g}$. This is, in fact, the correct answer. In general, dimensional analysis does not indicate the dimensionless prefactor, evidently 1 in this case. A highly readable introduction to dimensional analysis is found in Barenblatt (2003).

the tail. Ignoring prefactors, the maximum adsorption rate can be approximated
as

$$\phi \approx \frac{4\pi a k_b T}{6\pi \eta r_c} \tag{2.8}$$

for a spherical virus particle with radius r_c adsorbing to a host cell with radius
a in a fluid with viscosity η at temperature T, where k_B is the Boltzmann
constant. Given virus-host interactions taking place at room temperature ($T =$
293 K), in water ($\eta = 10^{-2}$ g/(cm× sec), where $k_B = 1.38 \times 10^{-16}$ erg/K, then

$$\phi \approx 3 \times 10^{-12} \frac{a}{r_c} \text{cm}^3/\text{sec}. \tag{2.9}$$

Viral ecologists tend to work in units where densities are measured per milliliter, and time is measured per hour. In such units, the maximum adsorption
rate can be written as

$$\phi \approx 4 \times 10^{-9} \frac{a}{r_c} \text{ml/hr}. \tag{2.10}$$

In Figure 2.9, the values of adsorption rate are shown as contours for
combinations of virus-host interactions in which hosts vary from 0.5 to 10 μm
in diameter and viruses vary from 20 to 200 nm in diameter. These values
of ϕ have an ecological interpretation in terms of the host-density-dependent
adsorption rate. If $\phi = 10^{-7}$ ml/hr, then the density of viruses interacting
with a cell population at a density of 10^7 cells/ml will decrease by a factor
of 2.78 in one hour due to adsorption alone. One of the earliest studies of
adsorption rates by Ellis and Delbrück (1939) estimated the adsorption rate
of one of the first "anti-*Escherichia coli* phage" to be $10^{-7.14}$ ml/hr at 25° C.
Given the volume of *E. coli* of approximately 1.5 μm^3, the effective diameter
is (coincidentally) 1.5 μm, such that for viruses with diameters from 50 to
100 nm, the anticipated diffusion-limited adsorption rate is between $10^{-6.8}$
to $10^{-7.2}$—close to the measured value. More generally, these upper limits
of diffusion-limited adsorption are not necessarily realized. They do provide
useful limits when considering models in which adsorption is unknown, yet
must be included in a model or in efforts to characterize the extent to which a
virus approaches diffusion-limited adsorption rates. The actual adsorption rate
in natural environments should also include binding of viruses to nontarget
cells, including debris, surfaces, and even other viruses.

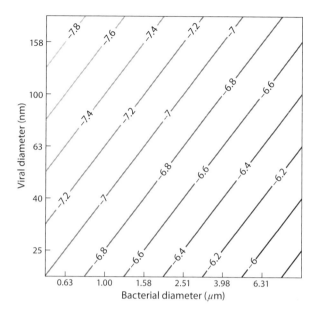

FIGURE 2.9. Maximum diffusion-limited adsorption rate in water as a function of viral and host size. The contours denote equivalent values of adsorption rate (ϕ), such that a value of -7 is equivalent to $\phi = 10^{-7}$ ml/hr. Note the log scale of x- and y-axes. The maximum adsorption rate is located in the bottom right corner, that is, where $\phi \approx 10^{-5.8}$ ml/hr, or 4.3×10^{-10} ml/sec.

2.6   SUMMARY

- Viruses have two key life history stages: inside and outside a host cell.
- Viral infections of microbes often lead to the death of host cells and the release of viral progeny.
- Viral infections can also lead to the integration of viral genomes with those of their hosts; induction of these genomes can result in subsequent lysis and release of viral progeny.
- Viruses are distinguished not only by the host they infect but also by the *quantitative* rates and levels at which these interactions take place.
- Viral life history traits reflect the combined interactions of viruses and hosts; that is, they are not encoded solely by viral genomes.
- Viral life history traits can vary by orders of magnitude, whether for time to lysis, burst size, probability of lysogeny, rate of induction, adsorption rate, or mortality rate.

- The viral life history traits introduced in this chapter will be key to establishing population and evolutionary models of virus-host interactions in subsequent chapters.
- The molecular and evolutionary basis of viral life history traits is of interest in its own right.

PART II

POPULATION AND EVOLUTIONARY DYNAMICS OF VIRUSES AND THEIR MICROBIAL HOSTS

CHAPTER THREE

Population Dynamics of Viruses and Microbes

3.1 ON MEASUREMENTS AND MODELS

In the mid-late 1970s, Bruce Levin and colleagues at Emory University embarked on a program to examine microbial and viral population dynamics. Their approach: to combine strains of bacteria and viruses in a flask, chemostat, or other vessel and to record the change in number of each population with time. Example experimental variables included resource input, strain type, and/or turnover time.

Their program appears straightforward in retrospect, but it is important to recognize three key innovative aspects: (i) the decision to measure the *dynamics* of the populations by repeated sampling over time; (ii) the integration of *mathematical models* with experimental work; (iii) the choice to study *multiple* interacting populations, rather than a single population. The seminal publications in the early stage of this program were those by Levin et al. (1977) and Chao et al. (1977). In contrast, microbiological studies at that time tended to measure the relative change in a population from the start to the finish of an experiment. Even today, many studies focus on a single population arising from a clonal isolate without the use of any mathematical models or theory whatsoever. The approaches could not have been more distinct.

The approach of biologists Bruce Levin and Lin Chao, in collaboration with mathematician Frank Stewart, built on general principles in ecological theory and specific models for population dynamics of viruses and hosts. The first modern model of virus-host population dynamics in which the hosts were themselves microorganisms was developed by Alan Campbell (1961). Campbell's model, like the work of Levin and colleagues, sought to describe the change in population densities arising from interactions among virulent phage and their bacterial hosts. The basic mechanisms, as laid out in Chapter 2, included the reproduction of host cells, the infection and lysis of host cells by viruses, and the decay of viruses. Campbell, like Levin and colleagues, incorporated the use of explicit delays between infection and lysis. Indeed, the system was more specific than many of the "general" models available at the time for the study of predator-prey interactions. The specificity helped reveal not only how populations might interact in principle but also how they interact in practice.

The Levin "school" continues to work at the interface of experiments and models. The approach was not only embraced at Emory but shared by many, particularly those involved in a new Gordon Research Conference in Microbial Population Biology that held its first meeting in 1985, with Bruce Levin as chair. Rather than recapitulating the entire history from the late 1970s to the present, I will illustrate through a series of models some of the *principles* introduced by Levin and colleagues, and subsequent authors, and will describe a few new twists as well. What Levin and colleagues found in those early papers, elaborated on by many since, is that viruses can rapidly decrease host populations—driving them well below densities found in the absence of viruses—and that host and viral populations can oscillate. They also found that population dynamics does not tell the whole story, but I'm getting ahead of myself—one story at a time. For now, I'll discuss just population dynamics, and in the next chapter, I'll continue with the rest of this story: what happens when populations evolve.

3.2   VIRUSES AND THE "CONTROL" OF MICROBIAL POPULATIONS

3.2.1   ECOLOGY AND CONTROL

How do interactions among organisms and between organisms and nutrients unfold in a complex environment? The study of ecology is preoccupied with this question. Unsurprisingly, there is no single answer, even if certain mechanisms do recur: top-down and bottom-up control. The difference between these two mechanisms can be illustrated with an example. Consider a laboratory flask containing a population of microorganisms and growth medium and from which predators and other pathogens of this microbial population have been excluded. One would expect the density of microbes to be limited by available nutrients in the medium. This is termed *bottom-up* control. Similarly, consider the scenario in which the resource is replete with nutrients, and a predator of the microbe is added. In this case, the density of microbes might be limited by the predators rather than by the available nutrients. This is termed *top-down* control. These paradigms provide useful limit cases for interpreting potential outcomes, even if the reality is that many systems exhibit responses that lie on the continuum between these two limits (Carpenter et al. 1985; Hunter and Price 1992; Pace et al. 1999).

How, then, do viruses affect microbes and microbial communities? The obvious answer is that they negatively affect individual microbial cells and, by extension, decrease the total number of microbes in an environment. Figure 3.1 includes the results of multiple experimental manipulations of host organisms with and without viruses. In each case, the density of hosts drops markedly with the addition of viruses. It would seem that viruses, like predators, have the potential to control a microbial population from the "top." Evaluating this potential requires consideration of the drivers, that is, the conditions that can lead to virus drawdown of host populations. Moreover, early dynamics do not always tell the whole story. The next few sections will introduce models that provide a conceptual and mathematical framework for understanding how viral infections change the dynamics of microbial populations and nutrients— potentially switching control from bottom up to top down.

3.2.2 MODELS OF VIRUS-HOST INTERACTIONS WITH EXPLICIT RESOURCES

The original experiments of Levin and colleagues (Levin et al. 1977; Chao et al. 1977) examined the effect of viruses on microbial population dynamics and concomitant changes in viral population dynamics. These interactions took place in a *chemostat*, a continuous-flow reactor in which fresh media with resource (nutrient) concentration J_0 is continually introduced at a rate ω (Figure 3.2). To maintain a constant volume, an equivalent amount of material is removed from the chemostat at the same rate. This material is a subsample of the chemostat contents and so can contain a mixture of resources, hosts, and viruses. The chemostat is usually stirred or otherwise mixed and can include sampling ports for evaluating the current state of the contents or for adding new populations or other media.

Inside the chemostat vessel, there are direct interactions between viruses (V) and hosts (N), direct interactions between hosts and the growth media (R), and likely many indirect interactions as a result of feedback processes. How much of the detail of what is occurring inside the chemostat vessel should be considered in the model? This is a recurring question that arises when biologists and theorists gather, and matters turn to a research problem of common interest. Sometimes, the biologist is the one asking, usually when the theorist has planned to omit some seemingly crucial element from the model. Sometimes, the theorist is the one asking, usually when the biologist has insisted on a more "realistic" representation. The tension is real, and there

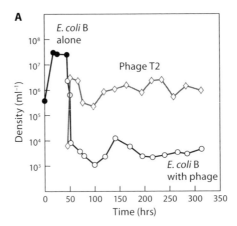

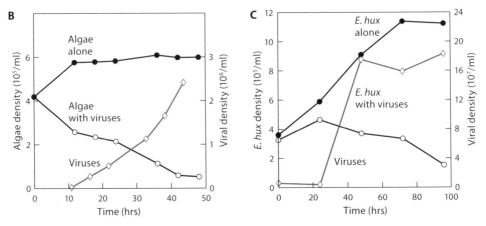

FIGURE 3.1. Viral drawdown of host populations. (A) Population dynamics of bacteriophage T2 infecting *E. coli* B in a chemostat. Phage were added at approximately 48 hr after inoculation of the chemostat with bacteria alone. (B) Population dynamics of viruses infecting the algae *Phaeocystis pouchetii*, in culture. Dynamics correspond to a control (without viruses) and a treatment (with viruses). The algae and viruses were isolated from Norwegian coastal waters. (C) Population dynamics of viruses Ehv86 infecting the algal host *Emiliania huxleyi* CCMP 374 (or Ehux374). Dynamics correspond to a control (without viruses) and a treatment (with viruses). *E. huxleyi* is a ubiquitous bloom-forming algae in marine surface waters. In all panels, the solid curve with filled circles denotes the density of hosts with no viruses. The solid curve with open circles denotes the density of hosts with viruses. Correspondingly, the solid curve with gray diamonds denotes the density of free viruses. Original data in Levin et al. (1977) (A), Jacobsen et al. (1996) (B), and Fulton et al. (2014) (C) estimated using PlotDigitizer 2.6.3.

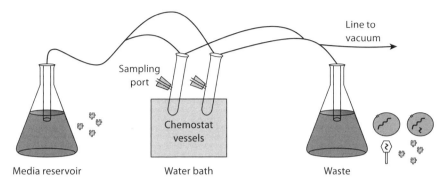

FIGURE 3.2. Schematic of a chemostat, adapted from the design of Bruce Levin and Omar Cornejo.

is no one correct answer. Here, the guiding credo is to "make everything as simple as possible, but not simpler". [1] The choice of a chemostat helps in one significant respect. Every entity in the chemostat is assumed to sample the same average environment. As a consequence, the representation of dynamics here begins by considering the change in the average concentrations $R(t)$, $N(t)$, and $V(t)$ over time rather than by describing the stochastic dynamics in space and time (see Schrag and Mittler (1996) for an extended treatment of spatial refuges that emerge in chemostat experiments).

The population dynamics of hosts, viruses, and resources can be modeled as a coupled system of ordinary differential equations. In such a system, the average population density and the resource concentration are treated explicitly (see Figure 3.3). Appendix B introduces the basic methodology for deriving so-called mean-field dynamics leading to coupled ordinary differntial equations (ODEs) from a set of Poisson processes. The present analysis will not include an explicit treatment of the infected state after infection and before lysis. This delay will be included and analyzed later in this chapter. Given these assumptions, the population dynamics can be written as

$$\frac{dR}{dt} = \overbrace{\omega J_0}^{\text{media inflow}} - \overbrace{f(R)N}^{\text{nutrient consumption}} - \overbrace{\omega R}^{\text{outflow}},$$

$$\frac{dN}{dt} = \overbrace{\epsilon f(R)N}^{\text{cell division}} - \overbrace{\phi N V}^{\text{infection and lysis}} - \overbrace{\omega N}^{\text{outflow}}, \qquad (3.1)$$

$$\frac{dV}{dt} = \overbrace{\beta \phi N V}^{\text{lysis}} - \overbrace{\phi N V}^{\text{infection}} - \overbrace{\omega V}^{\text{outflow}},$$

[1] This quotation is attributed to Albert Einstein, but its origins remain uncertain.

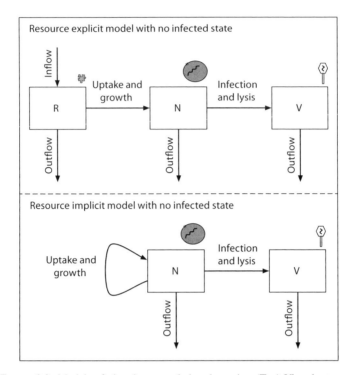

FIGURE 3.3. Models of virus-host population dynamics. (Top) Virus-host population dynamics with explicit resources and no infected state. (Bottom) Implicit resources and no infected state. The arrows denote the flow of material, whether nutrients, cells, or viruses.

where $f(R)$ denotes the uptake of resources by microbes (units of μg/(cells $\times$ hr), ϵ is the conversion factor between resources taken up and new cells produced (units of cells/μg), ϕ is the adsorption rate (units of ml/(cells $\times$ hr), and β is the burst size. The viral-associated life history traits were introduced in Chapter 2. The host-associated life history traits are implicitly defined in terms of the choice of the functional uptake response, $f(R)$. In modeling of microbial population dynamics, two choices are common (Smith and Waltman 1995; Hastings 1997; Case 2000):

$$\text{Type I}: \quad f(R) = cR, \tag{3.2}$$

$$\text{Type II}: \quad f(R) = \gamma \frac{R}{Q+R}. \tag{3.3}$$

The Type I response assumes that nutrient uptake increases with available nutrients with a slope c. The Type II response assumes that nutrient uptake increases with available nutrients when nutrients are at low concentrations and then saturates at a value of γ given high nutrient concentrations. The transition between high and low nutrient concentrations is denoted by the half-saturation constant, Q. These two responses can be made equivalent at low concentrations, by denoting $c \equiv \gamma / Q$. The interested reader should refer to Appendix B.1 for more details on the basis for and values of parameters associated with microbial growth. An alternative model (the Contois model) assumes that specific growth rate also decreases with increasing population size given the same resource availability (Contois 1959). Finally, the physical nature of the chemostat is included in the model by modeling the input of resources at a rate ωJ_0 and modeling the loss of resources, hosts, and viruses at rates ωR, ωN, and ωV, respectively. No explicit bacterial or viral mortality is considered here, because washout is assumed to be the dominant source of loss.

What are the expected dynamics of a resource-host-virus system in a chemostat? The effects of viruses on the population dynamics of this system can be understood by comparing the predicted equilibrium of the system with and without viruses. This comparison is inspired by analyses of consumer-resource dynamics (Murdoch et al. 2003) generally, and trophic cascades (Terborgh and Estes 2010) specifically. *Trophic cascades* are "reciprocal predator-prey effects that alter the abundance, biomass, or productivity of a community across more than one trophic link in the food web" (Denno and Lewis 2009). The term *trophic* implies that interactions occur among different functional classes of individuals, for example, predators and prey, herbivores and plants, or parasites and hosts. The term *cascade* implies that the effects of such interactions have secondary effects beyond the scale of the interaction. Although trophic cascades are often considered in a food web context, similar concepts apply to systems in which parasites infect and kill hosts.

The current treatment in the main text of this and subsequent chapters presumes the reader understands how to identify equilibria of nonlinear dynamical systems and how to analyze their stability (consult Appendix B.2). Here, the three equilibria can be identified by setting the derivatives of the system to 0. The values of the equilibria depend on the choice of functional responses. The derivation here utilizes the Type I functional response; however the switch from bottom-up to top-down control can be illustrated using either response (see the appendix of Weitz et al. (2005)). The equilibria for the system in Eq. 3.1 given the Type I response $f(R) = cR$ are

$$\text{Host and viral washout} : (J_0, 0, 0), \tag{3.4}$$

$$\text{Viral washout} : \left(\frac{\omega}{\epsilon c}, \epsilon J_0 - \omega/c, 0\right), \tag{3.5}$$

$$\text{Coexistence} : \left(\frac{J_0}{1 + \frac{c}{\tilde{\beta}\phi}}, \frac{\omega}{\tilde{\beta}\phi}, \frac{\epsilon c R^* - \omega}{\phi}\right), \tag{3.6}$$

where each set of values is in the form (R^*, N^*, V^*). In these solutions, $\tilde{\beta} = \beta - 1$ is the net viral production for successful lysis events given a burst size of β. These equilibrium solutions may not appear illuminating at first, but they do contain insights into what viruses can do to host populations and when such effects are most likely to be observed. Henceforth, the viral washout and coexistence equilibrium are denoted by the subscripts v and $\bar{v}$, respectively. R_v^* and $R_{\bar{v}}^*$ denote the resources at equilibrium with and without viruses, respectively.

What are the conditions that determine whether the input resource density and washout rate of the chemostat are sufficient to sustain host growth rather than host washout? The equilibrium density for hosts given the absence of viruses is $\epsilon J_0 - \omega/c$. Hosts can persist in a chemostat with available resources so long as their equilibrium density is positive, which occurs when $\epsilon J_0 c > \omega$. There are multiple ways to interpret this condition. As is apparent, the maximum growth rate of hosts is $\epsilon c J_0$, and the washout rate of hosts is ω. For hosts to grow, they must divide faster than they are washed out. This result can be derived a different way.

Observe that a host organism will remain in the chemostat for a time period of $1/\omega$, on average. If a small number of host cells was inoculated in an otherwise host-free chemostat whose resource density was J_0, then each of these host cells would produce $\epsilon c J_0$ number of new individuals per unit time. The average number of new individuals produced by a rare host during its lifetime in the chemostat would be

$$\mathcal{R}_0^{(n)} = \overbrace{\epsilon c J_0}^{\text{rate of host production}} \times \overbrace{\frac{1}{\omega}}^{\text{average time in chemostat}} > 1, \tag{3.7}$$

where $\mathcal{R}_0$ is termed the basic reproductive number, and the superscript $^{(n)}$ denotes it is associated with the host population. So long as a rare host produces more than one individual, then the host population should increase, on average. In this model, such increases of the host population are followed by eventual saturation at its equilibrium density—leading to a stable equilibrium. The same result holds for Type II functional responses, albeit the quantitative condition must be modified to reflect the saturating response (Weitz et al. 2005). As a rule, host growth is likely so long as the growth rate given a resource density of J_0 exceeds the washout rate. Because the washout rate can be adjusted experimentally, this condition can be realized so long as the washout rate is set below the maximum growth rate of the host. One key consequence of host persistence is that the new resource concentration at equilibrium *with* hosts, $\omega/(\epsilon c)$, must be *below* that of the resource concentration at equilibrium *without* hosts, J_0. A second key consequence is that the total number of hosts increases with increases in the resource concentration of the incoming media. In summary, there is a broad regime of flow and resource input conditions for which hosts can persist, drawing down resources below the concentration in the host-free scenario. This broad regime is one reason why chemostats are a useful experimental system for investigating microbial population and evolutionary dynamics.

What are the subsequent conditions under which viruses can coexist with their hosts without driving their hosts and consequently themselves to extinction? The logic is similar to that above, except now applied to the situation in which a rare virus is added to a chemostat containing resources and hosts, both at their equilibrium density. A rare virus comes into contact with hosts at a per-virion rate of $\phi N_{\bar{v}}^*$. The rate of virus removal from the chemostat due to washout is ω. The probability that the virus infects a host before being washed out is

$$\frac{\phi N_{\bar{v}}^*}{\phi N_{\bar{v}}^* + \omega}. \tag{3.8}$$

This fraction leverages the fact that event A occurs before event B with probability $v_A/(v_A + v_B)$, where v_A and v_B are the rates of each event, respectively (see Appendix B.4). Assuming that each viral infection is productive, then the average number of viruses produced by a rare virus is

$$\mathcal{R}_0^{(v)} = \overbrace{\frac{\beta \phi N_{\bar{v}}^*}{\phi N_{\bar{v}}^* + \omega}}^{\text{infection and lysis}} \tag{3.9}$$

The basic reproductive number of viruses, $\mathcal{R}_0^{(v)}$, must exceed 1 for the virus population to grow. The condition can be consolidated as

$$\tilde{\beta} \phi N_{\bar{v}}^* > \omega \tag{3.10}$$

This condition is analogous to that derived for hosts to grow in a chemostat, albeit here the hosts have become the resources, and the viruses have become the consumers! This criterion for viral growth can also be derived by considering the invasion dynamics of the virus population in Eq. 3.1. The per capita growth rate of viruses is $\tilde{\beta}\phi N_{\tilde{v}}^* - \omega$. The growth rate must be positive for viruses to invade a host-controlled chemostat, leading to the same condition as in Eq. 3.10.

How typical is viral growth? Consider a host population growing in a glucose-limited chemostat with washout rate $\omega = 0.1$ hr^{-1} and J_0 exceeding $100\,\mu$g/ml. Such conditions could maintain a host population of density $N_{\tilde{v}}^* \approx 10^8$/ml. The conditions for growth of a small virus population become $\tilde{\beta}\phi 10^9 > 1$. For virus burst sizes of approximately 100, the adsorption rate need only exceed 10^{-11} ml/hr, which is 0.0001 of the diffusion-limited adsorption rate (Chapter 2). Coexistence is likely even if adsorption is far less than the theoretical maximum. In other words, there is a range of parameters for which viruses are predicted to coexist with hosts in a chemostat (Figure 3.4 C). This result holds even if host growth is modeled as a saturating, rather than linear, response, and when other features of virus-host biology are included as well (Levin et al. 1977; Chao et al. 1977; Weitz et al. 2005; Smith and Thieme 2012). However, if adsorption is not high enough, or if the initial host density is too low, then a small virus population will not grow (Figure 3.4 B). In summary, the invasion of a virus population depends on the quantitative values of life history parameters and environmental conditions.

What other effects do viruses have on the community? This question can be addressed by comparing the predicted concentrations of resources and of hosts with and without viruses. A few definitions will be useful here. First, *resource enhancement* is defined as the ratio of resources at steady state with and without viruses. Second, *host drawdown* is defined as the ratio of hosts at steady state with and without viruses. At equilibrium, these two ratios can be derived for the chemostat model:

$$\text{Resource enhancement} = \frac{R_v^*}{R_{\tilde{v}}^*} = \frac{\epsilon c J_0}{\omega\left(1 + \frac{c}{\tilde{\beta}\phi}\right)}, \tag{3.11}$$

and

$$\text{Host drawdown} = \frac{N_v^*}{N_{\tilde{v}}^*} = \frac{\omega}{\tilde{\beta}\phi\,(\epsilon J_0 - \omega/c)}. \tag{3.12}$$

These algebraic forms do not look particularly revealing. Figure 3.5 includes results of 10^6 simulations of this model, with parameters randomly chosen

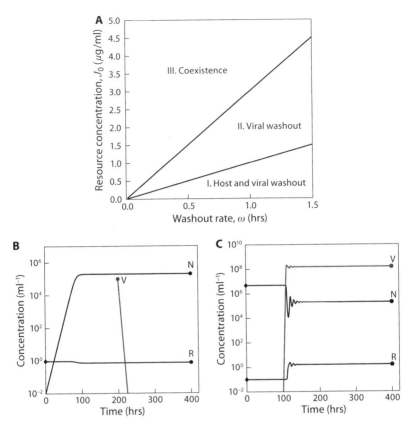

FIGURE 3.4. Qualitative outcomes of virus-microbe population dynamics for Eq. 3.1 given a Type I functional response. (A) Phase diagram of predicted ecological steady states as a function of changes to chemostat control parameters J_0 and ω. Additional parameters are $c = 10^{-6}$ ml/(cells$\times$hrs), $\epsilon = 10^6$ cells/μg, $\beta = 50$, and $\phi = 10^{-8}$ ml/(cells$\times$hrs). (B) Dynamics in phase II: small population of hosts invade, whereas large populations of viruses go extinct. Here, $\omega = 0.75$ and $J_0 = 0.95$. (C) Dynamics in phase III: viruses can invade a persistent population of hosts and resources. Here, $\omega = 0.1$ and $J_0 = 5$.

centered around the life history traits of *E. coli* B and bacteriophage T2. For each simulation, the resource and host levels are compared between the virus-free and the virus-present steady states. Viruses are not always predicted to persist, but whenever they do persist, then resources *always* increase, and hosts *always* decrease. This is the hallmark of a community that has shifted from bottom-up control, which implies that consumers are limited by resource availability, to top-down control, which implies that consumers are limited by predators or parasites. Is this luck? No—at least not in theory.

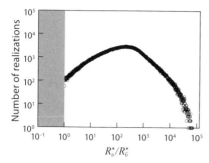 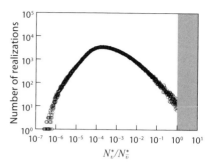

FIGURE 3.5. Ratios of resources and hosts with and without viruses. (Left) Resource enhancement. (Right) Host drawdown. Simulations utilize life history traits centered on those introduced in Chapter 2 and relevant chemostat parameters, $\omega = 0.1$, $c = 10^{-7}$, $J_0 = 10$, $\epsilon = 10^6$, $\phi = 10^{-8}$, $\beta = 50$. Latin Hypercube Sampling was utilized to sample parameter space one order of magnitude above and below these values. Sample results were binned in log-spaced bins.

The change in host density can be evaluated by comparing densities with and without viruses:

$$\text{Host drawdown} = \frac{\omega}{\tilde{\beta}\phi N_{\tilde{v}}^*}. \tag{3.13}$$

The condition for viruses to invade is that $\tilde{\beta}\phi N_{\tilde{v}}^* > \omega$ (Eq. 3.13). Therefore, hosts are drawn down to a density less than that without viruses whenever viruses invade and coexist with hosts. The change in resource concentration can be similarly evaluated:

$$\text{Resource enhancement} = \frac{\epsilon c J_0}{\omega + c N_v^*}. \tag{3.14}$$

The condition for coexistence in Eq. 3.6 implies the equivalence $\epsilon J_0 = \omega/c + N_{\tilde{v}}^*$. Therefore, the resource enhancement can be rewritten as

$$\text{Resource enhancement} = \frac{\omega + c N_{\tilde{v}}^*}{\omega + c N_v^*}. \tag{3.15}$$

As was just shown, $N_{\tilde{v}}^* > N_v^*$, so whenever viruses invade and coexist with hosts, the resource enhancement must be greater than 1. These points confirm the numerical results: viruses have the effect of decreasing host densities (see Figure 3.1) and increasing resource concentrations compared with those in virus-free environment. These two effects are theoretical hallmarks of a system under top-down control.

An additional feature of this system is that the virus density increases but the host density remains the same when more resources are added—a result that may appear counterintuitive. The logic is that the viruses reach a steady state because their washout rate balances their production rate on a per capita basis. Increasing the resource concentration of the inflow, J_0, does not change the per capita viral washout rate, ω. So, virus death remains unchanged, and therefore virus reproduction, which depends on N, must also remain unchanged, on a per capita basis. Host densities do not change if virus reproduction does not change. Adding more resources does increase the total host reproduction rate, and so there is a greater transfer from resources to hosts to viruses. The slope of the increase of virus density with resource increase depends on life history traits, but the qualitative response is generic, at least in this class of models. Deviations from such a trophic transfer relationship suggest other processes must be incorporated into the model structure.

3.2.3   MODELS OF VIRUS-HOST INTERACTIONS WITH IMPLICIT RESOURCES

Figure 3.3 includes an alternative representation of the dynamics of hosts $N(t)$ and viruses $V(t)$ in a chemostat with no explicit treatment of resource dynamics. As before, hosts should grow to a resource-limited density in the absence of viruses, yet now the growth depends on host density rather than resource availability. The transformation of a system from three variables (R, N, V) to two variables (N, V) simplifies analysis, but it can also lead to surprises. The quantitative details of dynamics can change even if the projected equilibrium states remain the same. Nonetheless, the viral effects at equilibrium in this particular reduced model can be derived from those in the explicit resource model in certain limits (see Appendix B.3). This implicit resource model is already widely used and will be the basis for further extensions considering additional complexity later in this monograph.

The population dynamics of N and V given implicit resources can be written as

$$\frac{dN}{dt} = \overbrace{rN(1 - N/K)}^{\text{logistic growth}} - \phi NV - \omega N,$$

$$\frac{dV}{dt} = \beta \phi NV - \phi NV - \omega V. \tag{3.16}$$

Here, r denotes the maximum growth rate, and K denotes the maximum population density. All other parameters have the same meaning as in the model with explicit resource dynamics. This model has three equilibria, written in the form (N^*, V^*):

$$\text{Host and viral washout}: \ (0, 0), \tag{3.17}$$

$$\text{Viral washout}: \ (K(1 - \omega/r), 0), \tag{3.18}$$

$$\text{Coexistence}: \ \left(\frac{\omega}{\tilde{\beta}\phi}, \frac{r}{\phi} \left(1 - \frac{\omega}{\tilde{\beta}\phi K} \right) - \frac{\omega}{\phi} \right), \tag{3.19}$$

where $\tilde{\beta} = \beta - 1$. This model presumes that the two dominant causes of host death are viral lysis and washout. As before, the same three outcomes are possible. In the absence of viruses, the condition that allows hosts to grow is that their maximum growth rate, exceeds the washout rate; that is, $r > \omega$.

What determines whether viruses can coexist with hosts? Consider the addition of a small population of viruses into an environment where there were only hosts, at their equilibrium. Will the viruses increase or decrease in number? The invasion of this viral population requires that the population increase from small to large density. On an individual virus particle basis, each virus can either decay or infect and lyse a host cell. The per-viral decay rate is ω. Initially, the per-viral infection and lysis rate is $\phi N(0) = \phi K(1 - \omega/r)$. As before, the average number of viruses produced by a rare virus is

$$\mathcal{R}_0^{(v)} = \beta \times \overbrace{\frac{\phi N_{\tilde{v}}^*}{\phi N_{\tilde{v}}^* + \omega}}^{\text{probability of lysis}}. \tag{3.20}$$

If $\mathcal{R}_0^{(v)} > 1$, then the virus population will increase. The condition for virus invasion is

$$\frac{\tilde{\beta}\phi N_{\tilde{v}}^*}{\omega} > 1, \tag{3.21}$$

which means that the initial multiplication rate of viruses must exceed the washout rate for viruses to increase in number. This is the same condition as was derived in the resource-explicit model. Increases in burst size, infection rate, and available hosts all promote viral invasion and subsequent coexistence. Increases in viral decay rate diminish the chance of viral invasion and subsequent coexistence. Formally, the same result can be derived by applying standard tools of local stability analysis (see Appendix B.2).

The implicit model of host-viral interactions retains the two key features of a top-down system, as explained previously. First, host densities always decrease whenever viruses invade and persist. The host drawdown level in this resource-implicit model is

$$\text{Host drawdown} = \frac{\omega}{\bar{\beta}\phi K(1 - \omega/r)}, \tag{3.22}$$

which must be less than 1—it is the inverse of the requirement for viruses to invade. Second, the growth rate increases with increases in the input resource concentration. In such a scenario, the host steady state will not change, whereas the viral steady-state density will increase. In summary, the resource-implicit and the resource-explicit model of virus-host population dynamics share many of the same core predictions.

Finally, it is worth offering another interpretation of the host density when controlled by viruses, $N_v^* = \frac{\omega}{\bar{\beta}\phi}$. At equilibrium, only one of the β viruses released from an infected cell can survive, on average, to infect another cell, releasing, yet again, β viruses. If more survived, then the population of viruses would increase over time. If fewer survived, then the population of viruses would decrease over time. The probability that the virus infects a host cell before being washed out is $\phi N_v^*/(\omega + \phi N_v^*)$. Equating these two probabilities yields

$$\frac{1}{\beta} = \frac{\phi N_v^*}{\omega + \phi N_v^*}. \tag{3.23}$$

Solving this equation for N_v^* yields the same result as before. The interpretation of this steady-state density is that each free virus has a $1/\beta$ chance of infecting a cell before dying at only this density. This result implies that variation in burst size may also destabilize population dynamics. Similar conclusions can be drawn from temporal variation in other life history traits.

3.2.4   VIRUSES WHOSE GENOME IS INHERITED DURING HOST DIVISION

The modification of host population dynamics due to viral infection is apparent when viruses must lyse their host cells to propagate. However, as was discussed in the prior chapter, many viruses can integrate their genome with that of their host's. This process is termed *lysogeny*. Other viruses can inject their genome into the host, and the genome, can persist stably during host cell divisions. This process is termed *pseudolysogeny*. The stable persistence of a virus genome

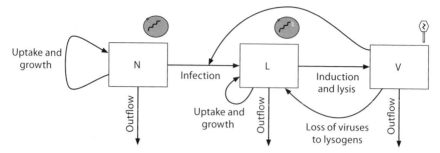

FIGURE 3.6. Model of virus-host population dynamics with implicit resources, where initial virus infection leads to lysogeny. The arrows denote the flow of material, whether cells or particles.

inside a host cell can lead to distinct cellular-level effects, such as (i) changes in the physiological state of the host, including modification of growth rate, or (ii) changes in the susceptibility of the host to subsequent infection. In such cases, viruses also modify host populations.

One model of virus-host interactions with implicit resources can be written given an assumption that the virus always integrates itself into the host genome, forming a lysogen. Here, the population density of lysogens is noted as L, and the dynamics are

$$\frac{dN}{dt} = \overbrace{rN\left(1 - \frac{N+L}{K}\right)}^{\text{density-limited growth}} -\phi NV - \omega N,$$

$$\frac{dL}{dt} = \overbrace{r_L L\left(1 - \frac{N+L}{K}\right)}^{\text{density-limited growth}} + \overbrace{\phi NV}^{\text{establishment of lysogeny}} - \overbrace{iL}^{\text{induction}} -\omega L, \quad (3.24)$$

$$\frac{dV}{dt} = \overbrace{\beta iL}^{\text{lysis}} -\phi NV - \overbrace{\chi \phi LV}^{\text{infection of lysogens}} -\omega V.$$

The rationale for this model is that both susceptible hosts and lysogens compete for resources during growth (see Figure 3.6). The uninfected host cell divides at a maximum rate r, and the lysogen divides at a maximum rate r_L. There is often an added transcriptional and translational burden associated with lysogeny, such that $r_L < r$. However, viruses also can carry genes that provide benefits to hosts, whether toxin genes or antibiotic resistance cassettes (Gandon and Vale 2014). Then, infected lysogens induce at a rate i per unit time, with each induction leading to lysis and release of β viruses.

Finally, viruses may infect lysogens, but there is evidence that lysogens are often resistant to infection by viruses of the same or related strains. This phenomenon is termed *superinfection exclusion*, *superimmunity* or even *homoimmunity* (see references in Labrie et al. (2010)). Translation of this concept to the model means that when superimmunity is surface mediated, then viruses do not adsorb to lysogens ($\chi = 0$), and when superimmunity is intracellularly mediated, then viruses infect but are subsequently cleared ($\chi = 1$). Intermediate regimes are also possible. More generally, the model in Eq. 3.24 has similarities to models of intraguild predation (Polis et al. 1989), in that both L and N compete for resources, but L acts to "predate" upon N via the infection of susceptible hosts by viruses released by lysogens.

Excluding the trivial steady state in which all populations go extinct, the steady states for this model include (1) a host-only steady state in which $N^* = K(1 - \omega/r)$ and $L^* = 0 = V^*$; (2) a lysogen-virus-only steady state in which $N^* = 0$, $L^* = K(1 - (\omega + i)/r_L)$ and $V^* = \beta i L^*/(\omega + \chi \phi L^*)$; (3) a coexistence steady state. Note that the closed-form equations for the coexistence steady state are not particularly illuminating. Nonetheless, it is possible to utilize some of their features to interpret the effects of lysogeny. As before, the establishment of hosts in the chemostat requires that $r > \omega$. But, what happens if lysogens and viruses grow and persist?

There are two possible outcomes. The first is that uninfected hosts are washed out from the chemostat permanently, owing to competitive exclusion by lysogens in the presence of viruses. Are there more or fewer lysogens in such an scenario? The ratio of lysogens to uninfected hosts between steady states (2) and (1) is

$$\frac{L_2^*}{N_1^*} = \frac{1 - \frac{\omega + i}{r_L}}{1 - \frac{\omega}{r}},$$

such that hosts (whether infected with a prophage or not) decrease in abundance so long as $\omega/(\omega + i) < r/r_L$. For $r_L < r$, this is always true. And when $r_L > r$ (i.e., viruses provide a benefit to hosts), it is still true so long as $i > \omega(r_L/r - 1)$; that is, induction is more costly than is the net increase in lysogen growth rate relative to uninfected host cells. When viruses lead to the local extinction of uninfected hosts in the chemostat, a broad set of conditions determines that there will be fewer cells than in the uninfected steady state.

The second possibility is that hosts, lysogens, and viruses coexist. In this case, the steady state satisfies the condition

$$r\left(1 - \frac{N^* + L^*}{K}\right) = \omega + \phi V^*, \tag{3.25}$$

such that that the total density of cells is equal to

$$N^* + L^* = K \left(1 - \frac{\omega}{r} - \frac{\phi V^*}{r} \right). \qquad (3.26)$$

This condition is derived by setting the first dynamical equation of Eq. 3.24 to 0. The total host cell density is necessarily less than the uninfected host cell density of $K(1 - \omega/r)$ in the absence of viruses. In summary, if viruses persist in the chemostat, then the total host cell density is less than that in the virus-free state. At least for this class of models, temperate viruses—like viruses that only lyse their host cell—are predicted to reduce host population densities below those of the virus-free state.

3.3 VIRUSES AND OSCILLATORY DYNAMICS

The interactions between viruses and their microbial hosts need not lead to a static equilibrium. Oscillations have been observed since the inception of quantitative studies of virus-host interactions (see Figure 5A of Levin et al. (1977)). Although the two sets of models described in the previous section reveal certain features of viral effects on a single microbial population, they do not necessarily predict others, including long-term oscillatory dynamics. Rather, the models predict damped oscillations leading to an equilibrium, as show in Figure 3.4C. What is the basis for these oscillations?

It is worthwhile to revisit the coexistence dynamics (see Figure 3.4C) by plotting the dynamics in the $N - V$ phase plane (see Figure 3.7). The arrows indicate the direction of dynamics with time; that is, the dynamics are damped counterclockwise cycles. These cycles imply that the host population peaks before the virus population does. Likewise, the subsequent host population minimum is followed by the viral population minimum, at which point the cycle repeats. Counterclockwise cycles are ubiquitous in studies of predator-prey dynamics. For example, they were identified in seminal studies of lynx-hare and mink-muskrat oscillations (Elton and Nicholson 1942). The rationale for the commonness of these cycles is that the predator abundance increases in response to abundant prey and subsequently drives down the prey population. With diminishing prey, the predator population decreases, and with fewer predators, the prey recovers. At that point the cycle repeats (Hastings 1997).

In classic predator-prey theory, the functional response of the predator determines whether a particular predator-prey system exhibits persistent, rather than transient, oscillations (Rosenzweig and MacArthur 1963). Here, the functional response of the virus is modeled as a "Type I" response; that is,

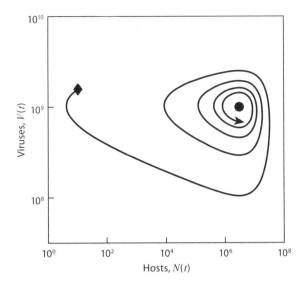

FIGURE 3.7. Counterclockwise phase-plane dynamics for virus-host interactions. Parameters are $\omega = 0.1$, $c = 10^{-8}$, $J_0 = 1$, $\epsilon = 10^8$, $\phi = 6.7 \times 10^{-10}$, and $\beta = 50$. Time begins at the location marked by the diamond and moves toward the asymptotic fixed point marked by the filled circle in the center of the counterclockwise spiral. Densities are per milliliter.

viral infection and lysis depends linearly on the density of hosts, $\sim \phi N$. Predator-prey systems with Type I functional responses tend not to exhibit persistent oscillations. If consumption rates saturate with increasing resource density, then other dynamics are possible. For example, the same system can exhibit oscillatory dynamics given a consumption rate with a Type II response proportional to $N/(C + N)$ (Rosenzweig and MacArthur 1963).

A system might oscillate for many other reasons besides oscillatory changes in environmental conditions. These changes include the presence of delays between the initiation of an event and its termination. Viruses lyse their host after a latent period, which varies with the host, virus, and environmental conditions. A working hypothesis, introduced in Chapter 2, is that the latent period should be similar to the doubling time of the host, because viral production requires host metabolism. The doubling time at the time of infection provides a strong indicator of metabolic throughput. The latent periods of T4 in mid-log *E. coli* can be as short as 20 minutes, whereas the latent period of cyanoviruses infecting nutrient-limited *Prochlorococcus* typically extend from 8 to 12 hours. The next sections investigate the consequences of incorporating the latent period in model dynamics.

3.3.1   MODELS OF VIRUS-HOST INTERACTIONS WITH AN INFECTED CLASS

Consider the model of virus-host dynamics with implicit resources in which infected hosts are treated explicitly (see Eq. 3.16). The latent period may be tightly regulated around a fixed value, or it may exhibit substantial variation. The model in this section assumes that infected individuals are lysed at a rate η; that is, the length of an infection is exponentially distributed with a mean infection period $1/\eta$:

$$\frac{dN}{dt} = rN\left(1 - \frac{N+I}{K}\right) - \phi NV - \omega N,$$

$$\frac{dI}{dt} = \overbrace{\phi NV}^{\text{establishment of infection}} - \overbrace{\eta I}^{\text{lysis}} - \omega I, \qquad (3.27)$$

$$\frac{dV}{dt} = \overbrace{\beta \eta I}^{\text{lysis}} - \phi NV - \omega V.$$

In this model, infected hosts also take up resources, competing with uninfected hosts. This is the rationale behind including the negative density dependence of uninfected and infected hosts on uninfected host growth. What are the possible outcomes within such a dynamical system?

It is worthwhile to consider two cases at the extreme limits: $\eta \to \infty$ and $\eta \to 0$. The first case corresponds to rapid lysis. One would expect the predicted dynamics to converge to that of the system with no infected state. To understand why, it is worth noting that the possible equilibrium points of the model in the $N - V$ plane must also satisfy $dI/dt = 0$. There is no change in the density of infected hosts so long as $I^* = \phi NV/(\eta + \omega)$. Given very rapid lysis, it is possible to approximate the dynamical system in Eq. 3.27 with only two variables, $N(t)$ and $V(t)$, by assuming that $I(t)$ rapidly converges to I^*. The formalism (and appropriateness) of such approximations is treated in the study of "fast-slow" dynamical systems see ecological examples in Cortez and Ellner (2010). Given such an approximation, the dynamics can be written as

$$\frac{dN}{dt} = rN\left(1 - \frac{N+I^*}{K}\right) - \phi NV - \omega N, \qquad (3.28)$$

$$\frac{dV}{dt} = \beta \eta I^* - \phi NV - \omega V. \qquad (3.29)$$

Replacement of I^* yields

$$\frac{dN}{dt} = rN \left(1 - \frac{N + \phi NV/(\eta + \omega)}{K} \right) - \phi NV - \omega N, \qquad (3.30)$$

$$\frac{dV}{dt} = \beta \phi NV \frac{\eta}{\eta + \omega} - \phi NV - \omega V. \qquad (3.31)$$

In the limit that $\eta \to \infty$, and removing all terms of order $1/\eta$, we obtain

$$\frac{dN}{dt} = rN \left(1 - \frac{N}{K} \right) - \phi NV - \omega N, \qquad (3.32)$$

$$\frac{dV}{dt} = \tilde{\beta} \phi NV - \omega V. \qquad (3.33)$$

This system is precisely that of Eq. 3.16. It should be expected that the dynamics of a model with very rapid lysis after infection will, indeed, approach that of a model without any explicit consideration of the infected state.

The second case corresponds to very long latent periods; that is, $\eta \to 0$. In this limit there is the danger, to the virus at least, that the latent period is so long that the average number of viruses produced falls below replacement levels. Recall that infected cells are also washed out of a chemostat. If the postinfection lysis rate is η and the washout rate is ω, then only a fraction, $\eta/(\eta + \omega)$, of infected cells actually lyse on average. These lysed cells produce β virions. Viruses should persist with hosts only so long as the average number of viruses produced by one virus in an otherwise virus-free population multiplied by the probability of infection exceeds 1. Is this, in fact, the case?

Consider the equilibrium solutions to Eq. 3.27, which can be solved algebraically to yield

$$N^* = \frac{\omega(\eta + \omega)}{\phi \left(\beta \eta - (\eta + \omega) \right)},$$

$$V^* = \frac{r \left(1 - N^*/K \right) - \omega}{\phi \left[1 + \frac{rN^*}{K(\eta + \omega)} \right]}, \qquad (3.34)$$

$$I^* = \frac{\phi N^* V^*}{\eta + \omega}.$$

The stability of this steady state to local perturbations can be determined based on the sign of the eigenvalues associated with the linearized

N–I–V system near this equilibrium (see Appendix B.2.3 for more details). From Appendix B.2, the virus-free equilibrium is predicted to be unstable when

$$\phi \tilde{K} \beta \eta > (\eta + \omega)(\phi \tilde{K} + \omega), \tag{3.35}$$

where $\tilde{K} = K(1 - \omega/r)$. This condition can alternatively be written as

$$\beta \underbrace{\frac{\eta}{\eta + \omega}}_{\text{viruses per infected host}} \times \underbrace{\frac{\phi \tilde{K}}{\phi \tilde{K} + \omega}}_{\text{probability of infection}} > 1, \tag{3.36}$$

precisely the condition posited earlier. The success of viruses depends on how well they exploit their hosts as well as on how well they survive extracellularly. Both of these variables contribute to the likelihood that a virus will invade and coexist with a host.

Recall that the mean latent period in this model is $1/\eta$. The condition for virus invasion can be rewritten as

$$\eta > \frac{\omega(\phi \tilde{K} + \omega)}{\phi \tilde{K} \tilde{\beta} - \omega}. \tag{3.37}$$

Thus, there should be a critical lysis rate, η_c, below which viruses will be unable to persist with hosts. Similarly, at very large values of η, the model reduces to the case in which there is no infected state, and so viruses and hosts should coexist. For intermediate lysis rates, something happens. That something is that all three populations—uninfected hosts, infected hosts, and viruses—oscillate. The oscillation is the steady state of the model for intermediate values of the lysis rate (see Figures 3.8 and 3.9). The peaks and valleys of the oscillations follow an expected ordering, just as predator-prey oscillations do. Consider a relatively small susceptible host population and a large virus and infected host population. Given the lack of hosts, new viruses will be produced but then will be washed out. As viruses and infected hosts decrease in abundance, susceptible hosts increase. This increase leads to new infections, which subsequently release viruses, leading to an increase in the virus population and the decline of the host population and, subsequently, that of the infected hosts—at which point the cycle continues.

It is evident that oscillations are possible for a wide range of η values. As was explained in Chapter 2, on life history traits, the latent period of a virus is often closely related to host physiology. Consequently, oscillations of

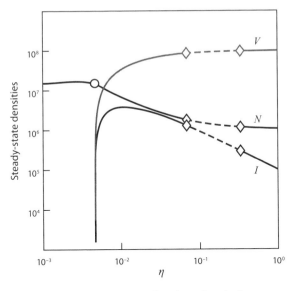

FIGURE 3.8. Steady-state densities as a function of η. At low $\eta < \eta_c$, viruses cannot persist, so that $N^* = \tilde{K}$. After a transcritical bifurcation (open circle), viruses and hosts coexist (with an infected host class). Later, the system undergoes a series of Hopf bifurcations (open diamonds). The steady-state dynamics are characterized by oscillatory dynamics for values of η between these critical values. For large values of η, the predicted steady states of N and V approach that of the model without an infected state. Life history parameters are $r = 0.16$, $K = 2.2 * 10^7, \phi = 10^{-9}, \beta = 50$, and $\omega = 0.05$. Note that the fixed point $N^* = \tilde{K}$ exists for $\eta > \eta_c$ but is unstable with respect to small perturbations of either I or V and is not shown.

biological interest might be expected to occur with a time scale set by $\eta \approx r$. In the preceding example, $r \approx 0.16/\text{hrs}$. But the oscillations that occur in this regime are characterized by periods *much greater* than the growth period and the latent period (see Figure 3.9).

Applied mathematicians Eduardo Beretta and Yang Kuang developed a theory of virus-host oscillations in 1980. Their theory, designed to consider oscillations in marine bacteria, is based on methods in dynamical systems for predicting when an otherwise stable equilibrium will become unstable, leading to oscillatory dynamics (Beretta and Kuang 1998). Such transitions are often mediated by small changes in parameters that lead to so-called Hopf bifurcations (Strogatz 1994). Here, the parameter of relevance is the average latent period, $1/\eta$. Technically, a Hopf bifurcation can be identified

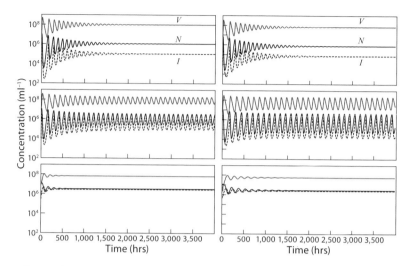

FIGURE 3.9. Viruses and their effect on population stability in models with continuous release (left) and discrete delays (right). (Left) Dynamics of N, I, and V given three postinfection continuous lysis rates, $\eta = 0.025$, 0.25, and 1 (bottom to top), with all other parameters as in Figure 3.8. (Right) Dynamics of N, I, and V given three postinfection delay periods, $\tau = 21.97$, 3.65, and 0.98 (bottom to top), with all other parameters as in Figure 3.8.

by finding the value of a critical parameter such that the eigenvalues with the largest real component are strictly imaginary and come in conjugate pairs. The magnitude of the imaginary component of these eigenvalues corresponds to the frequency of oscillations for parameters very close to the bifurcation point (see Appendix B.2). The condition for identifying such bifurcations is known as the *Routh-Hurwitz criterion* (Strogatz 1994). The conditions for bifurations are themselves generally not illuminating; however, they do help in calculating the period of oscillations. Using the Routh-Hurwitz criterion, Beretta and Kuang's prediction (eq. 55 of Beretta and Kuang (1998)) can be adapted to the present model, in which all populations are subject to a washout rate ω. In this case, the predicted oscillation period is

$$T = 2\pi \sqrt{\frac{\tilde{\eta} + \tilde{r} + (\tilde{m} + \tilde{\phi}\tilde{K})\zeta_0}{\tilde{m}\tilde{r}\tilde{\eta}(1 - \zeta_0)}}, \qquad (3.38)$$

where $\tilde{r} = r - \omega$, $\tilde{\eta} = \eta + \omega$, $\tilde{K} = K(r - \omega)/r$, $\tilde{m} = \omega$, and $\zeta_0 \approx 0.16$ is a fixed constant. Further assuming that $\eta \approx r$ and given that chemostat systems generally have washout rates that dominate viral mortality, the oscillation

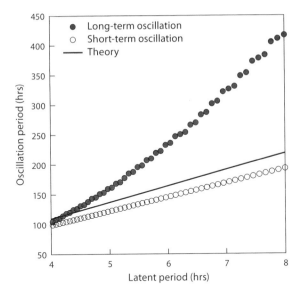

FIGURE 3.10. Oscillation period of fluctuating dynamics of $N(t)$, $I(t)$, and $V(t)$ as a function of average latent period, $1/\eta$. Short-term oscillations (open circles) were calculated from periods of small limit cycles associated with the unstable fixed point. Long-term oscillations (closed circles) were estimated by calculating the discrete Fourier transform over a period of 5000 to 2500 hours after a transient period of the same length. Then, the oscillation period was estimated by finding the frequency at which the power spectral density was maximal. In all cases, parameters were $\beta = 50$, $\phi = 10^{-9}$, $K = 2.24 \times 10^7$, and the following parameters varied along with latent period: $r = 0.63\eta$ and $\omega = 0.2\eta$. *Key point: The oscillation period induced by a delay can be much longer than the delay period; for example, oscillations range from 100 to 400 hours, more than 25 times longer than the latent period.*

period can be approximated as

$$T_{\text{chemostat}} \approx 2\pi \sqrt{\frac{2}{0.85r\omega}}. \tag{3.39}$$

Because $\omega < r$, it is apparent that the oscillation period will be greater than that characterized by the growth rate.

This phenomenon can be observed in Figure 3.10, in which the oscillation period is measured from the limit cycle arising from N–I–V dynamics given idealized host-phage systems in which $\eta \approx 1.6r$ and $\omega \approx 0.2r$; that is, latent periods are shorter than division times, but washout is significantly slower than both. In these simulations, the values of β and ϕ are held fixed at

50 and 10^{-9} mls/(cell $\times$ hrs), respectively. Then, the period of oscillations near the fixed point is estimated by finding the magnitude of the imaginary part of the pair of conjugate eigenvalues associated with the unstable fixed point. If this magnitude is λ_i, then the predicted period for small oscillations should be $T \approx 2\pi/\lambda_i$. These values can be compared against the theoretical predictions of Eq. 3.39 and are found to slightly overestimate the actual periods over this range by 11%-13%. A deviation is expected: predicted oscillation periods characterize the limit cycle only in the limit that the system parameters approach the Hopf bifurcation. In practice, the long-term oscillations need not have the same period as oscillations that move dynamics away from a fixed point (Figure 3.10). Nonetheless, the following prediction unites all these measurements: the realized oscillation is much longer than the latent period and increases with latent period. The same trend holds even when the latent period has a single value rather than an exponential distribution of values, as will be investigated in the next section.

3.3.2 MODELS OF VIRUS-HOST INTERACTIONS WITH AN EXPLICIT LATENT PERIOD

The prior model illustrates how delays between infection and lysis can, in certain parameter regimes, enable viruses to destabilize microbial population dynamics. Does the same mechanism apply if the delays are treated as finite and discrete rather than as continuous and exponentially distributed? There are a number of biological reasons why finite delays should be expected. It takes a minimum amount of time for new virus particles to be produced—this is the *eclipse period*. Then, even after the first intact virus particles appear in the cell, there may be additional delays before the cell is lysed, during production of viruses in the rise period (Chapter 2).

It would seem that modeling virus-host interactions with discrete delays is simpler and closer to the underlying biology. In fact, such models present a new class of problems. Here is one way of writing such a model:

$$\frac{dN}{dt} = rN\left(1 - \frac{N+I}{K}\right) - \phi NV - \omega N,$$

$$\frac{dI}{dt} = \phi NV - \overbrace{\phi N_\tau V_\tau e^{-\omega\tau}}^{\text{delayed lysis}} - \omega I, \tag{3.40}$$

$$\frac{dV}{dt} = \overbrace{\beta\phi N_\tau V_\tau e^{-\omega\tau}}^{\text{delayed lysis}} - \phi NV - \omega V.$$

The crucial difference between this and previous models introduced thus far is the use of subscripts τ associated with the state variables N and V. These subscripts imply that the value of that variable must be evaluated at a time prior to the current time; for example, $N_\tau \equiv N(t - \tau)$. The delayed event is lysis, which occurs at a fixed delay τ after infection. Equation 3.40 is also an example of a system of delay differential equations (DDEs). Analysis of DDEs merits monographs of its own (Smith 2011). As explained, not all infected cells end up releasing viruses inside the chemostat. The probability that an infected cell will remain inside the chemostat at a time τ after initial infection, rather than being washed out, is $e^{-\omega\tau}$. This is why the rate of lysis events at time t is equal to the rate of infection events at time $t - \tau$ multiplied by the probability that those infected cells survived in the chemostat for a period of time τ.

One might wonder whether models with continuous and fixed, yet finite, delays will make equivalent predictions. They might, but to make a fair comparison requires a few observations. In the prior model, β was the nominal burst size. This burst occurred only when cells were not washed out—the probability of which was $\eta/(\eta + \omega)$. In the present model, β is, again, the nominal burst size of infected cells. However, $e^{-\omega\tau}$ is the fraction of cells not washed out in the time τ between infection and lysis. To compare models with all other life history parameters fixed requires that $\eta/(\eta + \omega) = e^{-\omega\tau}$, or, alternatively,

$$\eta = \frac{\omega e^{-\omega\tau}}{1 - e^{-\omega\tau}} \qquad (3.41)$$

Juan Bonachela and Simon Levin implemented a similar comparison in their analysis of the evolutionary consequences of considering finite versus continuous delays between infection and lysis (Bonachela and Levin 2014).

What is the steady state of the model in Eq. 3.40? If there are no oscillations, then, at steady state, the value of a variable is the same as it was at any finite delay in the past. Fixed points can be identified by setting $N^* = N_\tau^*$ and $V^* = V_\tau^*$. These values are the same for the finite-delay model and the continuous-delay model so long as the effective burst sizes are equivalent, as in Eq. 3.41. However, the dynamics need not converge to a fixed point.

It is worth considering whether the dynamics remain similar between model formulations. For example, do intermediate delays destabilize the system of Eq. 3.40, just as rates of viral release that were neither too large nor too small destabilized the dynamics in the continuous-release model of Eq. 3.27? The short answer is yes. The long answer can be found in the applied mathematics literature (Beretta et al. 2002). Here, to illustrate this effect, we consider a numerical simulation of a discrete-delay model of virus-host infections given identical life history parameters as in a continuous-delay model. The finite delays are $\tau = 0.98$, $\tau = 3.65$, and $\tau = 21.97$, and the continuous lysis rates are $\eta = 1$, $\eta = 0.25$, and $\eta = 0.025$, respectively. Each comparison has the same effective burst size given that $\omega = 0.05$. Simulating a delay differential equation requires more information than simulating a differential equation without delays. Ask yourself, how many initial conditions are necessary to simulate the continuous-release versus the finite-delay model of host-viral interactions with an infected state? In the former, you need to know the initial density of uninfected hosts, infected hosts, and viruses. In the latter, you need to know the initial density of uninfected hosts, infected hosts, and viruses *at all times* $-\tau \leq t \leq 0$. Similarly, to predict the rate of change of densities at any given time in a model with delays, you need to know the density of variables at all times τ in the past. DDEs can be thought of as infinite-dimension dynamical systems, as they require history over a continuum. In practice, robust numerical simulations are available to simulate DDEs with a finite number of finite delays (e.g., DDE23 in MATLAB). Simulations confirm the same qualitative features as in the continuous-delay model (Figure 3.9). First, the steady state of the system is oscillatory for intermediate latent periods. Second, the structure of the cycle remains predator-prey-like with counterclockwise rotation in the phase plane. Finally, the duration of oscillations is much longer than the division time. In summary, the destabilization of microbial population dynamics by viruses is robust to changes in parameters and even the structure of the model representation.

3.4   LINKING MICROSCOPIC DETAILS WITH DYNAMICS

As is apparent, the choice of model structure can affect model predictions. In many instances, choices made to simplify a model will make no difference with respect to some predictions, but these same choices will have significant effects on others. The models already introduced to describe virus-host dynamics provide an excellent context for demonstrating this point. Recall the model proposed to describe virus-host population dynamics in which the infected host releases viruses at a rate η. In that model, viruses exclusively infected uninfected cells. Nonetheless, neither the ecological nor the physiological arguments for this assumption exclude the possibility that a virus may, with some low probability, infect a previously infected cell. Longer latent periods or spatial correlations in viral distributions relative to their hosts may further increase the probability of subsequent viral infections.

What would happen to model predictions if the model included an additional parameter—ρ—to represent the relative adsorption rate of viruses to infected hosts compared with their adsorption rate to uninfected hosts and if it was further assumed that the second infection does not change the productivity of the infection. The latter assumption implies that the number of viruses released and the time to release do not change for singly or multiply infected cells. In that case a variant of Eq. (3.27) can be written as

$$\frac{dN}{dt} = rN\left(1 - \frac{N+I}{K}\right) - \phi NV - \omega N,$$

$$\frac{dI}{dt} = \phi NV - \eta I - \omega I, \tag{3.42}$$

$$\frac{dV}{dt} = \beta \eta I - \phi NV - \overbrace{\rho \phi I V}^{\text{infection of infected cells}} - \omega V.$$

Here, $\rho \phi I V$ denotes the loss of infectious virus particles due to irreversible attachment to infected hosts. The previous model considered the case where $\rho = 0$. What happens when $\rho > 0$?

The full analysis of the model is facilitated by focusing attention on the equilibrium densities, irrespective of their stability. At steady state, $dI/dt = 0$, so that $I^* = \phi NV/(\eta + \omega)$, as before. It follows that at steady state

$$\frac{dN}{dt} = 0 = rN\left(1 - \frac{N + \phi NV/(\eta + \omega)}{K}\right) - \phi NV - \omega N, \tag{3.43}$$

$$\frac{dV}{dt} = 0 = \beta \phi NV \frac{\eta}{\eta + \omega} - \phi NV - \rho \phi NV \frac{\phi V}{\eta + \omega} - \omega V. \tag{3.44}$$

Note that including the possibility of multiple infections implies that viral decay (at steady state) no longer depends strictly linearly on V but, rather, nonlinearly on V. In fact, the larger V gets, the more important this detail becomes—consistent with the fact that multiple infections are more likely whenever V increases.

Appendix B.6 presents the complete algebraic solution for the equilibria of Eq. 3.42. However, the algebraic forms do not necessarily aid in building intuition for the effect of multiple infections on community structure. Instead, we consider a perturbative approach in which solutions are posited to be of the form $N^*(\rho) \approx N_0 + \rho N_1$, and $V^*(\rho) \approx V_0 + \rho V_1$, where the subscript denotes the order of ρ retained in the solution. The steady-state densities reported in Eq. 3.34 are the previously derived solutions for N_0 and V_0. N_1 and V_1 can be derived by expanding the $dV/dt = 0$ and $dN/dt = 0$ equations, respectively, retaining only those terms of order ρ. Terms of the order ρ^2 or higher can be ignored in the limit that $\rho \ll 1$. Solution of these equations yields

$$N_1 = \frac{\phi N_0 V_0}{\tilde{\beta}(\eta + \omega)}, \tag{3.45}$$

$$V_1 = \frac{-rN_1 \left(1 + \frac{\phi V_0}{\eta + \omega}\right)}{K\phi \left(1 + \frac{rN_0}{K(\eta + \omega)}\right)}, \tag{3.46}$$

where $\tilde{\beta} = \beta\eta/(\eta + \omega) - 1$, and by extension

$$I_1 = \frac{\phi (N_0 V_1 + N_1 V_0)}{\eta + \omega}. \tag{3.47}$$

Two key points from this perturbation analysis are, first, the sign of N_1 is positive, and the sign of V_1 is negative. This means that unproductive infections of infected cells lead to an increase in host density and a decrease in virus density. This is an intuitive conclusion. Second, the positive change in host density should be proportional to V_0. Recall that the virus density was proportional to r, the maximum growth rate. The maximum growth rate increases with increasing resource input density. A key secondary prediction is that multiple infections will move the system away from strict top-down control to partial top-down control. That is to say, the perturbation analysis implies that increases in resource input will lead to increases in both host and viral densities. The perturbation results agree with the exact solution of the model (see Figure B.4 in Appendix B.6).

What, then, is the effect of including this detail—that viruses can infect previously infected hosts—on the top-down control of the system? In Figure 3.11, the steady-state densities of N^* and V^* are shown as a function of r, the maximum growth rate of hosts. The solid line in Figure 3.11 (left panel) shows that as the maximum growth rate increases, the predicted host density at steady state is not expected to change as long as $\rho = 0$. This is a classic feature of models that exhibit top-down control, whether by predators or parasites. Moreover, the curve in Figure 3.11 (right panel) shows that as resources to the system increase, the predicted virus density at steady state is expected to increase. Again, in models of top-down control, the top predator (or parasite in this case) increases in abundance as resources are provided to its prey (or host).

At least some of the preceding features do change qualitatively when $\rho > 0$. To see how, examine the open circles in Figure 3.11 (left panel),

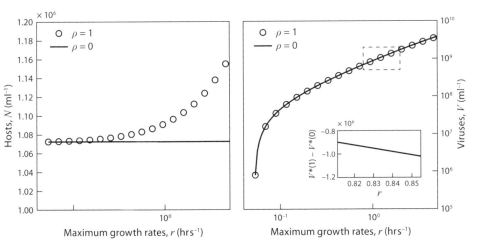

FIGURE 3.11. Effect of unproductive infection of infected hosts on the relationship between resource supply and population densities. The strength of the effect is measured in terms of $\rho\phi IV$, where $\rho = 1$ characterizes the relative adsorption of viruses to infected cells versus to uninfected cells. (Left) When $\rho = 1$, host density at steady state increases as resources are increased to the system (using the proxy of maximal growth rate). In contrast, when $\rho = 0$, the system is strictly top-down controlled, and N^* is a constant independent of r. (Right) Reinfections have only modest effects on viral densities. There is a decrease, but the decrease is less than the size of the points on a log scale. The inset shows the decrease in $V^*(\rho = 1)$ compared with $V^*(\rho = 0)$.

where $\rho = 1$. As is evident, the predicted host density at steady state is expected to *increase* with increasing maximum growth rate as long as $\rho > 0$. The predictions from this perturbation theory remain consistent with the results from the exact solutions. Ecologically, it would seem that reinfection does not alter the fact that viruses can limit host densities far below their carrying capacity. Yet, unlike in a strict top-down model of control, increases in resource supply that lead to increases in maximum growth rate do not flow directly to viruses. Instead, they flow partially to hosts. If one were to experimentally manipulate the resource supply and examine whether only virus density increased, the outcome would suggest of top-down control, but not strictly so. A similar phenomenon been explored in the context of other models in which virus productivity depends on resource availability and resource status of hosts (Weitz and Dushoff 2008; Menge and Weitz 2009; Wang and Goldenfeld 2010). This topic will recur in later chapters in discussions of the effect of viruses on host populations in more complex environments.

3.5   SUMMARY

- Models of virus-host interactions suggest that long-term persistence between a single virus population and a single host population is possible.
- Population dynamics models demonstrate that one potential effect of a viral infection is to shift a community from bottom-up to top-down control.
- One hallmark of top-down control is a decrease in host abundance with virus addition.
- A second hallmark of top-down control is an increase in resource concentration with virus addition.
- Virus-host population dynamics can exhibit oscillations. These oscillations are typified by phase lags between peaks in population abundances of host and virus, in which the host peak precedes the viral peak.
- The period of oscillations is often much longer than the latent period time between viral infection and lysis.
- The single-strain population dynamics models introduced here will form the base for investigating how viruses evolve, affect microbial community structure, and modify ecosystem functioning.
- The details of virus-host interactions—for example, whether a virus can reinfect a previously infected host—can crucially affect model predictions.

CHAPTER FOUR

Evolutionary Dynamics of Viruses or Microbes, but Not Both

4.1   VIRUSES AND THE NATURE OF MUTATION

The late nineteenth/early twentieth century was the start of a revolution in the biological sciences. Charles Darwin had posited a theory of evolution via natural selection, involving two key mechanisms: the generation of variation and the differential success of variants. It took Thomas Morgan and many others to show that change could indeed be heritable from one generation to the next. Nonetheless, many of the crucial mechanisms by which heritable change could be acquired were unknown, to say nothing of the molecule (DNA) that encodes the heritable nature of life. It was in this milieu that Salvador Luria and Max Delbrück published, in 1943, the results that would become the authoritative study on the nature of mutations within bacteria. Their first few sentences describe the crucial phenomena observed at the time:

> When a pure bacterial culture is attacked by a bacterial virus, the culture will clear after a few hours due to destruction of the sensitive cells by the virus. However, after further incubation for a few hours, or sometimes days, the culture will often become turbid again, due to the growth of a bacterial variant which is resistant to the action of the virus. This variant can be isolated and freed from the virus and will in many cases retain its resistance to the action of the virus even if subcultured through many generations in the absence of the virus. While the sensitive strain adsorbed the virus readily, the resistant variant will generally not show any affinity to it. (Luria and Delbrück 1943)

This observation has been repeated, many times since. Levin et al. (1977) described their study of phage-bacteria interactions as follows:

> Phage resistance in bacteria can occur as a single-locus mutation which would be very likely to arise in these cultures. In four of the six experimental

TABLE 4.1. Number of resistant colonies observed in Luria and Delbrück's 1943 experiment. This table includes a subset of the original data (Luria and Delbrück 1943). Each row represents a different experiment, and each column, a different replicate within that experiment. The number of replicates for each experiment was not fixed. *Key point: The number of resistant colonies observed within replicates of a given experiment varied significantly, with occasional observations of zero resistant colonies, and other observations of hundreds of resistant colonies.*

Experiment	Replica										
	1	2	3	4	5	6	7	8	9	10	11
1	10	18	125	10	14	27	3	17	17		
10	29	41	17	20	31	30	7	17			
11	30	10	40	45	183	12	173	23	57	51	
15	6	5	10	8	24	13	165	15	6	10	
16	1	0	3	0	0	5	0	5	0	6	107
17	1	0	0	7	0	303	0	0	3	48	1
21a	0	0	0	0	8	1	0	1	0	15	0
21b	38	28	35	107	13						

cultures ... *E. coli* B clones which were apparently resistant to the coexisting T2 populations did evolve.

The key question that Luria, a biologist, and Delbrück, a physicist, asked, and subsequently resolved, was whether bacteria became resistant *after* the interaction with a virus or whether resistant mutants had been present in the population *prior* to viral exposure.

To answer this question, Luria and Delbrück exposed a culture of bacteria to phage. The culture of bacteria originated with a single bacterium or a small number of susceptible bacteria that divided successively until they reached a very large population, often greater than 10^8 cells. A small volume of the culture of viruses was spread across an agar plate followed by a small volume of the bacterial culture. In this way, every bacterium on the surface could potentially interact with many viruses. The ancestral bacterium was known to be susceptible to lysis. Therefore, each bacterium on the surface of these agar plates should have been infected and lysed, leaving no trace of bacterial growth once the spread of viruses was complete. Instead, in every experiment, there was at least one replicate culture in which multiple resistant colonies were observed (Table 4.1). There was also substantial variation in the number of resistant colonies observed between replicate cultures. That variation included replicates that sometimes had many resistant colonies even when an independent replicate had none. Experiments 16 and 17 in Table 4.1

are two striking examples. One way to interpret this variation is that the experiments were flawed. Luria and Delbrück took a different approach.

The two leading hypotheses of the time were that mutation (i) occurred as a result of selection or (ii) arose independently of selection. How many resistant colonies would be expected given these two hypotheses, and how do the predictions compare with observations? First, consider the hypothesis that mutations are acquired as a result of selection. The experimental design, described previously, ensured that each bacterium on the surface of the agar plate almost certainly interacted with at least one virus. For a per bacterium probability of acquiring resistance, μ_a, the probability that these should be N' resistant colonies out of N_f total bacteria should follow a binomial distribution:

$$p(N') = \mu_a^{N'}(1 - \mu_a)^{N_f - N'}\binom{N_f}{N'}.$$ (4.1)

This binomial distribution follows from the assumption that resistance would be independently acquired within each bacterial cell with some small probability. Given that the final population was large ($N_f \gg 1$), and the probability of acquiring resistance was small ($\mu_a \ll 1$), the binomial distribution for the number of resistant colonies can be approximated as the Poisson distribution

$$p(N') = \frac{\lambda^{N'} e^{-\lambda}}{N'!},$$ (4.2)

where $\lambda = \mu_a N_f$. In a model of acquired mutation, the variation in the number of resistant colonies should be Poisson distributed, such that the variance is equal to the mean, and the standard deviation is equal to the square root of the mean. This model of acquired mutation is sketched in Figure 4.1A.

Second, consider the hypothesis that mutations are independent of selection. Resistant mutants present in the population prior to exposure to phage will reflect two processes: (i) a mutation at some time t_0 prior to the exposure time t_f and (ii) the growth of these resistant mutants in abundance. For example, if the first division of a bacterium led to one susceptible and one resistant bacterium, then all the descendants of the resistant mutant would themselves be resistant, such that at least $N_f/2$ of the population would be resistant, ignoring revertant mutations. This unlikely to happen given small, independent mutation rates μ_i. Therefore, a model of independent mutations behaves much like a slot machine or the lottery—winning is unlikely, but payoffs can be quite large when they do occur. The following continuous model represents the population dynamics of the number of susceptible hosts N and

A1. Growth of a bacterial population
from a single ancestor

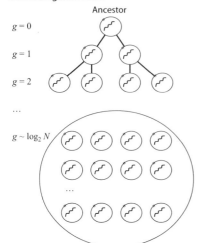

A2. Exposure of population to viruses

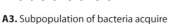

A3. Subpopulation of bacteria acquire
resistance and survive viral infection.

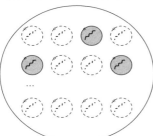

B1. Growth of a bacterial population
from a single ancestor

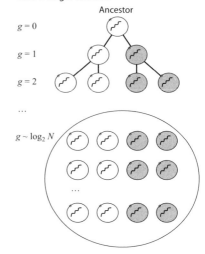

B2. Exposure of population to viruses

B3. Subset of resistant bacteria already present
and survive viral infection.

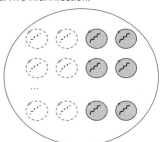

FIGURE 4.1. Alternative mechanisms of mutation: acquired (panels A1–A3) or independent (panels B1–B3). If mutations are induced, then the bacterial population (A1) does not contain any mutants prior to exposure to the virus (A2). Only after exposure are a random fraction, μ_a, of hosts "converted" into mutants, which survive the infection (A3). In contrast, in an independent mutation model, the bacterial population grows from wild type through a series of cell divisions. A resistant mutant arises with probability μ_i in a given division. If a resistant mutant arises during this proliferation process, then all the progeny of the mutant are also mutants (B1). After exposure to viruses (B2), only the resistant mutants survive (B3).

the number of resistant hosts N':

$$\frac{dN}{dt} = \overbrace{N}^{\text{growth of WT}} - \overbrace{\mu_i N}^{\text{mutation}}, \tag{4.3}$$

$$\frac{dN'}{dt} = \overbrace{\mu_i N}^{\text{new mutations}} + \overbrace{N'}^{\text{clonal growth of mutants}}, \tag{4.4}$$

where time is rescaled in terms of the growth rate r, which is assumed to be equivalent among strains. These equations can be solved analytically, noting that the host population grows as

$$N(t) = e^{(1-\mu_i)t}, \tag{4.5}$$

and recognizing that $N + N' = e^t$, the mutant population grows as

$$N'(t) = e^t \left(1 - e^{-\mu_i t}\right), \tag{4.6}$$

such that the population of resistant hosts is predicted to grow faster than that of the susceptible hosts! Further, given that $\mu_i \ll 1$ and $t_f \approx \log N_f$, then the number of resistant mutants expected after t_f in a continuously growing population should be

$$N'(t_f) \approx e^{t_f} \mu_i t_f = \mu_i N_f \log N_f. \tag{4.7}$$

The actual dynamics are not well approximated by a continuous population model, at least not at first. The reason is that continuous models are not well suited to describe dynamics when populations comprise one or very few individuals. There are unlikely to be any mutant individuals in the population until $\mu_i N \approx 1$. Instead, the average number of mutants measured in a series of C replicate experiments should be

$$N'(t) \approx \mu_i N_f (t_f - t_0), \tag{4.8}$$

where t_0 is the likely time for the first mutation.

Luria and Delbrück reasoned that t_0 in C replicates should satisfy

$$\mu_i \left(N_{t_0} - N_0\right) C \approx 1, \tag{4.9}$$

where N_0 is the small number of bacteria in the initial population. Because $N_{t_0} = N_f e^{t_0 - t_f} \gg N_0$, then

$$t_f - t_0 = \log \left(\mu_i N_f C \right), \tag{4.10}$$

such that the average number of resistant mutants is

$$N' = \mu_i N_f \log \left(\mu_i N_f C \right). \tag{4.11}$$

If the independent model of mutations is correct, then this equation can be used to estimate the mutation rate, given observations of N', N_f, and C. A solution for μ_i can be obtained with any standard nonlinear solver, given a suitable initial condition (e.g., $\mu_i \approx N'/N_f$). Irrespective of which model is correct, the data on the number of resistant colonies and total population sizes would simply yield two different estimates of the mutation rate. Given the original data in the study by Luria and Delbrück (1943), the mutation rates can be estimated to be $\mu_a = 1.7 \times 10^{-7}$, for the model of acquired mutation, and $\mu_i = 2.4 \times 10^{-8}$, for the model of independent mutations. These mutation rates are different, but the true mutation rate leading to resistance was not known a priori. How else can the data be used to distinguished these two models?

The difference between the two models lies in their predictions of variation. Consider first the model of acquired mutation as a consequence of interaction with a virus. In this case, the number of resistant colonies, N', should be Poisson distributed, with an average of $\bar{N}' = \mu_a N$. The standard deviation in the number of resistant colonies across replicates should scale as $\bar{N}'^{1/2}$, such that the coefficient of variation scales as $\bar{N}'^{-1/2}$. When the average number of mutant colonies exceeds approximately 50, the results between replicates should be similar, and there should rarely be very large numbers of resistant mutants or an absence of mutants.

In contrast, consider independent mutations emerging in a population that grows from a single individual to a population of N individuals. The mutants in the final population derive from single events that take place at a time τ between t_0 (the likely first mutant) and t_f (the final time). The expected number of events in a small interval $d\tau$ around time τ is $N(\tau)\mu_i d\tau$, which being Poisson distributed, implies that the variance in such events is also $N(\tau)\mu_i d\tau$. The total number of mutants in the final population is larger by a factor of $e^{t_f - \tau}$, because mutants that appear in the interval τ to $\tau + d\tau$ grow exponentially. The contribution to the total variance in the final population must be multiplied by this factor squared, $\left(e^{t_f - \tau} \right)^2$. The total variance is the sum of each independent contribution, such that the total variance is

$$\text{Var}(N') = \int_{t_0}^{t_f} d\tau \, \mu_i N(\tau) e^{2(t_f - \tau)} \tag{4.12}$$

$$= \int_{t_0}^{t_f} d\tau \, \mu_i N_f e^{-(t_f - \tau)} e^{2(t_f - \tau)} \tag{4.13}$$

$$= \int_{t_0}^{t_f} d\tau \, \mu_i N_f e^{(t_f - \tau)} \tag{4.14}$$

$$= \mu_i N_f \left(e^{(t_f - t_0)} - 1 \right). \tag{4.15}$$

Substitution in the approximation for $t_f - t_0$ (in Eq. 4.10) yields

$$\text{Var}(N') \approx C \left(\mu_i N_f \right)^2 \tag{4.16}$$

Here, the ratio of the variance to the average is $C\mu_i N_f / \log(C\mu_i N_f)$; that is, the variance is much greater than the mean, unlike in the case of Poisson fluctuations. *In summary: The variance in resistance mutants should be small if mutations are acquired after exposure to viruses, whereas the variance in resistance mutants should be large if mutations are independent of exposure to viruses.*

Based on the observed differences in the variation in resistant colonies, Luria and Delbrück concluded that mutations were independent of the selection pressure. This conclusion was cited along with other seminal contributions as the basis for the Nobel Prize in Physiology or Medicine awarded to Luria and Delbrück in 1969, which they shared with Alfred Hershey.

The literature on mutations arising in finite populations did not stop there. Luria and Delbrück's original paper continues to stimulate new theory and insights more than 70 years after the original proposal (Kessler and Levine 2013). Although the mathematical approaches to solving the model can be complex, the intuition behind the process is also accessible when numerically simulated. The key choice in developing a simulation of this process is to decide on the expected level of synchronicity in the cell division process. Two limiting cases are (i) a completely synchronized or (ii) a completely desynchronized population. In the former case, the population doubles in size via a series of approximately $\log_2 N$ cell divisions. For each division, one of the offspring is mutated with a probability μ. All daughters of mutant cells are themselves mutants. This model is equivalent to one in which the time between divisions is fixed at a. In the latter case, there are $N - 1$ events to consider, and at each event, one of the individuals in the population is selected to divide.

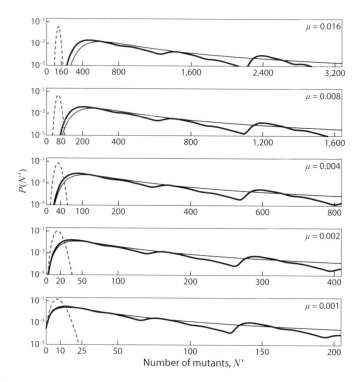

FIGURE 4.2. Simulated distributions under different scenarios for the probability of resistant mutants, $P(N')$, given a population that grows from a single susceptible cell to $N = 8192$ cells. Here, the mutational probability μ is the same for both the acquired model, $\mu_a = \mu$, and the independent model, $\mu_i = \mu$. The independent model is simulated using two limits of synchronization: (i) asynchronous cell division (smooth curve) and (ii) synchronous cell division (jagged curve). The dashed (narrow) distributions correspond to Poisson distributions of mutations arising from acquired mutations, with an expected value of $\mu_a N$. The scale of the x-axis changes in the five panels corresponding to simulations with increasing values of μ from bottom to top. In all cases, the acquired model predicts fewer mutants and less variation between replicates than does the independent model.

The offspring is mutated with a probability μ. This model is equivalent to one in which the time between divisions is exponentially distributed with average age of division a. Appendix C.1 provides further details on the simulation methods.

Simulations of either the synchronized or desynchronized models lead to results that are qualitatively similar to experiments (Figure 4.2). For independent mutation, the probability that there are N' mutants out of N_f

total individuals is broadly distributed, and the variance in the number of mutants scales nonlinearly with the expected number of mutants. For acquired mutation, the probability that there are N' mutants out of N_f total individuals is narrowly distributed, and the variance in the number of mutants scales with the expected number of mutants. This difference is consistent with the broadening of the mutant probability distribution as μ increases from the bottom to the top panel in Figure 4.2.

In summary, thinking carefully about the variation in the number of mutants, and not the average number of mutants, helped prove that mutations from susceptible to resistant occurred independently from the selective pressure.[1] However, Luria and Delbrück did not investigate the long-term population and evolutionary dynamics that emerge via virus-host interactions. Viruses can stimulate host evolution, by driving changes in the frequency of host genotypes. That topic—the other half of evolution via natural selection—is the subject of the next section.

Before addressing that subject, it is worth concluding this preliminary section with a question: what is evolution? The answer to this question would seem to be self-evident to nearly all biologists, but that is far from the case. Ask yourself this question, or, alternatively, ask a colleague. The answers will all too often fall under the category of "I can't quite define it, but I know it when I see it." Dictionary definitions are even worse and lead to greater confusion. The following definition will prove most useful here and will serve as our guide in this monograph:

Evolution: the change in the frequency of alleles/genotypes from one generation to the next.

According to this definition, one form of evolution is de novo changes in the frequencies of genotypes arising due to mutation. Another form of evolution is changes in the frequencies of existing genotypes due to adaptation or stochastic processes. With the preceding definition in mind, I will show how viruses can affect host evolution and, conversely, how hosts can affect viral evolution.

---

[1] The history of biology could have turned out quite differently had Luria and Delbrück used a different microbe: *Streptococcus thermophilus*, rather than *E. coli* (van der Oost et al. 2014). For *S. thermophilus*, the mutations are acquired subsequent to interaction with an invading virus. The reasons will be explored in the next chapter.

4.2   THE EFFECTS OF VIRUSES ON HOST EVOLUTION

4.2.1   RESISTANCE AND FITNESS

Resistance is clearly of *potential* benefit to a host. For example, consider a population that included two host genotypes—one rare resistant genotype and one common host genotype—that were then exposed to viruses. Initially, the resistant type would divide without loss due to infection, while the susceptible type would have an increased death rate owing to virus infection and lysis. Over time, the population of the resistant type would increase, and the population of the susceptible type would decrease, possibly to extinction. Is this the actual case? The framing of this question suggests a key difference between the setup described and the original experiment of Luria and Delbrück. The difference is that the change in the number of host genotypes can lead to subsequent changes in the number of viruses in the population. In turn, these changes modify the strength of the viral selective pressure. Even if the viral selection pressure is initially strong, it may not necessarily remain so. Adding a virus to a host population can lead to a selection pressure favoring the increase of a rare host genotype. But such increases are not inevitable. In contrast, not all resistant host genotypes will increase in abundance. Further, even if they do so, the addition of a virus can lead to the coexistence of multiple host genotypes that fluctuate in their relative abundance. In other words, viruses can stimulate host evolution.

A model of virus-host interactions with an explicit infected state was introduced as Eq. 3.27 in Chapter 3. This model included uninfected hosts (N), infected hosts (I), and free viruses (V). In this model, viruses are released with an exponentially distributed latent period whose average is $1/\eta = \tau$. To evaluate the effect of mutants on the population dynamics, this model can be extended to include a resistant host population (N') that cannot be infected by viruses. A full evaluation of such a model requires additional information, because mutations that confer resistance may also lead to other physiological effects on the host. Such dual effects of a single mutation are an example of *pleiotropy* (Lenski 1988b). The pleiotropic effect of relevance here is that the resistant host can be less efficient in taking up and/or utilizing resources, such that its growth rate, r', is less than that of the susceptible host; that is, $r' < r$. How valid is such an assumption?

The cost of host resistance is paramount in the study of virus-host interactions. The fact that resistance can evolve in the laboratory—often quite readily—raises the question, how can phage and other viruses of

microorganisms persist? Trade-offs between resistance and competition suggests a route forward. In the mid-to late 1980s, Richard Lenski, Brendan
Bohannan, and colleagues studied the interactions of *E. coli* B and bacteriophage T4 (Lenski and Levin 1985; Lenski 1988a,1988b; Bohannan and Lenski
1999, 2000). These studies focused, in part, on the strength and nature of
trade-offs in this virus-host system. The relative fitness of resistant mutants
compared with that of the susceptible wild type was measured in a competition
assay. The relative fitness of the mutant is defined as

$$W' = \log_2 \left[\frac{N'_f}{N'_0} \right] / \log_2 \left[\frac{N_f}{N_0} \right] \tag{4.17}$$

where N_0 and N'_0 are the initial ancestral and mutant host density and, similarly, N_f and N'_f are the final ancestral and mutant host density respectively,
after coculturing. The rationale behind this metric is that both mutant and host
will undergo distinct numbers of divisions during the incubation period, such
that N_f and N'_f can be written as $N_f = N_0 2^g$ and $N'_f = N'_0 2^{g'}$, respectively.
Equation 4.17 can then be rewritten as $W' \equiv g'/g$. There are many factors
that influence the change in population size. Nonetheless, the use of change in
total population serves to integrate these factors into a single index of relative
success particularly, given experiments of fixed duration.

What are the fitness costs of resistance, if any? Lenski (1988b) found
that mutant host isolates suffered a significant fitness disadvantage compared
with the ancestral susceptible host (Figure 4.3A). The decrease in growth rate
was further attributed to the nature of the mutation that conferred resistance.
Bacteriophage T4 adsorbs to the cell surface of *E. coli* at specific receptor sites
in the lipopolysaccharide (LPS) of the cell wall corresponding to the "first
glucose residue" of the LPS core (see discussion in Lenski (1988b)). Via locus
mapping methods available at the time, Lenski was able to attribute nearly all
mutations that conferred resistance to loci associated with LPS biosynthesis.
Modifying the structure of the LPS has the dual effect of decreasing the
adsorption rate of viruses through receptor modification and changing
the permeability of the LPS. This change in permeability leads to changes in
the profile of compounds that can enter the cell. Such changes negatively affect
cell physiology and fitness. Bohannan and colleagues repeatedly observed
"a trade-off in *E. coli* between resistance to bacteriophage and competitive
ability" (Bohannan et al. 2002). The resistance was extracellularly mediated,
and the loss of competitive ability manifested itself as a decrease in fitness
relative to the wild-type (susceptible) host. The strength of the fitness cost
varied depending on environment. Nonetheless, the basic mechanism has
become a paradigm in viral biology.

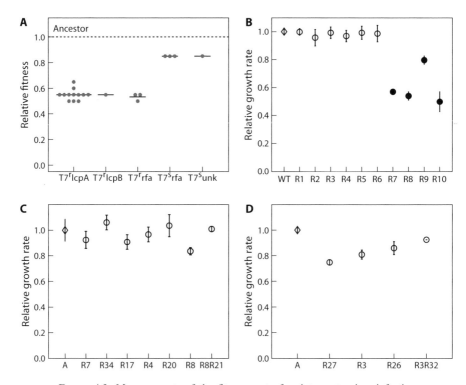

FIGURE 4.3. Measurements of the fitness cost of resistance to virus infection. (A) Relative fitness of *E. coli* resistant mutants interacting with bacteriophage T4. Each point denotes one of 20 strains. The solid bars denote the mean fitness per class of mutant. Analysis adapted from data originally presented in Lenski (1988b). (B) Relative growth rates of resistant strains of *Prochlorococcus* MED4 compared with ancestral type. Filled points denote those strains that have a significantly lower growth rate than the ancestor. Original data analysis and additional details are in Avrani et al. (2011). Relative growth rates of cyanobacteria *Synechococcus* WH7803 (C) and WH8101 (D) after passage through one or more viral selection experiments. Original data analysis and additional details are in Lennon et al. (2007).

The trade-off paradigm has its roots in the study of *E. coli* and associated phage, like many paradigms in viral biology, yet it does not apply universally to the diversity of virus-host systems in the environment (Breitbart 2012). In fact, attempts to replicate these landmark studies in other systems have led to contradictory findings. In multiple examples mutations that confer resistance to phage infection do not lead to directly measurable changes in fitness, for example, decreases in growth rates.

A prominent example is the analysis of cyanobacteria-cyanophage interactions (Lennon et al. 2007; Avrani et al. 2011). Cyanobacteria from the clades

Prochlorococcus and *Synechococcus* represent experimentally tractable and environmentally relevant marine photoautotrophs. Bacteria from these two clades are responsible for nearly 50% of all primary production in the global oceans. Moreover, many cyanophage are now available in culture, and their isolation and analysis has led to many novel findings, including the discovery of phage-encoded photosystem genes (Lindell et al. 2004; Sullivan et al. 2005; Clokie and Mann 2006; Lindell et al. 2007). Of relevance here is the use of cyanophage-cyanobacteria as an environmental model system to explore trade-offs between resistance and growth.

As a first example, Jay Lennon, Marcie Marston, and Jennifer Martiny isolated *Synechococcus* strains, grew them in culture, and used screening techniques similar to those of Luria and Delbrück to identify phage-resistant mutants (Lennon et al. 2007). They found multiple examples of resistant hosts that *did* have decreased growth rates, as well as multiple examples of resistant hosts that did *not* have decreased growth rates (Figures 4.3C and D). Similarly, Debbie Lindell's group has isolated many cyanophage that infect cyanobacteria from the *Prochlorococcus* clade, such as the host MED4 and the podovirus P-SSP7 (Avrani et al. 2011). Populations of cyanobacteria grown from a susceptible MED4 clone yielded multiple mutant isolates, all of which were resistant to infection by the original virus. In this group's experiments, nearly half these mutants had significantly lower growth rates (20%–50% lower), while other mutants suffered no growth cost of resistance (Figure 4.3B). Interestingly, Lindell's group hypothesized that the costs of resistance, when they exist, need not always be directly linked to a decrease in growth rate but, rather, to an increase in the susceptibility to other viruses. This notion—something like "whack a phage"—is emerging as an alternative paradigm (Avrani et al. 2012). Although there are pairs of strongly interacting virus-host populations, these populations rarely operate in isolation. The benefits of resistance may have as much to do with interactions with other organisms in the environment—for examples, bacteria, viruses, and microeukaryotes—as they do with interactions with small molecules like nutrients, signaling chemicals, and metabolites. The next section will address what happens when—despite these additional complexities—resistance to viral infection is, in fact, associated with a growth rate cost.

4.2.2 Conditions Favoring Invasion of a Resistant Host Mutant

Chapter 3 introduced a chemostat model of virus-host population dynamics with an infected state, given implicit resources (Eq. 3.27). This model can be extended to describe the interaction of a single virus population (V) with two host populations, one susceptible (N) and one resistant (N'), in which only infections of the susceptible host lead to an infected class I:

$$\frac{dN}{dt} = rN\left(1 - \frac{N + I + N'}{K}\right) - \phi NV - \omega N,$$

$$\frac{dN'}{dt} = r'N'\left(1 - \frac{N + I + N'}{K}\right) - \omega N',$$

$$\frac{dI}{dt} = \phi NV - \eta I - \omega I,$$

$$\frac{dV}{dt} = \beta \eta I - \phi NV - \omega V.$$

(4.18)

In this model the growth of hosts is limited by the total cell density, $(N + I + N')/K$, rather than only the susceptible and infected host densities, $(N + I)/K$, and viruses adsorb and infect susceptible hosts but do not adsorb to resistant hosts. If $N' = 0$ then this system can reach a steady state including positive densities of N, I, and V. But what happens if a small number of resistant hosts, $N' > 0$, are added to the system? The addition can be exogenous or endogenous, resulting from the mutation of a susceptible host into a resistant host, just as Luria and Delbrück showed (Luria and Delbrück 1943). The population density of resistant hosts is initially small—on the order of $1/S$, where S is the volume of the chemostat. Resistant individuals may subsequently divide, eventually giving rise to a large population of mutants. Mutations that confer resistance can also have a slower growth rate than that of susceptible hosts. These pleiotropic effects make it more challenging to answer the question, can the resistant population invade and, if so, with what probability?

To determine whether a mutant population can invade, it is essential to introduce the concept of the per capita growth rate, $\bar{r}$. For example, for a population that has a birth rate b and death rate d, the population dynamics can be written as $dx/dt = (b - d)x$, such that the per capita growth rate is $\bar{r} = b - d$. The sign of the per capita growth rate determines the fate of this population. The population grows exponentially if $b > d$ and decays exponentially if $b < d$. An alternative approach is to recognize that the "population" of mutants initially comprises a single or a few individuals. The probability that this single individual reproduces before dying is $b/b + d$; likewise, the probability that death occurs before reproduction is $d/b + d$.

TABLE 4.2. Equilibrium densities corresponding to the mutant-free environment (original), the coexistence equilibrium with both residents and mutants, a wild-type only equilibrium, and a mutant-only equilibrium.

Variable	Original	Coexistence	WT only	Mutant only
N	$\dfrac{\omega(\eta+\omega)}{\phi\,(\beta\eta-(\eta+\omega))}$	$\dfrac{\omega(\eta+\omega)}{\phi\,(\beta\eta-(\eta+\omega))}$	$K(1-\omega/r)$	0
I	$\dfrac{\phi N^* V^*}{\eta+\omega}$	$\dfrac{\phi N^* V^*}{\eta+\omega}$	0	0
V	$\dfrac{r\,(1-N^*/K)-\omega}{\phi\left[1+\dfrac{rN^*}{K(\eta+\omega)}\right]}$	$\dfrac{\omega}{\phi}\left(\dfrac{r}{r'}-1\right)$	0	0
N'	0	$K(1-\omega/r')-N^*-I^*$	0	$K(1-\omega/r')$

The total expected change in the population of N after one event is

$$\text{Net population change} = \frac{b}{b+d}\times 1 - \frac{d}{b+d}\times 1. \qquad (4.19)$$

For the net population change to be positive requires that $b > d$. However, this result is not sufficient to specify the probability that the population will increase in the long term. For example, even if a cell divides into two, there is a chance that both cells will be washed out before further divisions take place. A long-standing and well-known result from branching process theory addresses the dynamics of birth and death processes (see Sec. XVII.5 of Feller (1968)). This result can be used to show that a single individual is likely, via a stochastic process of discrete birth and death events, to grow to a very large number of individuals with probability $p = 1 - d/b$. Again, the probability of invasion is positive only when $b > d$ or when the per capita growth rate is positive, as it is in the deterministic model.

The preceding definitions require a specification of the per capita birth and death rates, yet such per capita rates are context dependent. The environment that resistant mutants experience is set by the wild-type hosts. Mathematically, this is equivalent to asserting that the initial conditions of the system are in a mutant-free equilibrium. The parameter dependence of (N^*, I^*, V^*) for this equilibrium are listed in Table 4.2. When mutants appear in the system, their density is far smaller than that of both susceptible hosts and infected hosts, that is, $N' \ll N^*$, and $N' \ll I^*$. In this limit, the mutant population dynamics can be approximated as

$$\frac{dN'}{dt} = r'N'\left(1 - \frac{N^* + I^* + N'}{K}\right) - \omega N', \qquad (4.20)$$

$$\approx N'\left[r'\left(1 - \frac{N^* + I^*}{K}\right) - \omega\right], \qquad (4.21)$$

where terms of order N'^2 have been ignored. Consequently, the per capita growth rate of the mutants is

$$\bar{r}' \equiv \frac{1}{N'}\frac{dN'}{dt} = r'\left(1 - \frac{N^* + I^*}{K}\right) - \omega. \qquad (4.22)$$

The critical value of the maximum division rate at which net growth becomes positive is

$$r'_c = \frac{\omega}{\left(1 - \frac{N^* + I^*}{K}\right)}. \qquad (4.23)$$

The mutants can invade when $r' > r'_c$ and cannot invade when $r' < r'_c$. For some parameter values, the equilibrium may actually be a fluctuating steady state. More generally, the critical maximum division rate for invasion should be

$$\tilde{r}'_c = \left\langle \frac{\omega}{\left(1 - \frac{N + I}{K}\right)} \right\rangle, \qquad (4.24)$$

where the *averages* are considered over the steady-state limit cycle rather than at fixed-point values (Klausmeier 2008). In practice, these values are often similar, but a reader interested in the cost of resistance for fluctuating dynamics should be cognizant of this difference.

Two limits will prove useful in deciphering the meaning of this critical value of the mutant maximum growth rate, r'_c. First, consider the case when the viruses of the resident population reduce that population to negligible amounts, that is, to a density much less than K. In that limit, $r'_c \to \omega$, such that any mutant whose density-independent growth rate exceeds the chemostat washout rate may invade, even if $r'_c \ll r$. Second, consider the case where the viruses of the resident population reduce that population barely below the original virus-free equilibrium; that is, $N^* + I^* \to K(1 - \omega/r)$. In this case, $r'_c \to r$, such that the maximum growth rate of mutants must approach that of the residents! In other words, theory predicts that when viruses are efficient in their exploitation of the resident host, then resistant mutants with small or large costs of resistance can invade. Returning to the second case, consider the limit when viruses are not efficient in exploiting the resident host. In that limit, only

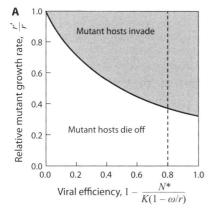
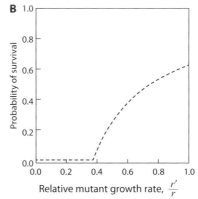

FIGURE 4.4. Invasion of mutant host when rare depends on the cost of resistance. (A) Alternative outcomes are invasion (shaded region) or extinction (white region) of mutant hosts. The likelihood of invasion is facilitated by increases in virus efficiency and decreases in the growth cost of resistance. The vertical dashed line is used to investigate the probability of mutant survival and invasion in the right panel. (B) Probability of mutant invasion given variation in the growth cost of resistance. These scenarios assumes a fixed value of virus efficiency, as denoted by the dashed line in the left panel. Common parameters are $r = 0.16$; $K = 2.2 \times 10^7$; $\eta = 1$; and the washout rate, $\omega = 0.05$. The variation in viral efficiency for the initial conditions is generated by varying the burst size from $\beta = 4$ to $\beta = 500$ given a fixed adsorption rate, $\phi = 10^{-9}$. For each value of β, a viral "efficiency" of host density drawdown is calculated relative to the equilibrium host density in the absence of viruses, $K(1 - \omega/r)$. Similarly, for each value of β, the minimum mutant growth rate, r', is calculated based on the invasion fitness criteria discussed in the text.

resistant mutants with relatively small costs of resistance can invade. Figure 4.4 illustrates how invasion success of resistant hosts depends on both the cost of resistance and the efficiency of exploitation of susceptible hosts by viruses. The numerical simulations confirm that mutant hosts with lower values of r' can invade a chemostat with viruses of higher efficiency. However if the fitness costs become too great, then mutants cannot invade, despite being resistant to viruses. As viruses become more efficient, an increasing range of mutants can invade (Figure 4.4A). For a fixed viral type, it is apparent that the probability of survival is zero until the mutant r' reaches r'_c (Figure 4.4B) and subsequently increases. The probability of invasion is

$$\text{P(invasion)} = 0 \qquad\qquad \text{when } r' < r'_c \qquad (4.25)$$

$$= 1 - \frac{\omega}{r'\left(1 - \frac{N^* + I^*}{K}\right)} \quad \text{when } r' \geq r'_c. \qquad (4.26)$$

This probability leverages the Feller result cited earlier in this chapter (see Sec. XVII.5 of Feller (1968)). This criterion assumes that deaths are due to washout and that the denominator is the birth rate. In summary, resistant mutants can invade insofar as the costs of resistance are not too great. The extent to which costs are prohibitive depends on the nature of the relationship between the virus and the susceptible host. The next section addresses what happens *after* the resistant host invades.

4.2.3   EVOLUTIONARY DYNAMICS OF HOSTS, MEDIATED BY VIRAL INFECTION

The question of what happens to a population of resistant and susceptible hosts has been addressed by a number of researchers, including Bruce Levin, Rich Lenski, Brendan Bohannan, Holly Wichman, Jim Bull, Paul Turner, and many others (Wichman et al. 1999; Turner and Chao 1999; Bull et al. 2006; Bull and Molineux 2007). Curiously, some of the earliest of such studies yielded dynamics that seemed rather inscrutable at the time. Only many years later were they understood to be part of a much more general phenomenon, exemplified by a chemostat study of *E. coli* populations mixed with bacteriophage T4 (Figure 4.5A). As is apparent, the total populations of bacteria and phage oscillate. Close examination reveals that the peak of the phage population either is concordant with or immediately follows the peak of the host population. This phenomenon is consistent with predictions of classic consumer-resource theory, in which resource peaks are followed by consumer population peaks. When the resource reaches its trough, the consumer decreases in abundance owing to resource depletion, and the cycle continues (Hastings 1997; Case 2000; Murdoch et al. 2003). The data seem consistent with this canonical population dynamics view of virus-host interactions, in which bacteria peaks precede viral peaks, until approximately 200 hours into the experiment. But at 200 hours something happens.

The something that happens is that the total bacterial population stops oscillating while increasing in density just as the viral population decreases in density and continues to oscillate. The viral oscillations have extended time lags between peaks that are nearly twice as long as before the change (Figure 4.5A). The experimentalists did not add anything to the chemostat and continued to measure the same thing: the dynamics of *E. coli* and bacteriophage T4. Yet, the dynamics of the system appeared to have shifted from classic Lotka-Volterra cycles where the bacteria peak precedes the virus peak (see Chapter 3) to what are now termed *cryptic* oscillations (Jones and Ellner 2007; Cortez and Ellner 2010). The label *cryptic* refers to the

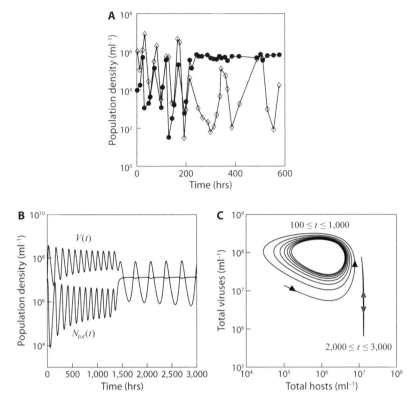

FIGURE 4.5. Cryptic viral oscillations in experiments and in theory. (A) Measured chemostat dynamics of total bacteria (*E. coli*) (filled circles) and free viruses (bacteriophage T4) (open diamonds). Note the apparent change in dynamics near $t \approx 200$. Points were extracted from Figure 4A of Bohannan and Lenski (1997) using PlotDigitizer 2.6.3. (B) Simulated dynamics of uninfected hosts and free viruses. At time $t = 1000$, a mutant host was introduced that was resistant to viral infection. The parameters for the simulation are $r = 0.16$, $r' = 0.12$, $K = 2.2 \times 10^7$, $\eta = 0.25$, $\beta = 50$, $\phi = 10^{-9}$, and $\omega = 0.05$. (C) Phase-plane portrait of dynamics of free viruses and uninfected hosts in the simulated model, focusing on two portions of dynamics (i) $100 \leq t \leq 1000$ and (ii) $2000 \leq t \leq 3000$. The numerical simulations are intended to recapitulate the qualitative phenomena of a transition to cryptic oscillations.

fact that a correlation or other inverse-problems-based approach would not be able to identify the relationship between the virus population and the host population after $t \approx 200$ hours. Still, the viruses continued to infect and reproduce. Otherwise, they would have been slowly eliminated from the chemostat via washout. The same type of phenomenon can be observed in numerical simulations of Eq. 4.18 for other life history and chemostat parameters. Figure 4.5B shows virus-host oscillations changing to cryptic oscillations where the viral population oscillations exhibit increased time lags and lower median densities. Figure 4.5C shows the same simulated dynamics in the phase plane. Here, the shift between Lotka-Volterra cycles and cryptic cycles is even more apparent. What may be surprising is that the simulations that yielded such cycles were already presented in Eqs. 4.18. This transition can occur for suitably chosen parameters, and the range of suitable parameters is quite broad. Those for the simulation that yielded Figures 4.5B–C are identical with those used in the $N–I–V$ model previously introduced, with one additional parameter associated with the N' population, r'.

What is the reason for such an apparently unusual response to the invasion of a chemostat by a resistant bacterium? The answer begins with the first step—the invasion of a chemostat comprising three populations: susceptible hosts, infected hosts, and free viruses. Each population is assumed to be at its equilibrium density (Table 4.2). Assuming that the conditions for invasion are satisfied, then the dynamics can—given stochastic escape from small mutant populations—be described using Eqs. 4.18. As the mutant population increases, the growth rate of hosts must decrease, leading to a decrease of susceptible hosts and of viruses. The mutant host population and the viruses need not drive the susceptible hosts and viruses to extinction. Owing to the assumed growth-rate cost of resistance $(r' < r)$, the susceptible hosts and viruses can continue to persist in the chemostat. Eventually, the system may converge to a steady state including a dominant population of resistant hosts and a subpopulation of susceptible hosts, whether infected or not, on which viruses persist (Table 4.2). This scenario was also suggested to be the mechanism of coexistence for cyanobacteria and cyanophage in some marine environments (Waterbury and Valois 1993). As in the $N–I–V$ model, this steady state need not be stable. Moreover, given that viruses do not infect the N' population, it may prove useful to examine the dynamics of the individual host populations rather than the total population.

Bohannan and Lenski did just that, by revisiting the dynamics of bacterio-phage T4 interacting with $E.\ coli$. They introduced a known resistant isolate at approximately 50 hours after the start of an experiment (Figure 4.6A). As should be apparent, the resistant bacteria grew in number and reached a nearly fixed level. In contrast, the original susceptible host population continued to oscillate along with the virus population. The peaks of virus density followed those of the host population, and the time between peaks increased. These results can be modeled, again using Eqs. 4.18, without any change in parameter values. The difference in interpretation is facilitated by distinguishing the

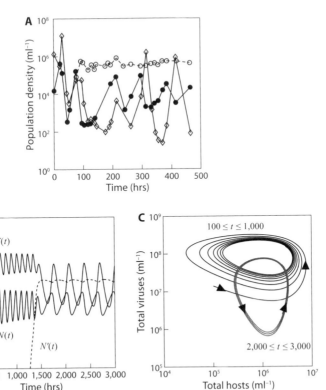

FIGURE 4.6. Cryptic host dynamics arising from virus-mediated host evolution. (A) Measured chemostat dynamics of total bacteria (*E. coli*) and free viruses (bacteriophage T4). A resistant host mutant was introduced to the chemostat vessel at $t \approx 50$ hrs (resistant bacteria, open circles; susceptible bacteria, filled circles; viruses, open diamonds). Points were extracted from Figure 2B of Bohannan and Lenski (1999) using PlotDigitizer 2.6.3. (B) The dynamics of uninfected hosts and free viruses. At time $t = 1000$, a mutant host arose that was resistant to viral infection. The parameters for the simulation are $r = 0.16$, $r' = 0.12$, $K = 2.2 \times 10^7$, $\eta = 0.25$, $\beta = 50$, $\phi = 10^{-9}$, and $\omega = 0.05$. (C) Phase-plane portrait of dynamics of free viruses and uninfected hosts, focusing on two portions of dynamics: (i) $100 \leq t \leq 1000$ and (ii) $2000 \leq t \leq 3000$.

separate host subpopulations. As can be seen in Figure 4.6B, the introduction of a resistant mutant gave rise to two apparent dynamics: a nearly stationary $N'(t)$ and predator-prey-like oscillations of $N(t)$ with $V(t)$. The nature of these oscillations is even more evident when viewed in the N–V phase plane (Figure 4.6C), where the separation in the dynamics occurred between $100 \leq t \leq 1000$ hours and between $2000 \leq t \leq 3000$ hours. The key take-away point here is that viruses can drive the evolution of host genotypes, that is, the fluctuations in the frequencies of N and N'.

How general is this effect? In other words, given the emergence of a mutant population, what combinations of parameters: $(r, r', \phi, \beta, \eta, K, \omega, J_0)$ lead (i) to invasion of the mutant and (ii) to viral-induced oscillations in host genotypes? The implicit condition for invasion was stated previously in Eq. 4.24, but the long-term outcomes include the possibility of extinction of the original virus-host pair as well as cryptic oscillations. Yoshida et al. (2007) claimed that cryptic oscillations could occur in a similar class of predator-prey models whenever (i) "defense was effective but not costly" and (ii) "coexistence steady state is locally unstable." These conditions were proved to apply in a narrow region of parameter space near bifurcation points, yet simulations suggest that they hold more generally. Numerical analysis of the current virus-host models confirms part of the conjecture of Yoshida et al. (2007), with one key exception. Small costs of resistance are not consistent with cryptic dynamics in the current model. The reason is that if there is not a sufficient cost for virus resistance, then the resistant host will invade and replace the susceptible host. This invasion also leads to the elimination of the virus. It remains an open question as to what global dynamics emerge—including destabilizing dynamics that approach extinction planes—when a virus, susceptible host, and resistant host are combined under chemostat-like conditions.

4.3   THE EFFECTS OF HOSTS ON VIRAL EVOLUTION

4.3.1   Viral Adaptation to Its Host

Virus-host interactions can also induce viral evolution. Viral mutants may differ in the hosts they can infect, as well as in their *quantitative* values of life history traits given the same host. These traits include the adsorption rate, latent period, burst size, and/or decay rate. Of course, the underlying genetic mutation may lead to changes in multiple traits simultaneously, resulting in so called pleiotropic effects. Just as discussed in the prior section, the success of mutants depends on the ecological context, so it seems natural to ask, what are the conditions that lead to invasion of a viral mutant into a community? Moreover, what is the expected change in viral life history traits given that viral mutants are continually generated as a result of infection? The population dynamics model from the previous chapter provides a foundation for formalizing these questions.

To begin, consider again a model of virus-host dynamics with implicit resources in a chemostat, as was introduced in Eq. 3.16:

$$\frac{dN}{dt} = rN(1 - N/K) - \phi NV - \omega N,$$

$$\frac{dV}{dt} = \beta\phi NV - \phi NV - \omega V.$$

(4.27)

This model presumed that every individual bacterium had the same life history traits, such that a single growth rate and carrying capacity was sufficient to describe the dynamics in the absence of viruses. Further, the model assumed that all viruses shared the same life history traits, such that their adsorption rate and burst size were identical. Both newly produced host cells and virus particles might differ from their parents in their life history traits. These offspring are termed *mutants*, and the parents are termed *residents*.

To evaluate the fate of viral mutants in a population requires a few simplifying assumptions. The first is to assume that the population has reached an equilibrium. At equilibrium in this model, recall that

$$N^* = \frac{\omega}{\tilde{\beta}\phi},$$

$$V^* = \frac{r(1 - N^*/K) - \omega}{\phi},$$

(4.28)

where $\tilde{\beta} = \beta - 1$. These densities imply that there are $\beta\phi N^*V^*$ new viruses produced per unit time. There should also be $\mu\beta\phi N^*V^*$ mutant viruses produced per unit time. The second simplifying assumption has to do with how small μ is. Mutation probabilities measured per nucleotide per replication vary with virus type. RNA viruses tend to have much higher mutation rates than do dsDNA viruses, yet even dsDNA viruses have higher mutation rates than do their hosts (Duffy et al. 2008; Sanjuán et al. 2010). Here, the relevant mutation rate is that of a change in a viral genome sequence that leads to a change in a particular life history trait. Even if the mutation probability is on the order of 10^{-6}, the rate of mutations that change a particular life history trait can be far lower. As a working hypothesis, assume that μ is so small that it is sufficient to consider, at most, two subpopulations at any given time: that of the wild type and that of the mutant. With these assumptions in mind, consider one mutant virus particle whose life history traits are β' and ϕ'. What is the likely fate of this founding mutant virus?

As it turns out, the fate depends on the life history traits of the mutant virus relative to those of the wild type. This should not be surprising, given that the concept of relative fitness of hosts was previously introduced in this chapter.

The starting point for any analysis of invasion is to establish the context for invasion. Consider resident hosts and viruses at densities N^* and V^*, respectively, as in Eq. 4.28. The rate of adsorption of a rare mutant virus to an individual host cell in the environment is $N^*\phi'$. Mutant viruses are washed out of the system at a rate ω. Therefore, the mutant is washed out before infecting a host with probability $\omega/(N^*\phi' + \omega)$. In turn, the probability that a mutant virus will infect a host before being washed out of the chemostat is $N^*\phi'/(N^*\phi' + \omega)$, producing β' new virus particles. Given the probabilities of these two alternative fates, the mutant virus will produce $\beta' N^*\phi'/(N^*\phi' + \omega)$ new viruses on average. A population of virus mutants can be established if

$$\beta' \frac{N^*\phi'}{N^*\phi' + \omega} > 1. \tag{4.29}$$

Recall that N^* is the host density, whose value is determined by the life history traits of the wild-type virus. Substitution of $N^* = \omega/(\tilde{\beta}\phi)$ gives the condition for the growth of a population of virus mutants as

$$\tilde{\beta}'\phi' > \tilde{\beta}\phi. \tag{4.30}$$

This same condition for the invasion of a viral mutant can be established via a local stability analysis of the three-dimensional dynamical system involving N, V and V' (Appendix C.2). In summary, viral mutants with larger values of the product of their burst sizes and adsorption rates are likely to establish themselves in the chemostat.

After the mutants are established, the dynamics can be described in terms of the continuous dynamics of three populations:

$$\frac{dN}{dt} = rN\left(1 - N/K\right) - \phi NV - \phi' NV' - \omega N,$$

$$\frac{dV}{dt} = \beta\phi NV - \phi NV - \omega V, \tag{4.31}$$

$$\frac{dV'}{dt} = \beta'\phi' NV' - \phi' NV' - \omega V'.$$

Here, $V'(t=0) \ll V^*$, as the density of mutant viruses is initially far less than that of resident viruses. There are a few ways to think about this expanded system. From a strictly mathematical point of view, the dynamics of this three-dimensional system can be considered as a perturbation from the equilibrium, $(N^*, V^*, 0)$. From a biological point of view, the question is whether the population of mutant viruses continues to increase or decrease.

In this system, the per capita growth rate of the mutant viral population can be written as

$$\bar{r}_{V'} \equiv \frac{1}{V'} \frac{dV'}{dt} = \beta' \phi' N - \phi' N - \omega. \tag{4.32}$$

The per capita growth of the population immediately upon introduction of the mutant virus can be approximated by assuming that $N \approx N^*$. The rationale for this choice is that the mutant virus is invading an environment in which densities are determined by the resident types. The instantaneous per capita growth rate is expected to be

$$\bar{r}_{V'} = \frac{\beta' \phi' \omega}{\tilde{\beta} \phi} - \frac{\phi' \omega}{\tilde{\beta} \phi} - \omega \tag{4.33}$$

$$= \omega \left(\frac{\beta' \phi'}{\tilde{\beta} \phi} - \frac{\phi'}{\tilde{\beta} \phi} - 1 \right) \tag{4.34}$$

$$= \frac{\omega}{\tilde{\beta} \phi} \left(\tilde{\beta}' \phi' - \tilde{\beta} \phi \right). \tag{4.35}$$

In other words, $\bar{r}_{V'} > 0$ when $\tilde{\beta}' \phi' > \tilde{\beta} \phi$. When viral life history traits favor establishment of a viral mutant, then the number of such mutants in the population is expected to grow over time. What about the resident type? At least initially, its per capita growth rate can be written as

$$\bar{r}_V \equiv \frac{1}{V} \frac{dV}{dt} = \beta \phi N^* - \phi N^* - \omega. \tag{4.36}$$

At equilibrium, $\bar{r}_V = 0$ given $N^* = \omega / (\tilde{\beta} \phi)$. This makes sense because the per capita growth rate of the resident virus should be zero at equilibrium. Altogether, this analysis implies that a mutant virus whose life history traits favor establishment should tend to subsequently increase in number relative to the residents.

Figure 4.7A shows an extreme example in which a mutant virus type with burst size 100 invades an environment with a single resident virus whose burst size is 25. Not only does the mutant population increase in number, it replaces the resident population. In the process, the frequency of mutants increases from rare to common and eventually to 1. In addition, the average burst size of viruses in the population increases from $\beta = 25$ to $\beta' = 100$. That the mutant virus replaces the resident virus is consistent with the recognition that both viruses are competing for the same host. The resident virus must interact with at least $N^* = \omega / (\tilde{\beta} \phi)$ hosts for it to persist. The mutant virus

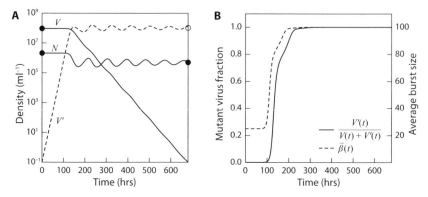

FIGURE 4.7. Invasion of mutant virus, followed by replacement of the resident virus by the mutant virus. (A) Dynamics of N, V, and V'. Points at $t = 0$ are the equilibrium densities of N^* and V^*. Points at the end of the dynamics are the new equilibrium densities of N^* and V'^* after invasion. (B) Dynamics of the relative fraction of mutants and the average burst size. Parameters of the initial model are $r = 0.16$, $K = 2.2 \times 10^7$, $\phi = 10^{-9}$, $\omega = 0.05$, and $\beta = 25$. The mutant virus has life history traits $\beta' = 100$ and $\phi' = 10^{-9}$.

can persist even when host density drops to $N^* = \omega/(\tilde{\beta}'\phi)$. Since $\beta' > \beta$, then $\tilde{\beta}' > \tilde{\beta}$; the mutant virus has reduced the resource for the resident virus below a critical threshold. From an equilibrium perspective, there is no simultaneous solution to Eqs. 4.31 in which $N^* > 0$ and both $V^* > 0$ and $V'^* > 0$. The equilibrium $(N^*, V^*, 0)$ is unstable with respect to perturbations in V'. In contrast, the equilibrium $(N'^*, 0, V'^*)$ is stable with respect to perturbations in V. In other words, although V' can invade V, V cannot invade V'. The reasons are precisely as described earlier, since the conditions favoring V' invasion of V would have to be satisfied when the roles are reversed, but they cannot be, at least not given the structure of the present model.

Is this behavior generic? For many ecological models, including this one, *invasion implies replacement.* Let us unpack this phrase. Invasion means that the per capita growth rate of the mutant is positive when the mutant is rare. The term *replacement* implies that when the invasion fitness of the mutant is positive then the population of the mutant type increases and that of the resident type decreases toward extinction. The mutant then becomes the resident. If such events happen consecutively, then one can imagine a sequence of replacements of residents by mutants, who themselves become the residents and then get replaced by even better adapted mutants, and so on. This sequence of events represents the adaptation of a population to its environment. In the case of viruses, the environment includes its host or hosts.

Although the model structure has been intentionally simplified, this simplification has enabled the examination of adaptation of viral life history traits via the mechanism of a series of adaptive sweeps. Sequences of such adaptive sweeps can be formalized in terms of a theory of adaptive dynamics describing how phenotypic traits change over evolutionary time.

4.3.2   FROM ADAPTIVE SWEEPS TO ADAPTIVE DYNAMICS

The theory of adaptive dynamics considers how phenotypes evolve in response to repeated invasions of rare mutants. It was formalized in a series of papers in the mid-to late 1990s (Dieckmann and Law 1996; Geritz et al. 1997, 1998; Dieckmann and Doebeli 1999). Adaptive dynamics builds on the following assumptions regarding the slow evolutionary dynamics in which phenotypes change over time in response to the ecology of the system: (i) mutations are rare; (ii) mutations have small effect; (iii) invasion implies replacement, except at branching points; (iv) equilibrium dynamics converge to a fixed point; (v) populations are assumed to be monomorphic. All these conditions can be relaxed in generalizations of the theory. The first assumption implies a crucial feature of this theory: ecological dynamics are very fast compared with evolutionary dynamics. Obviously, there are limitations to the theory, and, as with all theories, it has its critiques and debates (Waxman and Gavrilets 2005; Geritz and Gyllenberg 2005). Nonetheless, the slow-evolution limit is particularly useful in thinking about how ecology affects the evolutionary change of life history traits. The theory of adaptive dynamics has been derived and treated extensively elsewhere (Dercole and Rinaldi 2008). Here, the focus is on predicting the evolutionary dynamics of the average viral phenotypic trait.

A few preliminaries are in order before such predictions can be made. These preliminaries are of a general nature, consistent with the terminology of the field. To begin, consider a monomorphic population with N individuals each of which has the same life history trait x. Occasionally, a rare mutant type is produced with a distinct life history trait x'. Will the next mutant to invade and replace the resident have a trait $x' > x$ or $x' < x$? To invade, the next mutant should have a positive invasion fitness when rare, for example, $\bar{r}_{V'} > 0$ in the context of viral mutant invasion. In adaptive dynamics, $s_{x'}(x)$ is defined as the invasion fitness of a rare mutant with phenotype x' in an environment set by a population of individuals each with phenotype x.[2] The phrase "set by a

---

[2] The traditional adaptive dynamics derivations use the formalism $s_y(x)$ for the invasion of a rare mutant with phenotype y invading an environment set by a population each with phenotype x. Here we use the x' notation for the mutant, given two coevolving populations.

population" denotes that other features of the environment, such as resource availability, will reach an equilibrium partially determined by the trait value x. In adaptive dynamics, mutants are assumed to have small trait differences from the resident. Therefore, the relevant invasion fitness is for values of $x' = x + \delta$, where δ is some small deviation in trait space. The invasion fitness for mutants whose traits are similar to those of residents can be approximated via a Taylor expansion as

$$s_{x'}(x) \approx s_{x'}(x)|_{x'=x} + \left.\frac{\partial s_{x'}(x)}{\partial x'}\right|_{x'=x} \delta + \mathcal{O}(\delta^2) \tag{4.37}$$

$$\approx \cancel{s_{x'}(x)|_{x'=x}} + \left.\frac{\partial s_{x'}(x)}{\partial x'}\right|_{x'=x} \delta + \cancel{\mathcal{O}(\delta^2)} \tag{4.38}$$

$$= \left.\frac{\partial s_{x'}(x)}{\partial x'}\right|_{y=x} \delta, \tag{4.39}$$

where $s_x(x) = 0$, as an invader with the same phenotype would neither increase nor decrease in population. In words, the direction of trait evolution depends on the *derivative* of the invasion fitness with respect to trait values of the mutants.

Evolution should favor an increase in the average trait of the population if mutants with phenotype $x' > x$ have positive invasion fitness values, and mutants with phenotype $x' < x$ have negative invasion fitness values. Evolution should lead to decreases in the average trait of the population if mutants with phenotype $x' > x$ have negative invasion fitness values, and mutants with phenotype $x' < x$ have positive invasion fitness values. Figure 4.8 illustrates this concept. An additional insight from adaptive dynamics is that not only the direction but the strength of adaptation is determined by the environment that the resident population sets. Once the mutant invades, it replaces the resident and is then subject to other invasions. In this way, the average trait value can change over evolutionary time, as measured in units of invasion events or time between events. Alternative outcomes that arise in this framework, including evolutionary branching, will be discussed in Chapter 5.

The process of invasion and replacement lead to stochastic evolutionary trajectories in phenotypic traits. This stochasticity is in addition to that arising from demographic stochasticity of ecological dynamics, that is, the discrete processes of birth and death of individuals. The prior example illustrated how a mutant virus with burst size 100 could invade and replace a resident virus with burst size 25. However, there can be variation in the time of first appearance of the mutant or, alternatively, the burst size associated with the first mutant that can outcompete the resident. If the monomorphic trait of the population is x, then the population dynamics can be written as

$$dN(x, t)/dt = f(x, N, E), \qquad (4.40)$$

such that the equilibrium of the population N^* for a given trait value x occurs so long as $f(x, N^*, E) = 0$ in an environment E. Further, let $P(x, t)$ be the probability that the population has the monomorphic trait value x at time t. The time t refers to evolutionary time, again assuming that ecological dynamics are very rapid compared with evolutionary dynamics. Over evolutionary timescales, the average trait value can be written as

$$\hat{x}(t) = \int dx\, x\, P(x, t). \qquad (4.41)$$

The dynamics of the average trait value can be written as

$$\frac{d\hat{x}}{dt} = \int dx\, x\, \frac{dP(x, t)}{dt}. \qquad (4.42)$$

Predicting the evolutionary dynamics of traits requires a theory of the change in probability, $P(x, t)$, that a population has a particular trait value x at a particular moment in time.

To predict the long-term evolutionary dynamics of a trait under selection, the possible stochastic outcomes of individual evolutionary trajectories must be defined. The probability that individuals of the resident population have a particular trait value x at time t follows the master equation (van Kampen 2001)

$$\frac{d}{dt} P(x, t) = \int dx' \left[w(x|x')P(x', t) - w(x'|x)P(x, t) \right], \qquad (4.43)$$

where $w(x'|x)$ is the probability per unit time of the transition $x \to x'$. The transition probability rate, $w(x'|x) = \mathcal{M}(x', x)\mathcal{D}(x', x)$, is the product of the mutation generation rate, $\mathcal{M}(x', x)$, and the probability, $\mathcal{D}(x', x)$, that mutants survive to outcompete the resident type. According to the steps outlined in Appendix C.3, the dynamics of the mean resident trait are

$$\frac{d\hat{x}(t)}{dt} = \frac{1}{2}\mu\sigma_\mu^2 N^*(\hat{x}) \frac{\partial s_{x'}(\hat{x})}{\partial x'} \big|_{x'=\hat{x}}, \qquad (4.44)$$

where μ is the mutational probability per birth, and σ_μ^2 is the variation of the mutational distribution. This is a generic equation for the dynamics of the mean phenotypic trait, known as the *canonical equation of adaptive dynamics*

(Dieckmann and Law 1996). From an ecological view the two key terms are N^*, the steady-state density of the population, and $\partial s_{x'}(\hat{x})/\partial x'|_{x'=\hat{x}}$, the selective derivative, which determines the direction of phenotypic evolution. Moreover, the population size also determines the magnitude of phenotypic change, because it sets the scale at which mutants are generated. The simplest model of evolution arising from this theory of adaptive dynamics is akin to "hill climbing," albeit where the hill is constantly being modified by the walker. This process need not continue inevitably. For example, the evolution in traits may lead to the convergence of the trait to a value x^* such that mutants with phenotypes $x' > x^*$ and $x' < x^*$ have negative invasion fitness (Figure 4.8). This is an example of convergence to an evolutionarily stable strategy in that alternative mutants cannot invade when rare (Smith 1973; Geritz et al. 1998). The fitness of individuals with trait x^* is said to have reached a maximum in the environment.[3] These preliminaries enable the examination of the evolution of a particular virus life history trait: the latent period.

4.3.3   EVOLUTION OF THE VIRAL LATENT PERIOD

There are many possible traits to consider when thinking about viral evolution, for example, the target host, efficiency of entry, or efficiency of host utilization. The evolution of latent period has the advantage of both theoretical and experimental bases of support (Wang et al. 1996; Abedon et al. 2001; Wang 2006; Shao and Wang 2008). What should be the "optimal" latent period for a given virus when infecting a fixed host? The rationale for building up one's intuition that latent period might evolve to some intermediate value can be illustrated with the following example.

Consider the infection of a single host by a virus whose latent period is τ and whose burst size is β. How many viruses will the infection produce, on average? (*Hint*: The answer is not β). Even when forms of cellular stochasticity and variation in viral life history traits are neglected, the average *realized* number of virions produced per infection is the burst size multiplied by the probability that the virus is able to complete its infection cycle.

Hosts live in a fluctuating environment. If, for example, the mortality/ washout rate of hosts is ω, then the probability that the virus will complete its infection cycle is $e^{-\omega\tau}$. The average number of viruses produced per infection given a mortality rate ω is $\bar{\beta} = \beta e^{-\omega\tau}$. There is a trade-off between β and τ, such that a virus must extend its latent period to yield more progeny

---

[3] The next chapter discusses what happens when the fitness of individuals with trait x^* converges, instead, to a minimum.

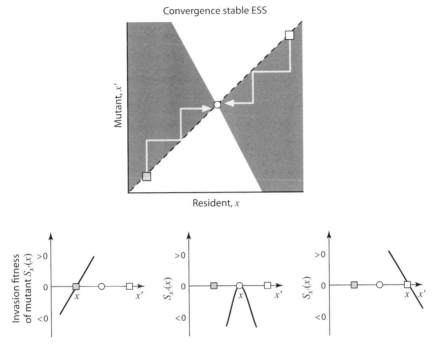

FIGURE 4.8. Schematic of ecologically mediated evolutionary dynamics, that is, adaptive dynamics. (Top) A pairwise invasibility plot, in which the sign of the invasion fitness, $s_{x'}(x)$, is shown for combinations of resident, x, and mutant, x', trait values. Shaded areas denote positive invasion fitness, and white areas denote negative invasion fitness. The squares denote possible starting points for an evolutionary dynamics. The circle denotes the endpoint, an *evolutionarily stable strategy* (ESS). The orthogonal arrows denote a possible sequence of invasion and replacement events in trait space toward the ESS. (Bottom) Local structure of the fitness landscape, as quantified in terms of the invasion fitness. Each of the panels denotes $s_{x'}(x)$, on the y-axis, as a function of x', on the x-axis. The parameter x is, from left, the shaded square, the open circle, and the open square. The leftmost panel denotes the fact that mutants x' with traits larger than the resident x have positive invasion fitness. The rightmost panel denotes the fact that mutants x' with traits smaller than the resident x have positive invasion fitness. The center panel denotes the fact that no mutant type can invade the resident, because all have negative invasion fitness. *Key point*: Evolution in this scenario over the long term will converge to the ESS.

(see Chapter 2). The optimal latent period for a single infection should be the value of τ^* that satisfies

$$\left. \frac{\partial \rho(\tau - \tau_e)e^{-\omega\tau}}{\partial \tau} \right|_{\tau=\tau^*} = 0, \tag{4.45}$$

where ρ is the intracellular virus production per unit time after an eclipse period τ_e. The solution to this equation is $\tau^* = \tau_e + 1/\omega$. As mortality risks to the host cell increase, then it is better—for the virus—to lyse early, whereas as mortality risks decrease, it is better—for the virus—to lyse later. For very low mortality risks, limitations on intracellular resources will eventually limit the optimal τ^*, ensuring that it is neither too low nor too high. The drawback to this approach is that the logic regarding optimality considers only a single replication cycle and does not take into account the potential feedback between viral infection and host availability. The long-term fitness of viruses is the relevant factor in considering the evolution of traits.

A significant step in this direction was taken by Ing-Nang Wang, who combined experiment and theory to characterize the extent to which intermediate latent periods are more fit than rapid or delayed latent periods (Wang 2006). The measure of fitness defined by Wang was $1/T \log (V_T/V_0)$, that is, the per capita growth rate of the viral population. In this setup, viruses and bacteria were combined in a batch culture experiment, given an initial growth medium. The latent period of virus strains differed, such that those viruses with shorter latent periods had smaller burst sizes, and those with longer latent periods had larger burst sizes. It was hypothesized that viral fitness would reach its maximum at intermediate latent periods (Figure 4.9), the rationale being that early latent periods lead to many more opportunities for reinfection. This increases the effective number of generations of viruses of that strain available to propagate on hosts. The cost of early lysis is that each replication produces fewer viruses. In contrast, long latent periods result in more viruses but at the cost of a reduced number of opportunities for each virus to find a new host and repeat the cycle. An additional subtlety here is that the bacteria grew during the experiment at a rate of approximately two divisions per hour. Viruses in the batch culture at later times adsorb to host cells more rapidly than at the beginning, because more host cells are available.

The N–I–V model with an explicit time delay τ from Chapter 2 can be adapted, at least over short times, for simulating batch culture experiments by setting $\omega = 0$, which designates the absence of inflow and outflow. Here, numerical simulations of in silico batch cultures were implemented using host life history traits as estimated by Wang (2006) and using in silico viral mutants whose traits (τ, β) follow the relationship $\beta = 7.6(\tau - \tau_e)$, where $\tau_e \approx 0.5$ hr. The experimental fitness values of viral mutants and model fitness values of in silico mutants display the same qualitative and even quantitative features (Figure 4.9). The strains with intermediate latent periods are most fit, reflecting the balance between the benefits of early release, with more cycles, and that of later release, with more output per cycle.

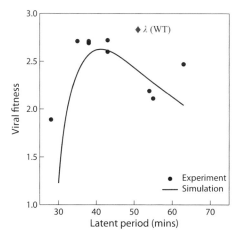

FIGURE 4.9. Fitness peaks at intermediate latent periods. Experimental data are from Wang (2006). Each black circle represents a distinct phage mutant with a distinct S gene. The fitness of each virus is calculated as $1/5 \log V_5/V_0$, where V_t is the virus density in the medium at time t, with the fitness of the WT strain with the wild-type S gene shown as a diamond. Theoretical estimates of fitness were calculated here as a result of simulated dynamics of the N–I–V model of Eq. 3.40, described in Chapter 3. Simulations included life history parameters from Wang (2006): $r = 2$, $K = 7.35 \times 10^9$, $\phi = 10^{-8}$, $\omega = 0$. Further, in silico viral mutants were evaluated over the range $28.6 < \tau < 63$ min given the predicted empirical trade-off between latent period and burst size, $\beta = 7.6(\tau - \tau_e)$ (see Table 2 of Wang (2006)). There are no free parameters in the fit.

The preceding-analysis focused on a single or limited number of cycles. The examination of viral latent period evolution can be extended to multiple invasion events using the theory of adaptive dynamics. To do so, recall again the model of N–I–V dynamics in a chemostat from Chapter 3, albeit with exponentially distributed delays between infection and lysis.[4] A viral mutant is characterized by its life history traits, η' and β', assuming no change in the adsorption rate ϕ. In this case the dynamics of the host-resident virus–mutant virus system can be written as

$$\frac{dN}{dt} = rN\left(1 - \frac{N + I + I'}{K}\right) - \phi NV - \phi NV' - \omega N, \qquad (4.46)$$

$$\frac{dI}{dt} = \phi NV - \eta I - \omega I, \qquad (4.47)$$

---

[4] The interested reader should refer to Bonachela and Levin (2014) for a treatment of the different inferences drawn from evolutionary dynamics where the latent period is modeled as fixed versus continuous.

$$\frac{dV}{dt} = \beta \eta I - \phi N V - \omega V, \tag{4.48}$$

$$\frac{dI'}{dt} = \phi N V' - \eta' I' - \omega I', \tag{4.49}$$

$$\frac{dV'}{dt} = \beta' \eta' I' - \phi N V' - \omega V'. \tag{4.50}$$

The mutant virus is invading an equilibrium set, in part, by the life history traits of the resident virus. Given that I' and V' dynamics depend only on each other and the value of N, the invasion of the mutant can be calculated based on the stability of the subsystem

$$\begin{aligned} \frac{dI'}{dt} &= \phi N^* V' - \eta' I' - \omega I' \\ \frac{dV'}{dt} &= \beta' \eta' I' - \phi N^* V' - \omega V' \end{aligned} \tag{4.51}$$

near the equilibrium

$$N^* = \frac{\omega(\eta + \omega)}{\phi \left(\beta \eta - (\eta + \omega) \right)} \tag{4.52}$$

(see Eq. 3.34 of Chapter 3). Here we assume that the original equilibrium is a fixed point, as in Bonachela and Levin (2014). Analysis of the general case and its influence on fluctuating coexistence in evolutionary dynamics is warranted.

After standard methods for linearizing this system near its fixed point are applied, the eigenvalues of Eqs. 4.51 are positive whenever

$$\frac{\omega(\eta + \omega)}{\phi \left(\beta \eta - (\eta + \omega) \right)} > \frac{\omega(\eta' + \omega)}{\phi \left(\beta' \eta' - (\eta' + \omega) \right)} \tag{4.53}$$

or, alternatively, whenever $N^* > N'^*$. The mutant virus with a different average latent period and burst size can invade a resident virus if it draws down the host population to a lower equilibrium than that set by the resident virus. This is the same principle as for the simpler model that ignored details of latency. This result can be anticipated so long as invasion implies replacement; that is, whichever consumer can draw its resources to a lower level represents the most fit type in the system. Consider that the possible mutants include a maximum value of $\beta(\eta)$ that saturates as $\eta \to 0$. In this case, η should evolve toward a single, optimal η^*, which for most smooth trade-off functions will represent the global optimum, or what a game theorist would call an *evolutionarily stable strategy* (ESS) (Smith 1973).

This evolutionary process can be simulated using any of the previously introduced population dynamics models as a basis: the N–V model or N–I–V model with finite or exponentially distributed delays. In the interest of presenting the simplest model, consider a simulation of the N–V model assuming that $\beta = \rho(\tau - \tau_e)e^{-\omega\tau}$ for fixed $\rho = 7.6$ viruses/min, and $\tau_e = 28.6$ min. This trade-off curve was measured in Wang (2006) for infections of *E. coli* variants by phage λ variants. The preference for early or late lysis depends on the risk to long latent periods. This risk is embodied in the washout rate. To examine the evolution toward intermediate latent periods, consider the scenario where $\omega = 1$. In this case, the latent period is predicted to converge toward an equilibrium and to remain there, uninvasible by other mutants that are restricted to the trade-off curve. The canonical equation of adaptive dynamics (Eq. 4.44) can then be compared with stochastic simulations of a "slow" evolutionary process.

In brief, the stochastic simulations proceed as follows. First, the ecological equilibrium is determined for a given resident virus with latent period τ. Then, a random mutant virus is generated with latent period τ' deviating from the resident by a normal distribution with mean 0 and standard deviation σ_μ. Whether the mutant invades is determined by calculating Feller's rule for branching random walks, $p = 1 - d/b$, where d is the death rate and b is the birth rate as calculated based on the mutant life history trait and the resident equilibrium. If it invades, then the mutant becomes the resident, and the process continues. Extended details of the simulation approach are explained in Appendix C.4. The results shown in Figure 4.10 demonstrate the evolutionary dynamics of the latent period phenotype over time, in an ecological context. Two key features emerge. The evolutionary dynamics converge to the optimal value of τ^*, irrespective of whether the initial resident has a latent period below or above this optimum. And the adaptive dynamics theory agrees, with no further fitting parameters, with the stochastic trajectory.

In closing, the present examination of viral latent periods suggests that viral life history traits are evolvable, both in theory and in practice. The "optimal" values are reached by a process of random mutations and natural selection that depend on the environment, including the life history traits of both viruses and hosts. But hosts can also evolve, and so a complete analysis of virus and host evolution requires consideration of *co*evolutionary dynamics.

4.4   SUMMARY

- Preexisting variation in host phenotypes include variants with different levels of susceptibility to viruses, including complete resistance.

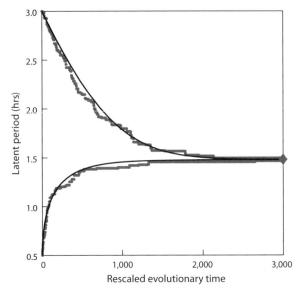

FIGURE 4.10. Evolutionary dynamics of viral latent period converge to an evolutionarily stable strategy. Stochastic simulations (gray circles) are compared with the canonical equation of adaptive dynamics (black line). Parameters common to both models are $\mu = 10^{-7}$, $\sigma_\mu = 1/60$, $K = 7.5 \times 10^9$, $\omega = 1$, $r = 2$, $\phi = 10^{-8}$, $\rho = 7.6 \times 60$, and $\tau_e = 28.6/60$, where time is in hours, and densities are in units of mililiters. The predicted optimum is denoted by the large gray diamond, $\tau^* = 1.48$. Details of the simulation methods are available in Appendix C.4.

- Formative studies of the basis of the mutation rate relied upon virus-host interactions and the possibility of the evolution of resistance to infection.
- Viruses represent a strong selective pressure and can induce evolution among hosts.
- Host evolution, as induced by viruses, includes novel forms of ecological dynamics, including cryptic dynamics.
- Infection of hosts represents a strong selective pressure for viruses.
- Viruses that differ in their life history traits vary in their fitness and can invade and replace existing viral strains.
- The latent period represents a model trait for the further study of the evolution of intermediate phenotypes.
- Evolution among other traits is also possible, including who infects whom.

CHAPTER FIVE

Coevolutionary Dynamics
of Viruses and Microbes

5.1   FROM SENSITIVITY RELATIONS TO COEVOLUTION

In 1945, Salvador Luria examined interactions arising between multiple strains of bacteriophage and bacteria (Luria 1945). The result was what we would now term an *infection network* but what Luria called "sensitivity relations" (Figure 5.1). Despite the change in terminology, the concept remains the same. Luria exposed the *E. coli* strain B to the virus strains α and γ. Luria then isolated 11 bacterial variants and four phage variants that evolved through repeated exposure of hosts to viruses. The bacteria differed in their degree of susceptibility. Some bacteria could be infected by all four viruses, but others could be infected by only one, two, or three viral types and, in one case, none of the viral types. Similarly, the viruses differed in their infectivity; viruses infected anywhere from three to nine hosts.

At the time, virus-host interactions were thought to be mediated by physiochemical interactions between the viral capsid and the host surface. The site of contact was called a "receptor" (Burnet 1930), a term and idea still in use. Surface receptors include outer membrane porins, carbohydrates, and, occasionally, lipids within the cell surface (Labrie et al. 2010). Mutations in a single protein or in pathways involved in metabolic production of lipids and carbohydrates may have the effect of changing host surface structure and, consequently, susceptibility to infection. In parallel, mutations in viral genes may affect the capsid and tail structure of virus particles or, possibly, the lipid membrane inside the capsid. According to Luria: "mutation to virus-resistance may be compensated for by an independent complementary mutation of the virus" (Luria and Delbrück 1943). We now know that defense mechanisms include both extracellular and intracellular mechanisms (Labrie et al. 2010). Similarly, counterdefense mechanisms may affect the structure, genotype, or intracellular regulation of the virus. For example, the CRISPR system involves interactions that unfold inside the cellular cytoplasm (Barrangou et al. 2007;

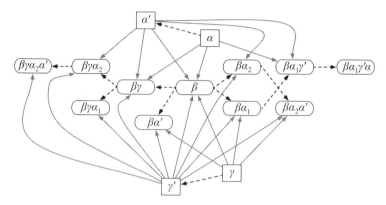

FIGURE 5.1. Phage-bacteria infection network adapted from Luria (1945). Bacteria are denoted by ovals, and phage by squares. The original caption reads: "The sensitivity relations between mutant bacterial strains and virus mutants. Broken arrows indicate mutations. Solid arrows indicate activity of a virus on a bacterial strain."

Marraffini and Sontheimer 2010; van der Oost et al. 2014). Nonetheless, the twin concepts of mutation and countermutation or, more generally, defense and counterdefense, remain the basis for coevolutionary dynamics unfolding in virus-host systems.

This early example demonstrates the potential for both viruses and hosts to change who they interact with via mutation and selection. One of the conclusions of Luria's study was that bacteria could evolve resistance to all co-occurring phage (see strain $B\alpha_1\gamma'\alpha$ in Figure 5.1). This conclusion—if broadly applicable—would imply that phage would have limited opportunities to coevolve along with hosts in natural environments. This idea continued to predominate 40 years later. Richard Lenski and Bruce Levin documented an example in their collaborative work (Lenski and Levin 1985). They inoculated 0.15 ml chemostats with *E. coli* B strains (derived from Luria's original isolation) along with phages T2, T4, T5, and T7, respectively. For each combination, bacterial density declined initially due to viral lysis. This decline was followed by an apparent recovery of the bacterial population to densities expected under resource-limited conditions (e.g., the recovery process took approximately 100 hours in the case of replicates with T4). The dominant host strain in the recovered population was determined to be phage resistant. Nonetheless, a background level of phage persisted, by infecting and lysing a small subpopulation of susceptible hosts. The total phage abundances were 1/1000 of the total bacterial abundances. These results led Lenski and Levin to conclude that "the coevolutionary potential of virulent phage is less than that of their bacterial hosts." It seemed that coevolution could occur but could be short-circuited by the host.

In 2002, Angus Buckling and Paul Rainey proposed a different host-virus system for studying coevolution in the laboratory. They used *Pseudomonas fluorescens* SBW25 and its associated phage, SBW25Φ2 (Buckling and Rainey 2002a, 2002b). In their experiments using 6 ml static flasks, inoculated with both bacteria and phage, Buckling and Rainey transferred 60 μL of culture to fresh media every 2 days. The resulting environment was spatially structured and included gradients in oxygen and other metabolic by-products. Previous studies had identified spatial heterogeneity as a potential mechanism by which viruses could persist with hosts during experiments long enough to enable coevolutionary dynamics to take place (Schrag and Mittler 1996). In Buckling and Rainey's experiments, both bacteria and phage communities persisted, first over approximately 80 generations (Buckling and Rainey 2002b) and then over 400 (Buckling and Rainey 2002a). In the process, the sensitivity of evolved bacteria and phage changed, just as Luria, Lenski, and Levin had observed, but the coevolutionary dynamics did not terminate with a bacteria resistant to all phage. Buckling and Rainey concluded that "phage are not fundamentally constrained in their ability to coevolve with bacteria" (Buckling and Rainey 2002a). The following year, Mizoguchi et al. (2003) made similar observations of persistent coevolutionary dynamics between *E. coli* and phage PP01, this time in chemostats. These experiments used nutrient-rich broth in an effort to simulate gut-like environments. The results included persistent phage communities in which total phage densities exceeded bacterial densities 100 hours after co-inoculation. Marston et al. (2012) observed similar results in cyanophage-cyanobacteria chemostats. A recent compilation of other coevolution experiments contains even more examples of persistent coevolution using phage-host systems (Dennehy 2012)). Altogether, a new paradigm has emerged and with it a question: what can explain the long-term persistence, rather than the termination, of coevolution of phage and bacterial hosts?

Addressing this question requires stepping back and formalizing some of the terms that are core to the study of coevolutionary dynamics. Recall that evolution refers to the heritable change in the frequency of genotypes over time. Similarly, coevolution refers to the heritable change in the frequency of genotypes of two distinct population types over time. These populations may be plants and pollinators, predators and prey, or viruses and their microbial hosts. The term *coevolutionary dynamics* is conventionally applied when the following conditions are met: (i) two types of populations, usually corresponding to two different species, have subpopulation variation; and (ii) the frequencies of the subpopulations within each type vary with time. Conditions for long-term coevolutionary dynamics, even in constant environmental conditions, often require that there be frequency- or density-dependent selection, such that no single type can dominate over time. Instead, the fitness landscape changes with the emergence of different types. The change in abundances of and even in identities of types provides opportunities for new invasions (van Valen 1973; Dieckmann and Doebeli 1999; Nowak 2006).

The use of the term *coevolutionary dynamics* proves somewhat problematic when considering viruses and microbes. First, the definition of a bacterial species, to say nothing of a virus species, is fraught with controversy (Fraser et al. 2009). Instead of entering this debate, the convention we use here is to distinguish types based on whether they have distinct phenotypes for which the phenotypic variation has a genetic basis. This convention excludes consideration of dynamics in which a population displays phase variations unrelated to their genotypes. An example of such phase variation occurs in bacterial persisters that transition into a dormant-like state in which they are often recalcitrant to antibiotics and even viral infection (Pearl et al. 2008). Second, there is often a presumption that evolutionary dynamics are slow, while ecological dynamics are fast. In other words, the total population densities change relatively quickly, while the relative frequencies of genotypes in a population change relatively slowly. This presumption has dominated the study of ecology, and certainly theoretical ecology, in the past. As is increasingly being realized, "rapid" evolutionary dynamics are ongoing in many predator-prey (Yoshida et al. 2007; Hiltunen et al. 2014), host-pathogen (Duffy and Sivars-Becker 2007), and certainly virus-microbe systems (Bohannan and Lenski 1999, 2000). In the rapid limit, the frequency of virus and host types and the total populations of viruses and hosts change concurrently.

In this chapter, I will examine how a single pair of viruses and hosts can diversify into two coexisting virus and host types via a combination of mutation and selection. Then, I will explore the effect of coevolutionary dynamics on ecology, that is, how characteristic features of virus-host cycles can be modified when the viruses and hosts possess subtype variations, each with its own distinct life history traits. Next, I will explore the effect of ecology on coevolutionary dynamics, that is, how virus-host interactions provide the context for selection of previously rare types. In some instances, such interactions may give rise to ongoing changes in the dominant members of a complex virus and host community, and even to the transition from low-diversity to high-diversity communities.

5.2   TOWARD "NOVEL" COEVOLUTION: ON THE PROBABILITY OF COMPENSATING MUTATIONS

5.2.1   COEVOLUTIONARY POKER: STARTING WITH ONE PAIR

Chapter 4 presented evidence concerning mechanisms by which hosts mutate to become resistant to infection by a virus population. But whether resistant hosts will increase in relative abundance depends on other factors. If resistance comes at too great a cost, then resistant hosts may die out. If resistance comes at too small a cost, then resistant hosts may replace susceptible hosts, driving them and the original virus population to extinction. For intermediate costs of resistance, the evolution of host resistance can lead to oscillations in the abundances of the two host types, one susceptible and one resistant. Other possible outcomes include the possibility that a virus mutant can emerge. If the virus mutant possesses counterresistance, then it will be able to infect the resistant host. Further, if these viral counterdefense mutants can become substantially abundant, a selective benefit may arise for the emergence of a second host mutant, perhaps resistant to both viral types. Coevolutionary dynamics may unfold in this way.

To begin, consider a host-virus system at equilibrium with densities of H^* hosts and V^* viruses. Further, assume that viruses are removed from the system at a rate of m per unit time, such that the density of new virus particles generated per unit time must be mV^*. What happens to a single resistant host cell of type H' that appears in the system at time $t = 0$? For example:

- What is the probability that a counterdefense viral mutant, V', is already present in the community at $t = 0$?
- If a counterdefense viral mutant is present, what is the probability it will infect and lyse the single host mutant?

Answering these questions requires more information regarding the host range of viral mutants, V' (see Figure 5.2). *Host range* is defined as the set of hosts that a given virus can infect. One possibility is that viruses of type V' can infect and lyse hosts of type H' but cannot infect and lyse hosts of type H. In that case, there is no selective benefit for V' to invade in the absence of H'. Other assumptions may also yield the same conclusion, for example, if there are significant trade-offs in the life history traits of V'. Because births must balance deaths at equilibrium, then resident viruses should be produced at a

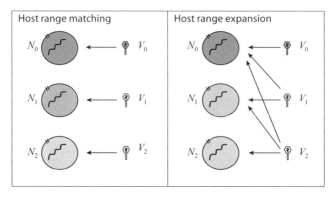

FIGURE 5.2. Schematic of infectivity between hosts—the resident host (N_0) and mutants (N_1 and N_2)—and viruses—the resident (V_0) and mutants (V_1 and V_2). Solid lines correspond to strong infection, and the absence of lines corresponds to no infection. The left panel illustrates a model in which each virus specializes on one host. The right panel illustrates a model of host-range expansion in which the mutant virus can infect the resident and subsequent mutant(s).

rate of mV^*. A fraction μ of these will be counterresistance mutants. Here, the value of μ specifically refers to the probability that a new virus has a distinct infection phenotype compared with the resident virus, enabling it to infect H' but not H. In the absence of H', the viral mutant dynamics follow

$$\frac{dV'}{dt} = \mu m V^* - m V', \tag{5.1}$$

assuming that mutant viruses decay at the same rate as the resident. Thus, there should be a background density of approximately $V'^* = \mu V^*$ mutant viruses on average.

The problem of predicting whether at least one mutant virus is present depends on the size of the environment or experiment of interest. The units of V and of V' are particles per unit volume, for example, particles/ml. In a sufficiently large volume, there will always be at least one mutant. Whereas using densities to track population and evolutionary dynamics is often appropriate, it is decidedly less so when one is interested in the probability of a single event and not a fractional event! There should be approximately $\mu V^* S$ viral mutants in a finite volume of size S. In practice, V^* is often on the order of 10^7/ml. Mutation rates involving a nucleotide substitution for a virus range from $\approx 10^{-7}$ for dsDNA viruses to $\approx 10^{-4}$ for ssRNA viruses (Figure 2.6 of Chapter 2). However, the phenotype mutation rate, μ, relevant here may be far smaller. The preferred way to estimate this effective phenotype mutation rate is to grow resident viruses from single isolate on a susceptible host. This process should generate a small background of counterdefense mutants. Then, the virus population should be grown again on a host that they cannot infect. The number and variance of measured plaques among replicated plates can be used to estimate the phenotype mutation rate (as per the Luria-Delbrück fluctuation test, introduced in Chapter 4). For convenience, assume that the mutation rate is $\mu = 10^{-9}$, that is, approximately 100-fold less than nucleotide mutation rates for dsDNA viruses. Given this value, and a resident virus density of 10^7/ml, flasks of approximately 1000 ml will likely contain around 10 virus mutants.

What, then, is the probability that a rare resistant host will divide before being infected and lysed by one of the mutant counterresistant viruses? The answer to this question provides intuition as to whether a resistant host is likely to grow in number from rare to abundant. Assuming that the sample is well mixed, then mutant viruses will adsorb to mutant hosts at a rate ϕ. The probability that the host will be infected by at least one virus is just 1 minus the probability that it is not infected by any viruses. However, because the number of mutant viruses is likely small, one must calculate this probability under all scenarios where the number of viruses, $k = 0, 1, 2 \ldots$, given an expected number $v = \mu V^* S$:

$$
P(\text{doubling}) = \sum_{k=0}^{\infty} \overbrace{\left(\frac{e^{-v} v^k}{k!} \right)}^{\text{prob. of } k \text{ viruses}} \times \overbrace{\left(\frac{r}{r + \phi k / S} \right)}^{\text{prob. of division given } k \text{ viruses}}, \tag{5.2}
$$

where r is the growth rate of hosts. Given $\phi \mu V^* \ll r$, then this probability can be approximated as

$$
P(\text{doubling}) \approx 1 - \phi \mu V^* / r. \tag{5.3}
$$

Alternatively, $1 - P(\text{doubling})$ is approximately

$$
P(\text{lysis of rare resistant host by mutant viruses}) \approx \phi \mu V^* / r. \tag{5.4}
$$

This equation can also be anticipated by noting that when two events take place, one rare—in this case infection and lysis at a rate $\phi \mu V^*$—and one common—in this case division at a rate r—then the probability of the rare event is just the rate of the rare event divided by that of the common event. A baseline estimate can be established by assuming $\phi \approx 10^{-9}$ ml/hr, $V^* \approx 10^7$/ml, $r \approx 1$ hr, and $\mu \approx 10^{-9}$. The result is on the order of 10^{-11}—a very small number. In words, a rare resistant host is much more likely to divide than to be infected and lysed by the small number of mutant viruses in the population.

In summary, the chance that a viral mutant might be *somewhere* increases with system size, to the point that in a large enough volume a mutant will be present somewhere. However, the chance that one of those viral mutants will interact with the resistant host is extremely unlikely. This first analysis should suggest that a resistant host type H' is likely to start dividing and increasing in number for quite some time before any of the background viral mutants infect and lyse any of the H' population. There is a sweet spot given combinations of volumes and mutation rates, such that counterresistant viruses are likely present yet are unlikely to have any affect on the dynamics of resistant hosts— at least at first.

5.2.2   COEVOLUTIONARY POKER: FROM ONE PAIR TO TWO PAIRS

The previous section established a general set of conditions for the proliferation of a mutant host population even when a very small number of mutant viruses are present that can, in principle but not in practice, infect and lyse the hosts. The next obvious question is, will the mutant viruses start to propagate on the growing population of resistant mutants before it is too late? The timing depends on whether viruses released via lysis of the original susceptible host population N drop below a critical level. These viruses are the source of counterresistant mutants in the absence of infection of mutant hosts. To answer this question it helps to consider a dynamic model. A model of interactions among wild-type hosts (N), wild-type viruses (V), and mutant hosts (N') was introduced in Chapter 4, namely,

$$\frac{dN}{dt} = rN\left(1 - \frac{N + N'}{K}\right) - \phi NV - \omega N, \tag{5.5}$$

$$\frac{dN'}{dt} = r'N'\left(1 - \frac{N + N'}{K'}\right) - \omega N', \tag{5.6}$$

$$\frac{dV}{dt} = \beta\phi NV - \phi NV - \omega V. \tag{5.7}$$

As described in Chapter 4, the population of resistant mutants increases so long as the cost of resistance is not too great. Resistance may also be partial and still be beneficial. The specific condition in the present model is that

$$r' > \frac{\omega}{1 - \frac{\omega}{\bar{\beta}\phi K'}}, \tag{5.8}$$

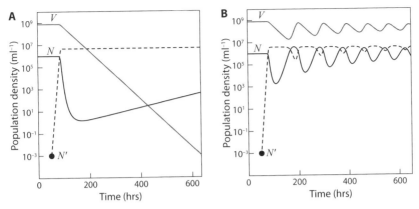

FIGURE 5.3. Invasion of a mutant host that is resistant to viral infection with a growth-rate cost (A) or carrying-capacity cost (B) of resistance. A single mutant N' is introduced at $t = 50$ (circle) and the population of N' increases (dashed line). For small growth costs (A), there is local extinction of viruses V (gray line) including temporary decline followed by rebound of resident hosts N (black line). For small carrying-capacity costs (B), numbers of resident viruses and hosts decrease but then oscillate. Common parameters for these simulations are $r = 1$, $\omega = 0.05$, $\phi = 10^{-9}$, $K = 5 \times 10^{6}$, and $\tilde{\beta} = 50$. Mutant host parameters are $r' = 0.8$ (A) and $K' = 4 \times 10^{6}$ (B). Here, $S = 10^{3}$ ml, such that $N'(t = 50) = 1/S = 10^{-3}$.

where $\tilde{\beta} = \beta - 1$. Here, because the resistance is complete, the invasion criterion depends on viral efficiency. Moreover, large costs of resistance may prevent the invasion of a resistant mutant. But in practice, the benefit of resistance can be quite substantial. For example, given $\omega = 0.05$, $\phi = 10^{-9}$, $\tilde{\beta} = 50$, and $K' = 5 \times 10^{6}$, the mutant host must grow at a rate of $r' > 0.0625$/hr. If the resident divides once per hour, then it will take significant costs to bacterial growth to override the selective benefit of resistance. When invasion does occur, the emerging mutant host population can lead to the decline and local extinction of the resident virus population (Figure 5.3A). Invasion can also lead to oscillations and coexistence among all three populations (Figure 5.3B).

Virus mutants may already be present in the population. As noted earlier the background level of counterresistance mutants should be on the order of μV^*. There are three limits to consider, $\mu V^* S \ll 1$, $\mu V^* S \gg 1$, and $\mu V^* S \approx 1$. The first case suggests that in a finite volume, mutant viruses will not be present and so are unlikely to affect system dynamics. The second and third cases, however, present new questions and challenges. Consider the following dynamic model in which there are wild-type hosts (N), resistant hosts (N'), wild-type viruses (V), and counterresistant viruses (V'):

$$\frac{dN}{dt} = rN\left(1 - \frac{N+N'}{K}\right) - \phi NV - \omega N,$$

$$\frac{dN'}{dt} = r'N'\left(1 - \frac{N+N'}{K'}\right) - \phi'N'V' - \omega N',$$

$$\frac{dV}{dt} = \beta\phi NV(1-\mu) + \mu\beta'\phi'N'V' - \phi NV - \omega V,$$

$$\frac{dV'}{dt} = \mu\beta\phi NV + \beta'\phi'N'V'(1-\mu) - \phi'N'V' - \omega V'.$$

(5.9)

In this model, V can infect N but not N', and V' can infect N' but not N. The variable μ is the phenotypic mutation rate of viruses where V' mutants are generated at a rate μ from V viruses, and vice versa. If $N' = 0$, then there there will be a background level of $V'^* = \mu V^*$ viral mutant types. And, as is evident from the previous analysis, these mutant viruses will not come into contact with the mutant hosts, at least at first. But, if $N' > 0$, then as the mutant host population increases, so too does the probability of contact and infection by counterresistant virus mutants. The basic reproductive number of mutant viruses (Chapter 4) is approximately

$$\mathcal{R}_0^{V'} = \beta\frac{\phi'N'}{\omega + \phi'N'}.$$

(5.10)

Therefore, when the mutant host density exceeds a critical level, $N_c' = \omega/\left(\tilde{\beta}\phi'\right)$, the mutant viruses will begin to grow in number, because $\mathcal{R}_0^{V'} > 1$. In a finite population one also must consider whether viruses are present at all. The combined conditions for virus presence and increase are

$$\beta\frac{\phi N'(t)}{m + \phi N'(t)} > 1 \quad \text{and} \quad V'(t)S > 1;$$

(5.11)

that is, viruses should be present and increase in number when present.

The dynamics of invasion by N' and V' individuals is as follows. In the absence of significant N'–V' interactions, $N'(t)$ increases and $V'(t)$ decreases, because slightly fewer new virions are being produced by N–V interactions. There is a time window when the preceding condition holds, thereby rescuing the V' population and permitting coevolutionary dynamics via a stepwise process. This argument contrasts with models of an infinite population.

In such a case, viral mutants are continually generated, and eventually the viral mutant population moves from rare to common once an N' mutant appears—irrespective of the value of μ or S. Consider the case in which the life history traits of hosts and viruses are the same, with the exception that V exclusively infects N, and V' exclusively infects N'. The point at which viral mutants begin to increase in number coincides with the rise of N' to N^*. This is likely to occur because $N^* < K$, and the mutant host population will grow toward its resource-limited capacity rather than virus-limited equilibrium. When does the rescue occur? The approximate time for the N' population to grow from a single individual to a population of size $N'_c = m/(\tilde{\beta}\phi)$ is $t'_c = (1/r) \log N_c$. If none of the mutant viruses infect a mutant host during this period, then the total number of mutant viruses remaining from this initial cohort will be

$$\mu V^* S e^{-\frac{\omega \log N'_C}{r}}$$

or

$$\mu V^* S \left(\frac{\omega}{\tilde{\beta}\phi}\right)^{-\omega/r}.$$

Consider an example in which $V^* = 10^9$, $\mu = 10^{-9}$, $r = 1$, $\omega = 0.05$, $\phi = 10^{-9}$, $\tilde{\beta} = 50$, and $S = 100$. Given these conditions, 50 of the approximately 100 mutant viruses that are present when the first mutant host appears will still be present in the population when the mutant host population crosses the threshold for mutant virus propagation. Additional mutant viruses will also be produced during the intervening rise. In other words, if $\mu V^* S \gg 1$, then a viral mutant is likely to be somewhere in the population once the host mutants are abundant, leading to spread of the viral mutants. Consequently, the rise of the mutant virus population can rescue the resident virus (Figure 5.4).

This rescue is apparent when considering the steady-state solution to this two-host/two-virus system, in which each virus specializes on a distinct host. Although the mutation rate is essential for understanding the dynamics of invasion, it is relatively unimportant for characterizing the steady state. In that case, the two hosts and two viruses should approach densities of

$$N^* \approx \frac{\omega}{\tilde{\beta}\phi} \quad \text{and} \quad N'^* \approx \frac{\omega}{\tilde{\beta}'\phi'}, \tag{5.12}$$

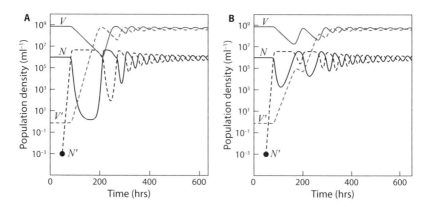

FIGURE 5.4. Invasion of a mutant host resistant to viral infection with a growth-rate cost (A) or carrying-capacity cost (B) of resistance. A single mutant N' is introduced at $t = 50$ (soild circle), and the population of N' increases (dashed line). In both circumstances, the small subpopulation of mutant viruses that can invade the mutant host increases, rescuing (at least in the case of the growth-cost scenario) the resident viruses. Fluctuating coexistence is the norm with two viruses specializing on each of the two hosts. Common parameters for these simulations are $r = 1$, $\omega = 0.05$, $\phi = 10^{-9}$, $K = 5 \times 10^6$, and $\tilde{\beta} = 50$. Mutant host parameters are $r' = 0.8$ (A) and $K' = 4 \times 10^6$ (B). Here, $S = 10^3$ ml, such that $N'(t = 50) = 1/S = 10^{-3}$.

as consistent with top-down control. The viruses should approach densities of

$$V^* = \frac{1}{\phi} \left[r \left(1 - \frac{N^* + N'^*}{K} \right) - \omega \right], \tag{5.13}$$

$$V'^* = \frac{1}{\phi'} \left[r' \left(1 - \frac{N^* + N'^*}{K'} \right) - \omega \right]. \tag{5.14}$$

Even if the number of mutant viruses is finite, but small, such invasions may still be possible because there is a time window over which additional viruses are being produced. The coexistence of two parasites each specializing on distinct prey has long been known to permit species coexistence, as can be inferred from general arguments regarding the number of species that can persist on a set of limiting resources (Levin 1970). Similar conclusions were also reached independently by Janzen and Connell in hypothesizing a mechanism by which host-specific predators promote diversity in forest communities (Janzen 1970; Connell 1970).

5.3   THE EFFECT OF COEVOLUTION ON HOST AND
VIRAL POPULATION DYNAMICS

5.3.1   COEVOLUTION CAN REVERSE THE STRUCTURE OF
PREDATOR-PREY CYCLES

It is evident that hosts and viruses can evolve resistance and counterresistance. Further, it also evident that there are broad regimes of ecological and evolutionary conditions under which a system may increase in complexity, that is, from a single host-virus pair to two hosts and two viruses. This is still vastly more simple than the extent of diversity found in natural environments. In fact, there are many more mutant types in the background of such "simple" systems. Nonetheless, thinking carefully about such simple settings provides a means to recognize not only how ecology can affect coevolution but how coevolution can affect ecology. The notion that ecology affects coevolution means that density-dependent selection leads to changes in the fitness of populations. Such changes can facilitate invasion and extinction through changes in genotype and associated life history traits. This is one area in which ecology affects coevolution and is why resistant hosts can invade a population of susceptible hosts so long as the costs of resistance are not too great and why counterresistant viruses can then increase in abundance in response to the changing landscape of hosts. The notion that coevolution affects ecology means that changes in the frequency of genotypes can lead to concomitant changes in the total population of host and viruses in a community. Such changes can be qualitative and can even fundamentally alter the apparent classic signal of virus-host oscillations.

Consider a two-host/two-virus model of population dynamics, similar to that in Eq. 5.9, but with each virus able to infect both hosts. This structure is similar to two-predator/two-prey communities (Cortez and Weitz 2014). Both sets of hosts and viruses have distinct life history traits. Setting $\mu = 0$ makes it possible to ignore subsequent introduction of variants into the system. The dynamics in that case may be written as

$$\frac{dN_1}{dt} = r_1 N_1 \left(1 - \frac{N_1 + N_2}{K} \right) - \phi_{11} N_1 V_1 - \phi_{12} N_1 V_2 - \omega N_1,$$

$$\frac{dN_2}{dt} = r_2 N_2 \left(1 - \frac{N_1 + N_2}{K} \right) - \phi_{21} N_2 V_1 - \phi_{22} N_2 V_2 - \omega N_2,$$

$$\frac{dV_1}{dt} = \tilde{\beta}_{11}\phi_{11}N_1V_1 + \tilde{\beta}_{21}\phi_{21}N_2V_1 - m_1V_1 - \omega V_1,$$

$$\frac{dV_2}{dt} = \tilde{\beta}_{12}\phi_{12}N_1V_2 + \tilde{\beta}_{22}\phi_{22}N_2V_2 - m_2V_2 - \omega V_2. \tag{5.15}$$

Here, the decay of viruses includes that due to washout and an additional first-order virus decay. The additional decay may be due to the intrinsic stability of viruses or other factors that dilute the concentration of free viruses. The bacteria vary in their vulnerability to infection and, in turn, have different growth rates. Similarly, the viruses can vary in their infectivity. The ensuing dynamics depend on the life history traits of the viruses and hosts. Consider a scenario in which the two hosts differ in their growth rates but also in their susceptibility to infection, such that host 1 grows more slowly than host 2 but is harder to infect. Concurrently, virus 2 has a higher decay rate than does virus 1 and a higher adsorption rate ϕ on both hosts. There is a trade-off between infectivity-associated life history traits and other life history traits.

Simulations of Eqs. 5.15 exhibit oscillatory dynamics for some parameter combinations with trade-offs. What is unusual is that the oscillations can be atypical for consumer-resource models. In Lotka-Volterra–like dynamics, the predator population peaks after the prey population; the same principle applies to interactions between a single virus and a single host population (Chapter 3). After the predator peaks, the prey population declines owing to predation, followed by a predator decline due to lack of prey. Then, prey recover, and the cycle repeats. Simulations of Eqs. 5.15 reveal that the total virus population peaks *before* the total host population (Figure 5.5A). Then, the virus population declines to its lowest value, then the host declines to its lower value, and finally, the virus population peaks again, and the cycle repeats. The oscillations are reversed with respect to the canonical Lotka-Volterra system. Indeed, without further information about the identity of the distinct curves, one might imagine that the host was exploiting the virus, rather than the other way around. But further information is available. During the time over which the total population dynamics are oscillating, the strain abundances are also changing (Figure 5.5B).

The mechanism by which a two-host/two-virus system can exhibit reversed peak ordering requires examining the changes of each of the four strains across one cycle (Figure 5.6). Consider a time point corresponding to peak virus density (upper left of phase-plane dynamics). At this point, the system is dominated by high-vulnerability hosts and low-offense viruses. Selection favors the increase of hosts that have lower vulnerability, because the benefits

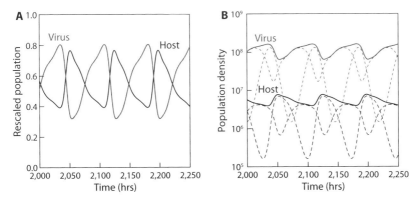

FIGURE 5.5. Virus peaks precede host peaks in a two-host/two-virus model. (A) Dynamics of total virus (gray line) and host (black line) populations. The total populations are rescaled so that the curves denote $V(t)/(\beta K)$ and $N(t)/K$, where $V(t) = V_1 + V_2$, and $N = N_1 + N_2$, respectively. (B) Strain and total population dynamics, without rescaling. Parameters for this model are $r_1 = 1.28$, $r_2 = 2.6$, $K = 10^7$, $\phi_{11} = 2.3 \times 10^{-9}$, $\phi_{12} = 6.35 \times 10^{-9}$, $\phi_{21} = 9.75 \times 10^{-9}$, $\phi_{22} = 1.04 \times 10^{-8}$, $\tilde{\beta} = 20$, $m_1 = 0.64$, $m_2 = 0.9$, and $\omega = 0.01$.

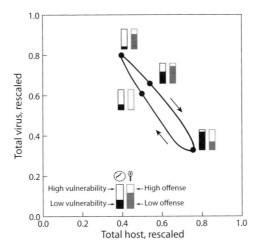

FIGURE 5.6. Phase plane of virus-host dynamics including relative frequency of genotypes. The core principles of this analysis, including illustrations for distinct parameter sets were introduced in Cortez and Weitz (2014). (Main) Phase-plane dynamics in which the four points denote times in the trajectory when the genotype frequencies are evaluated. The arrows denote the direction of dynamics forward in time. The simulation parameters are identical with those in Figure 5.5. (Legend) The two bars denote the relative frequencies of host and viral genotypes, respectively.

of avoiding viral infection and lysis outweigh the costs of a lower intrinsic growth rate. Not only do hosts with lower vulnerability invade, they also lead to an increase in total host density and a decrease in total virus density. As hosts increase in abundance, there is a selective advantage for high-offense viruses to increase (see bottom right of phase-plane dynamics). Therefore, when the total host population reaches its peak, the system is then dominated by low-vulnerability hosts and high-offense viruses, the reverse of the relative abundance in the virus peak. The host abundances then decrease given the large number of potential hosts who are now vulnerable to virus infection and lysis. Concurrently, there is a selective advantage for high-vulnerability hosts to emerge, given that they do not pay as great a cost in growth rate because of the large number of co-occurring high-offense viruses (see midleft time point of phase-plane dynamics). Finally, given a decreasing number of available hosts, the high-offense viruses are replaced by low-offense viruses. The virus population increases, in part, because low-offense viruses also have a lower mortality rate. At this point the cycle repeats.

In summary, models of coevolutionary dynamics present a mechanism by which the change in relative abundance of strains can also affect the structure of the overall ecological dynamics. The effect includes a potential reversal of the ordering of total virus and host peaks. The mechanism is applicable to the study of consumer-resource and exploiter-victim systems, including extensions to other functional responses for growth, interactions, and decay (Cortez and Weitz 2014). Thus far, the common element identified among models that predict such reversals is that there should be a trade-off between life history traits in the absence of interactions, with life history traits associated with the strength of the interaction across trophic levels. In the present case, the trade-off is between growth and susceptibility, for hosts, and between infectivity and mortality, for viruses. Experimental evidence for the reversal of canonical phase relations between viruses and hosts is considered next.

5.3.2   EXPERIMENTAL EVIDENCE FOR REVERSED CYCLES, INDUCED BY COEVOLUTIONARY DYNAMICS OF HOSTS AND VIRUSES

Bruce Levin and his research group have been instrumental in studies of the population and evolutionary dynamics of bacteria and phage. Their seminal contributions were covered in Chapters 3 and 4. Although early studies in the Levin group focused on *E. coli* as the focal host, the range of organisms has expanded to include many others, notably *Vibrio cholerae*, the causative agent of cholera. Outbreaks of the disease occur frequently in the Indian subcontinent and elsewhere when there is restricted access to clean water. The outbreak of cholera in Haiti following the devastating earthquake of 2010 is one such example (Piarroux et al. 2011). Cholera is usually treatable given access to medical care and clean water. Infected patients often die when such treatment is not possible (Colwell and Huq 1994).

V. cholerae and its phage are commonly found in marine environments, and relationship between the two is multifaceted. The primary virulence factor of cholera is the CT toxin (Waldor and Mekalanos 1996). This toxin is secreted by the bacterium, enters a group of human cells called enterocytes, where it enzymatically induces the release of water and ions from cells—leading to a cascade of symptoms including severe diarrhea. John Mekalanos and his group identified this toxin to be of viral origin. The gene responsible for CT is located in a prophage region and can be found in temperate phage (Waldor and Mekalanos 1996). Viral-borne toxins are not unique to cholera; for example, the shigella toxin is borne by lysogenic phage that facilitate enhanced virulence in infected bacteria (Jackson et al. 1987). But even if it seems that phages "help" their bacterial hosts propagate on human and other eukaryotic hosts, there are many bacteriophage that can infect and kill *V. cholerae* (Faruque et al. 2005a, 2005b).

It is within this context that Yan Wei and Bruce Levin examined *V. cholerae*–phage dynamics in the laboratory (Wei et al. 2011). Wei and colleagues observed that *V. cholerae* and phage could persist for long periods in a chemostat, exceeding 750 hours, the equivalent of approximately 1200 cell generations. During this period, bacterial cell density remained at approximately 2×10^6/ml rather than at the virus-free level of 2×10^8/ml. Virus densities in this experiment were often one to two orders of magnitude greater than that of bacteria. These features are indicative of a bacterial community that is, in effect, controlled by viruses rather than limited by resources. A subsequent experiment revealed similar features (Figure 5.7). However, a close examination of the time-series data revealed something unusual.

Discussions of virus-host dynamics often presume that virus density tends to peak *after* host density, leading to a subsequent host trough, followed by a viral trough. With viruses at their lowest density, the host density can increase, and the cycle repeats. This boom-and-bust cycle, as explained in Chapter 3, is akin to the structure of oscillations in predator-prey dynamics (Lotka 1925; Volterra 1926). One way to identify such cycles is to examine

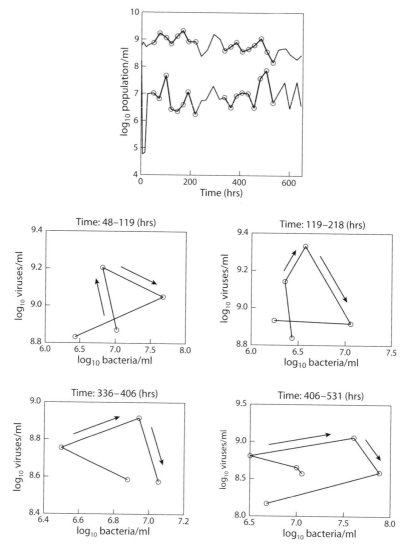

FIGURE 5.7. Population dynamics of cholera phage and *V. cholerae* exhibit clockwise cycles. Analysis adapted from the original presentation in Cortez and Weitz (2014). (Top) Original data from Wei et al. (2011), courtesy of Bruce Levin, plotted as time series on semilog axes. Circles denote those segments that are evaluated in the phase planes in the middle and bottom panels. (Middle and Bottom) Phase-plane plots of continuous time segments of viral versus. bacterial densities exhibiting candidate clockwise dynamics (arrows denote direction of flow over time). *Key point: Virus peaks appear to precede host peaks, in contrast with the canonical observation that host peaks precede virus peaks, as in Lotka-Volterra dynamics.*

the dynamics in the phase plane. Classic Lotka-Volterra cycles appear to rotate counterclockwise when the prey (or host) density is plotted on the x-axis, and the predator (or virus) density is plotted on the y-axis. In contrast, when portions of the data from Wei et al. (2011) are plotted in the phase plane they appear to follow a clockwise trajectory (Figure 5.7). These clockwise cycles can be seen in the intervals 48–119 hrs, 119–218 hrs, 336–406 hrs, and 406–531 hrs. Interpreted classically, it would seem that hosts eat viruses. Moreover, given a single host type and single virus type in each respective population, then there is no way for such inverted cycles to occur, in theory. Two questions arise here: what statistical evidence is there to support the identification of clockwise cycles, and what possible mechanisms can explain the emergence of such cycles?

Analysis of the orientation of cycles in time-series data usually focuses on predicting whether a cycle is likely to occur. For example, cyclical dynamics are predicted when a predator-prey system is initialized close to an equilibrium associated with a Hopf bifurcation (Abrams 2000). In two-dimensional systems, the orientation of the cycle, whether clockwise or counterclockwise, can also be evaluated near the equilibrium. One may think of such methods, broadly, as focusing on the forward problem, that is, predicting cyclical dynamics when the governing equations are known. Far less commonly examined is the inverse problem: identifying a cycle in time-series data when the governing dynamical equations are not known.

One such approach is to look at "streamlines" in time-series data (Holmengen and Seip 2009; Holmengen et al. 2009). The streamline method was originally used for analysis of fluid streamlines as a means of detecting vortexes (Sadarjoen and Post 1999). The underlying idea is that a fluid tracer positioned near a vortex center will wind around the vortex center, returning nearly identically to where it was released. The path of the tracer is termed a *streamline*. In the case of time-series data, consider the set of points in the phase plane $\mathbf{p}_i = (x_i, y_i)$, for example, where x_i and y_i are measurements of host and virus population levels, respectively, at times t_i. Further, consider an interval over which a cycle has been putatively identified, (i_0, i_f), such that the time over which a cycle occurs is $[t_{i_0}, t_{i_f}]$. The winding angle over this interval, α_w, is defined as

$$\alpha_w \equiv \sum_{i=i_0+1}^{i_f-1} \angle(\mathbf{p}_{i-1}, \mathbf{p}_i, \mathbf{p}_{i+1}), \qquad (5.16)$$

where $\mathbf{p}_i$ are the points of the streamline and $\angle(\mathbf{p}_1, \mathbf{p}_2, \mathbf{p}_3)$ measures the signed angle between the two line segments defined by $\mathbf{p}_1$–$\mathbf{p}_2$ and

$\mathbf{p}_2$–$\mathbf{p}_3$, respectively. Here, the convention is that clockwise rotation is positive, and counterclockwise rotation is negative. Further, the distance between the initial and final points is defined in terms of the Euclidean distance in phase space:

$$d_w \equiv \sqrt{\left(x_{i_f} - x_{i_0}\right)^2 + \left(y_{i_f} - y_{i_0}\right)^2}. \tag{5.17}$$

In this way every candidate interval can be described in terms of the summary statistics (α_w, d_w). An interval during which a clockwise cycle occurs should be characterized by $\alpha_{\circlearrowright} \approx -2\pi$ and $d_{\circlearrowright} = 0$, whereas an interval during which a counterclockwise cycle occurs should be characterized by $\alpha_{\circlearrowleft} \approx 2\pi$ and $d_{\circlearrowleft} = 0.$[1] The cycle characteristics of the four putatively identified clockwise cycles are (Cortez and Weitz 2014)

48–119 hrs: $(\alpha_w = -5.1, d_w = 0.59)$,

119–218 hrs: $(\alpha_w = -5.0, d_w = 0.21)$,

336–406 hrs: $(\alpha_w = -4.0, d_w = 0.17)$,

406–531 hrs: $(\alpha_w = -4.9, d_w = 0.54)$.

In all cases, the features seem close to that of an idealized clockwise cycle. But perhaps such values could have been the result of random correlations in time series? A null model is required to evaluate this possibility.

Consider an ensemble of random time series that preserve the statistical properties of point-to-point correlations of the original data, as far as is possible given the sampling intensity. In such an ensemble, the deviations in both the prey and predator time series can be reshuffled randomly for each time series within the ensemble. These randomly generated time series may exhibit spurious winding and may appear to return to the original point—and provide a basis for evaluating the statistical significance of the observed values of α_w and d_w. For each time series in the ensemble, the same cycle statistics were calculated and then compared with the statistics associated with the actual data. Let $f_{\alpha,d}$ be the fraction of the random time series whose winding angle was both closer to -2π of a pure clockwise cycle than the data and whose start-to-end distance was closer to 0 than the data. The results for this

---

[1] The sign convention follows the "right-hand rule," despite the fact that terminology might suggest that clockwise cycles would be better suited to having a positive sign.

analysis were: ($f_\alpha = 0.018$, $f_{\alpha,d} = 0.009$), ($f_\alpha = 0.058$, $f_{\alpha,d} = 0.0026$), ($f_\alpha = 0.0066$, $f_{\alpha,d} = 0.0053$), and ($f_\alpha = 0.089$, $f_{\alpha,d} = 0.017$) for the four putative cycles (Cortez and Weitz 2014). Such features are strongly unlikely to have occurred by chance in a time series of this length and with similar point-to-point deviations.

Wei and colleagues observed the unusual nature of the observed fluctuations in population dynamics in their empirical study (Wei et al. 2011). When plating out their phage populations, they observed distinct plaque morphologies from which they isolated "B" (big) and "T" (turbid) *V. cholerae* phage. These phage differed in their apparent infectivity of the host population in plaque assays. Similarly, isolation of bacteria from the chemostat experiments led to the identification of distinct bacterial types, including the wild type susceptible to both viruses; bacterial types that were resistant to B and to T phage, respectively; and even a minority of cells that were resistant to both B and T phage. The double-resistant mutants were approximately 50% less fit when grown without phage; therefore, they had a significant growth cost to resistance. The take-away point from this analysis is that both the host and virus populations comprised distinct subpopulations with distinct life history traits. It would seem that such distinct subpopulations enabled long-term coexistence and, perhaps, the apparent reversal of canonical predator-prey dynamics.

Both theory and available experiments suggest that coevolution can affect, and indeed fundamentally alter, ecological dynamics. This finding has a rather important consequence. In practice, we do not know which viruses infect which hosts in natural environments (Weitz et al. 2013; Deng et al. 2012). Nonetheless, the basis of some methods for identifying putative partners is to leverage canonical signatures of population dynamics. For example, if the density of viruses, as estimated by some marker gene, has peaks that tend to follow the density of some environmental host, as estimated by a marker gene, this might indicate an infectivity relationship between the two marked populations (Needham et al. 2013). Such a hypothesis may indeed be valid. But if the markers used to identify or group populations of hosts and viruses include subtype variation, as they most certainly do, then many interacting partners will not have the desired signature (Hiltunen et al. 2014). Further elaboration of the effects of coevolution on ecology is needed, with a focus on microscale diversity and trade-offs (Wichman et al. 1999; Forde et al. 2008; Poullain et al. 2008; Paterson et al. 2010; Meyer et al. 2012). The results will further progress toward understanding and predicting the dynamics of viruses and hosts in complex natural environments.

5.4   ECOLOGICAL EFFECTS ON THE COEVOLUTIONARY
DYNAMICS OF TYPES AND TRAITS

5.4.1   A SIMPLE MODEL OF DISCRETELY STRUCTURED COEVOLUTION

The previous section considered dynamics arising from interactions between two hosts and two viruses. As was shown, the relative abundances of all four types change as a result of interactions in nonstructured environments. Such models can reveal the way that evolution affects and even changes ecological dynamics, leading to clockwise rather than counterclockwise cycles. However, there is a limit to such models. Because they consider a small set of populations from the outset, they cannot explain a generative mechanism for new diversity. A more realistic model would include the possibility that new types could arise, not once, not twice, but as often as a viable mutant offspring was produced, which would then constitute the founding individual of a lineage. The subsequent survival of that lineage would depend on the ecological context.

To begin to explore dynamics in such a direction, consider the following illustrative chemostat model using the implicit-resource formulation introduced in Chapter 3:

$$
\frac{dN_i}{dt} = \overbrace{r_i N_i \left(1 - \frac{\sum N}{K_i}\right)(1 - \mu_N) + \mu_N \sum_{i' \neq i}^{S} \theta_N(i, i') r_{i'} N_{i'} \left(1 - \frac{\sum N}{K_{i'}}\right)}^{\text{growth}} \overbrace{\phantom{+ \mu_N \sum_{i'}^{S}}}^{\text{mutations}} -
$$

$$
\overbrace{\sum_{j=1}^{S'} \phi_{ij} N_i V_j}^{\text{viral lysis}} - \overbrace{\omega N_i}^{\text{washout}} , \tag{5.18}
$$

$$
\frac{dV_j}{dt} = \overbrace{\sum_{i=1}^{S} \beta_{ij} \phi_{ij} N_i V_j (1 - \mu_V)}^{\text{viral lysis}} + \overbrace{\mu_V \sum_{j' \neq j}^{S'} \sum_{i=1}^{S} \theta_V(j, j') \beta_{ij'} \phi_{ij'} N_i V_{j'}}^{\text{viral mutation}}
$$

$$
- \overbrace{\omega V_j}^{\text{washout}} . \tag{5.19}
$$

There are S host types and S' virus types in this model. The mutation rate, per host division, is μ_N, and the mutation rate, per virion replication, is μ_V. The population of host type i increases owing to reproduction events in which the host does not mutate (the first term in Eq. 5.18) and owing to reproduction events in which a host type i' reproduces and mutates to host type i (the second term in Eq. 5.18). The population of host type i decreases owing to viral lysis and washout. Similarly, the population of virus type j increases owing to lysis events in which the virus does not mutate (the first term in Eq. 5.19) and owing to lysis events in which a virus type j' releases a burst of virions including a mutant of type j (the second term in Eq. 5.19). The actual populations may be small and ideally modeled as a population of a discrete number of individuals. This is one reason why the current model is meant to be illustrative.[2]

To move from the general form of the model to a specific implementation requires choices, not only with respect to life history parameters but to the nature of the state space and of the mutational connections therein. The indices i and j are the strain type of hosts and viruses, respectively. Further, the functions $\theta_N(i, i')$ and $\theta_V(j, j')$ are the probability that a mutation changes host i to i' and virus j to j', respectively. Hosts and viruses are further constrained by the requirement that $\sum_{i' \neq i} \theta_N(i, i') = 1$ and $\sum_{j' \neq j} \theta_V(j, j') = 1$. The mutation connections between strain types represent a consolidation of a complex genotype space and genotype-to-phenotype maps. For example, Katia Koelle and colleagues have proposed a model in which influenza genotypes can be described as embedded within two-dimensional landscapes (Koelle et al. 2006, 2010). In this model, mutated viruses are considered to evolve in sequence space through a neutral network of genotypes. Occasionally, mutants evolve to a new neutral network, with a distinct set of epidemiologically relevant phenotypic characteristics, like interactions with the immune system. However, extensive study of a specific virus-cell interaction is required to propose, to say nothing of to analyze, such models. As an alternative, consider models in which the state space, connections between them, and genotype-to-phenotype map are simplified. In this way, distinct models can be evaluated by combining different assumptions. Using such models, it is possible to examine minimal assumptions necessary for persistent coevolution and to generate hypotheses regarding the factors underlying such coevolution.

5.4.2 DYNAMICS ARISING FROM SIMPLE MODELS OF STRUCTURED COEVOLUTION

The system described in Eqs. 5.18–5.19 provides one route for examining the potential for ecological interactions that mediate the selection for de novo

---

[2] Models adopting cutoffs to reduce artifacts caused by tracking small populations are treated in the next section.

generated mutants. The specification of such a model requires assumptions that may vary between models and between host-virus systems. Each of the following classes of assumptions represents a target for the empirical investigation of virus-microbe coevolution.

Ordering of host and viral types: The first set of assumptions concern the underlying state space of genotypes, each with distinct phenotypes. For a set of S genotypes, an ordered set is one in which mutations are restricted to occur between "adjacent" types; that is, type 1 can mutate to become type 2, type 2 can mutate to become types 1 and 3, type 3 can mutate to become types 2 and 4, and so on. Ignoring back mutations leads to a scenario in which type 1 can mutate to become type 2, type 2 can mutate to become type 3, and so on. Given increasing S, it would take $\sim S$ number of mutations to transform one type to any other type, on average. In contrast, an example of a disordered set of genotypes is one in which mutations transform type i to another type j with a fixed probability. The transitions between types resemble that of a random graph. Given increasing S, it would take $\sim \log S$ number of mutations to transform one type to any other type, on average. The structure of the space radically changes the extent to which one type is "close" or "far" with respect to other types (Burch and Chao 2000). *In the numerical simulations in this section, host types and virus types are ordered from 1 to S. Mutations of host type i and virus type j can generate new types of type i + 1 and j + 1, respectively.*

Infectivity between host and virus types: Selective pressures imposed by viral lysis of hosts determine, in part, the trajectory of coevolutionary dynamics. In particular, the $S \times S'$ matrix of adsorption ϕ_{ij} is key to determining trait dynamics, in both an ecological and an evolutionary sense. One possible scenario is one in which there is a single virus that can infect and lyse each host. For example, $\phi_{ij} = \phi_0$ if $i = j$, and $\phi_{ij} = 0$ otherwise. This situation coincides with the example of host-range matching in Figure 5.2. Alternatively, another scenario is one in which each virus infects the viruses of its ancestral type. For an ordered set of types, this corresponds to the case in which $\phi_{ij} = \phi_0$ if $i \leq j$, and 0 otherwise. This situation coincides with the host-range expansion case in Figure 5.2. *In the numerical simulations in this section, viruses of type j can infect and lyse hosts of type i ≤ j with adsorption rate φ and cannot infect hosts of type i > j.*

Life history trade-offs: Chapter 4 presented evidence for the existence of trade-offs between life history of hosts and their susceptibility to infection, along with evidence for the existence of trade-offs between life history traits of viruses and their host range. Such trade-offs are not inevitable, and certainly not universal. Moreover, trade-offs are often examined in the context of low-diversity communities; for example, the growth rate of a mutant host or of a mutant virus is examined relative to that of an ancestral type. There is far less evidence of these trade-offs for interactions between environmental viruses and their microbial hosts. One possibility is that the number and structure of interactions between hosts and viruses are statistically decoupled from life history traits. Another possibility is that there are strict trade-offs between order and life history traits. Two examples of trade-offs in the case of host-range expansion are (i) growth-rate-resistance trade-off, in which r_i is a decreasing function of host type i; and (ii) infectivity-efficiency tradeoff, in which $\phi_{ij} = \phi_j$, so long as $i \leq j$, and 0 otherwise, where ϕ_j is a decreasing function of viral type j. Alternatives are possible between these limits. *In the numerical simulations in this section hosts differ in their carrying capacity, such that $K_{i'} < K_i$ if $i' > i$; that is, host type 1 can grow to a higher density than any other type. Viruses differ in their host-independent burst sizes, such that $\beta_{j'} < \beta_j$ if $j' > j$; that is, virus type 1 has the largest burst size of all types.*

Dynamics of the model proceed given the model described in Eqs. 5.18–5.19, with its assumptions and initial conditions. The initial conditions considered here are that $N_1 > 0$, $V_1 > 0$, and $N_i = 0$ and $V_j = 0$ for $i > 1$ and $j > 1$, with only a single population of viruses and hosts initially present. The initial conditions mean to replicate experimental scenarios in which a chemostat is inoculated with a low-diversity community derived from a single clonal population of hosts and a single clonal population of viruses. As was explained in Chapter 4, these "clonal" populations may already contain small subpopulations of variants, whereas in the present model, small subpopulations of variants are introduced via the approximation of the mutation process.

The dynamics that ensue include changes in population abundances of existing types, generation of novel types, and changes in phenotypes. Given the initial conditions described, the changes in the populations of host type 1 and virus type 1 resemble that of top-down controlled systems, exhibiting counterclockwise oscillations in the phase plane in which host peaks precede virus peaks. Mutation generates small subpopulations of additional types. Hosts of type 2 and above are not susceptible to lysis by the dominant virus type 1. Moreover, host type 2 has the most advantageous of life history traits of those that cannot be infected by virus type 1. Host type 2 begins to increase in abundance faster than all the other subpopulations. When this density exceeds a critical level, then virus type 2 can grow in abundance, if present in the system. The growth of virus type 2 is based on a combination of successful lytic infections of the previously dominant type and this new type (Eq. 5.10 presented the critical host threshold in a single host-pair system). In parallel with the growth of host type 2, viruses of type 1 begins to decrease in abundance, as they cannot infect host type 2. Figure 5.8A shows the phase-space dynamics of the total population, beginning with a single dominant virus-host pair. The dynamics exhibit Lotka-Volterra–like oscillations, albeit the center of the cycle shifts owing to the change in the traits of the types that dominate the system. The inset of Figure 5.8 (top) shows the fraction of the total host population comprising the numerically dominant type. Similar results are obtained in the case of virus dynamics. Despite tracking the dynamics of $\sum V$ versus $\sum N$, such phase-space plots can be interpreted as tracking the relationship between the dominant virus and host types.

Which pairs dominate over time? Figure 5.8B presents the changing combinations of dominant host and phage types. Beginning with host type $i = 1$ and viral type $j = 1$, a sequence of shifts between dominant types occurs, first, with a change in the dominant host type from i to $i + 1$ and then, second, with a change in the dominant viral type from $j = i$ to $j = i + 1$ (Figure 5.8B). The sequence of the first of such steps shows how the system changes from a community dominated by host type $i = 1$ and viral type $j = 1$ to one dominated by the combination $i = 15$ and $j = 15$. Recall that given the definitions of life history traits in the model, the change in types also suggests that there is a change over time in the average phenotypic traits of hosts and viruses. Figure 5.8C shows the change in the average carrying capacity of hosts and in the average burst size of viruses. Both these traits decrease over time, despite the fact that improvements in both of these traits are usually associated with increases in fitness. The difference here is that the selective advantage, to hosts, of decreasing death via lysis is more of a benefit than the loss induced by the increase in competition from other hosts. Thus, hosts of type $i + 1$ can invade and replace a resident host of type i. The same logic applies to viruses: access to new hosts is more of a benefit to a virus than the cost of a modest decrease in burst size.

This model, given the initial assumptions, leads to escalation dynamics, similar to that of an arms race. The term *arms race*, in the present context, denotes an escalation in defensive and offensive traits associated with both hosts and viruses, respectively. The newly evolved virus types have the ability to infect ever-more hosts. The newly evolved host types have the ability to resist against infection by more viruses. In this arms race, hosts lead and viruses track. But hosts do not escape virus infection, at least for the parameters and model assumptions. The coevolutionary dynamics also

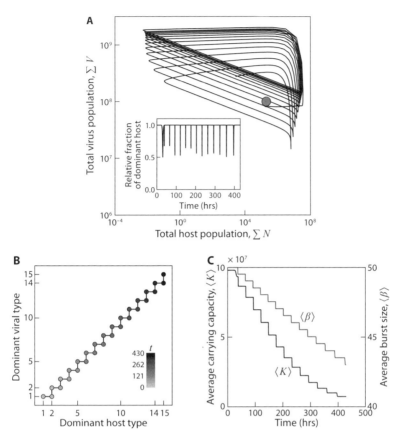

FIGURE 5.8. Stepwise coevolutionary dynamics among viruses and hosts. (A) Phase-plane dynamics of total virus and total host populations. The initial condition is denoted with a solid gray circle. The inset denotes the relative fraction of the dominant host, which is nearly 1 at all times, even as the identity of the dominant host changes. (B) Dynamics of the numerically dominant host and viral type, where the axes denote the strain identification, i and j. The shading in each combination of host and viral type denotes the time at which this combination first appears (see shaded legend). (C) Dynamics of the mean carrying capacity, $\langle K \rangle$, and mean burst size, $\langle \beta \rangle$. The mean carrying capacity is defined as $\langle K \rangle = \sum_{i=1}^{S} N_i K_i / \sum_{i=1}^{S} N_i$. The mean burst size is defined as $\langle \beta \rangle = \sum_{j=1}^{S'} V_j \beta_j / \sum_{j=1}^{S'} V_j$. Parameters are specified in the text.

affects the total population abundances, progressively diminishing the large oscillations in both host and viral densities. Although an ecological model would predict that viruses could control host densities, the coevolutionary model predicts something a bit different. For example, an ecological model of a single virus-host pair would lead to top-down limitation of hosts by viruses at a density 1/500th of the carrying capacity with the current model parameters, whereas the observed dynamics include the emergent property that the dominant host population exceeds at least 10% of their carrying capacity for 15% of the simulation. This occurs because host evolution leads to a temporal niche in which hosts have partially diminished lysis pressure. In summary, coevolutionary dynamics is a mechanism that has the potential to shift a system from strict top-down control to partial top-down control. This stepwise coevolutionary dynamics is idealized, and greater complexity and variability can, and does, unfold in natural systems and in other models. Specifically, the current model allows for arbitrarily low densities of types and does not include the introduction of novel hosts and viruses.

In considering coevolutionary dynamics arising from this model, it is worth recalling a debate that took place more than 30 years ago. The debate was centered on a question, as posed by Rodin and Ratner: "Do phage and bacteria undergo endless cycles of defense and counter-defense?" (Rodin and Ratner 1983a, 1983b). Their answer was yes (in theory). But to make such a conclusion Rodin and Ratner assumed extreme specificity in virus-host interactions, such that a virus type could infect only one host type, and a host type was infected by only one virus type. Moreover, they assumed that the emergence of a new virus-host pair would always eliminate the previous virus-host pair. In response, Richard Lenski argued, instead, that the answer best supported by current evidence was no. (Lenski 1984) claimed that there exists "a fundamental asymmetry in the coevolutionary potential of bacteria and phage." The reasons cited include the fact that replacement of one pair by another is a dynamical process, and the invasion of a new pair does not always eliminate the previous pair (as was described earlier in this chapter). Further, the experimental phage-host systems focused on at that time exhibited repeated evolution of host mutants that were completely resistant to lysis by co-occurring phage. But there is now increasing recognition that coevolutionary dynamics can be sustained for extended periods in the laboratory (Dennehy 2012). The next section addresses recent theoretical efforts to consider greater realism in evaluating mechanisms that might drive long-term coevolutionary dynamics.

5.4.3   COEVOLUTION DYNAMICS WITHIN COMPLEX MODELS: FROM ESCALATIONS TO DIVERSIFICATION

Coevolutionary dynamics of viruses and hosts include changes in the relative frequencies of both viruses and hosts (Wei et al. 2011) and in the de novo emergence of novel host and virus types with distinct life history traits, including distinct ranges of cross infection (review in Dennehy (2012)). It is less clear whether the mechanisms identified in simple models correspond to those emerging in systems of significantly greater complexity. There are, in fact, a number of proposals for alternative models of coevolutionary dynamics between viruses and microbes. These models focus on distinct features of coevolution, including (i) the balance between top-down and bottom-up selection (Weitz et al. 2005; Williams 2013); (ii) the effect of spatial structure on coevolution (Haerter et al. 2011; Haerter and Sneppen 2012); (iii) the impact of details of trade-offs on the potential for coevolution-induced diversification (Forde et al. 2008); and (iv) the possible effect of intracellular defense mechanisms on coevolutionary dynamics (Levin 2010; Childs et al. 2012; Weinberger et al. 2012). Rather than claim that such models are exhaustive or, as yet, definitive, one way to review their findings is to consider what mechanisms and/or principles are retained or modified in contexts in which an effort has been made to link models with measurements.

The first series of models in this chapter showed how a pair of hosts and viruses can be invaded by a second pair. Two of the "complex" models (Weitz et al. 2005; Williams 2013), listed in the preceding paragragh, provide a roadmap for how coevolution can drive diversification and lead to an increase in the number of types, well beyond two pairs. These two models are formalized based on a common set of assumptions. They both describe the dynamics of an initially low-diversity virus-host community interacting in a chemostat in which the abundances of resources, hosts, and viruses are tracked. Viruses and host populations have a single phenotype trait, y_i and x_i, respectively, from which other life history traits can be derived. Interactions in these models assume that there exists an optimal host trait value for the environment, x^*, such that hosts would evolve toward that trait value in the absence of viruses. Moreover, viruses are presumed to specialize in the model, such that there exists an optimal virus type, $y^*(x)$, for each host with trait x. However, viruses also can infect similar hosts, that is, those whose trait x' is near x. This initial component of the dynamics can be described using the language of adaptive dynamics.

Recall that the long-term evolution of latent period was used as an example to show how a trait could evolve toward a fitness maximum (Figure 4.8). That is, a mutant with any other latent period would not be able to invade, when rare, and replace the resident type which had evolved to the optimal latent period. However, it is also possible for evolution to converge to a fitness minimum (Dieckmann and Doebeli 1999; Doebeli and Dieckmann 2000). These points of convergence are termed *branching points* (Figure 5.9).

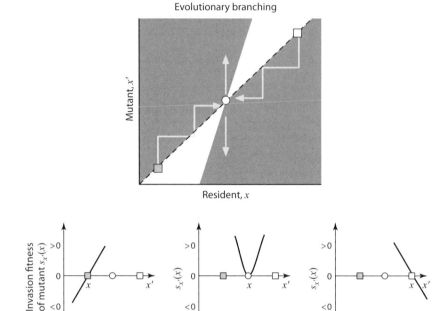

FIGURE 5.9. Schematic of ecologically mediated speciation in evolutionary dynamics. (Top) A pairwise invasibility plot, in which the sign of the invasion fitness, $s_{x'}(x)$, is shown for combinations of resident x and mutant x' trait values. Shaded areas denote positive invasion fitness, and white areas denote negative invasion fitness. The squares denote possible starting points for an evolutionary dynamics. The circle denotes a branching point, that is, the trait value associated with a diversification event. The orthogonal arrows denote a possible sequence of invasion and replacement events in trait space toward the branching point. (Bottom) Local structure of the fitness landscape, as quantified in terms of the invasion fitness. Each of the panels denotes $s_{x'}(x)$ on the ordinate axis as a function of x' on the abscissa. The parameter x is, from the left, the shaded square, the open circle, and the open square. The leftmost panel denotes that mutants x' with traits larger than the resident x have positive invasion fitness. The rightmost panel denotes that mutants x' with traits smaller than the resident x have positive invasion fitness. The center panel denotes that nearby mutant types can invade the resident, because all have positive invasion fitness. *Key point: Evolution in this scenario over the long term will converge to the branching point and then lead to subsequent diversification.*

At branching points, selection acts disruptively—to borrow the language of population genetics—so that coexistence by more than one population is possible. The study of adaptive dynamics has shown the robustness of this process as a mechanism of adaptive diversification (Doebeli 2011). In the

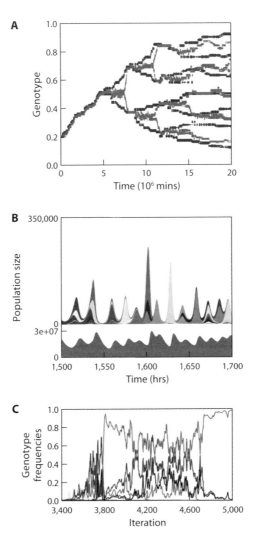

FIGURE 5.10. Simulated coevolutionary dynamics among viruses and their microbial hosts. (A) Dynamics of trait diversification in a chemostat model from low to high diversity. Darker and lighter trajectories denote host and viral genotypes, respectively. Reprinted with modifications to annotations from Figure 2c of Williams (2013), with permission. (B) Dynamics in a CRISPR model with a continuous ODE framework. The top panel shows the dynamics of multiple host populations whose density and diversity (shades denote distinct types) change with time. The bottom panel shows the dynamics of the total density of viruses. Reprinted with modifications to annotations from Figure 4 of Childs et al. (2012), with permission. (C) Dynamics in a CRISPR model with a population genetics framework. Relative frequencies of distinct host strains, which differ in their

present context, assume that hosts incur growth costs in evolving to minimize lysis by viruses. Therefore, two types of host populations may be able to coexist, one that is even more resistant to viruses and one that is permissive to viruses but that reduces the growth cost. Figure 5.9 provides a schematic of the process for the evolution of monomorphic populations with a single trait. Analysis of coevolutionary branching is based on similar principles but involves additional complications (Doebeli and Dieckmann 2000).

This initial convergence of populations toward branching points is exhibited in the models of both Weitz et al. (2005) and Beckett and Williams (2013). The dynamics, in both cases, can become increasingly diverse given suitably high levels of specialization on the part of viruses and suitably strong enough tradeoffs on the part of hosts. Figure 5.10A shows an example of such diversification, with a typical quasi-stationary state seen at the end of the dynamics. The principle of both models is that two hosts can coexist given a single virus, one that is better at obtaining resources, the other better at avoiding infection. However, given two hosts, two viruses can coexist, each specializing on a given host. With two viruses, three hosts can coexist, and so on. In these models, strains appear stochastically and are removed when their densities decrease below a critical value. This also help address artifactual dynamics that arise when infinitesimally small populations are tracked, in models like that of Eq. 5.18–5.19. In summary, coevolution enables the system to increase its diversity progressively, beyond that predicted for a single-resource environment.

However, not all models agree that adaptive diversification is inevitable. Another class of coevolutionary model has recently been introduced, coincident with the discovery of a novel mode of bacterial resistance to virus infection, termed CRISPR defense, where CRISPR stands for Clustered Regularly Interspaced Short Palindromic Repeats (van der Oost et al. 2014). The characteristic genomic signature of the CRISPR system was identified first by bioinformaticians who identified regions within bacterial genomes characterized by palindromic repeats. These repeats, from a handful to hundreds, were separated by sequences of seemingly viral or plasmid origin, so-called spacers (Bolotin et al. 2005). This genomic structure was easily identified in other bacteria and archaea genomes, yet its function remained a mystery.

FIGURE 5.10 (*Continued*)

genotypes owing to acquisition of virus protospacers. This section of dynamics includes two examples of nearly complete sweeps by a single strain, as well as strain interference. Reprinted with modifications to annotations from Figure 4B of Weinberger et al. (2012), with permission.

In 2007, Rodolphe Barrangou and colleagues provided what was to become the definitive explanation for this genomic organization (Barrangou et al. 2007). Their subject organism was the Gram-positive *Streptococcus thermophilus*, a key microbe in the dairy industry. Barrangou and colleagues confirmed that the spacers were indeed of viral and plasmid origin, and could be acquired from a virus. More important, they demonstrated that the CRISPR locus, including repeats, spacers, and upstream functional genes, constituted a novel type of intracellular defense mechanism against phage infection and plasmid entry. Molecular studies have since revealed the processes involved (e.g., Hale et al. 2009; Haurwitz et al. 2010; Deltcheva et al. 2011; Jackson et al. 2014): (i) acquisition of a small piece of foreign DNA, which is inserted into the CRISPR locus; (ii) transcription of spacers from the CRISPR locus; (iii) guided identification of foreign DNA so long as the foreign DNA contains a (nearly) exact match to the integrated spacer; (iv) elimination of the foreign DNA upon identification of a successful match. The CRISPR system constitutes a form of adaptive immune defense in bacteria and archaea (Sorek et al. 2008; Horvath and Barrangou 2010). The system has many novel properties, including that bacteria can evolve by a form of Lamarckian evolution in which mutation is directed, rather than random (Koonin and Wolf 2009). The term *Lamarckian* here refers to the fact that hosts incorporate viral genomic content into their own genome as a direct result of interactions with an invading virus. Hence, their behavior and interactions shape directed changes in their genomes. The directed modification of the bacterial genome subsequent to attack by phage is then passed on to progeny cells.

How does coevolution unfold when hosts are evolving via Lamarckian evolution and viruses are evolving via Darwinian evolution? One hypothesis is provided by a set of recent models, which differ in some of their postulates but all try to integrate the Lamarckian feature of the CRISPR system into their core assumptions. In a continuous dynamical system model (Childs et al. 2012), hosts possess a limited number of spacers and are nearly perfectly immune to viruses that contain a matching "protospacer," that is, the same DNA sequence as contained in the host CRISPR locus. Moreover, in this model, viruses mutate at random, but hosts directly mutate spacers in their CRISPR locus with a small probability after surviving infection by a virus. Host diversity increases as a result of directed acquisition of genetic elements from viruses that provide enhanced immunity to virus infection. Notably, hosts that invade may arise from single clones or even cohorts of clones that are distinct genotypically but similar phenotypically, that is, have similar immune profiles with respect to the virus community. This situation is termed *distributed immunity* (Childs et al. 2014). But as a result of virus mutation, the effectiveness of spacers decreases over time, leading to a coevolutionary dynamic in which the diversity of viruses and hosts fluctuates in time, even as the genotypes of viruses and hosts continue to diverge (Figure 5.10B).

Similar results were found in another model of CRISPR-induced immunity that utilized a population genetics framework. In this model total population sizes were fixed but dynamics were stochastic, including that of births, deaths, and appearances and dynamics of mutants (Weinberger et al. 2012). Diversification increased progressively until individual host types with a locally optimal combination of spacers evolved. Then, these types increased in number, nearly sweeping through the population either to fixation or near-fixation. However, owing to the change in the relative abundances of viruses, now being removed directly as a result of CRISPR immunity, the instantaneous fitness of the previously dominant strain decreased. Then, a subsequent diversification event took place (Figure 5.10C).

The continuous and discrete models of CRISPR-induced coevolution differ in their implementation. Nonetheless, they agree that coevolution can lead to fluctuating dynamics. In such fluctuating dynamics the host population is often most well adapted to concurrent or very recent viruses. The hosts are poorly adapted to prior and future virus communities, because they tend not to contain the set of spacers that were dominant far into the past or will be in the future. This is a form of time-shift comparison. Time-shift experiments are becoming an increasingly important tool for characterizing the mode of coevolution between viruses and host communities (Gómez and Buckling 2011; Buckling and Brockhurst 2012; Koskella and Meaden 2013).

Detailed examination of the functional interactions in these two classes of models raises an important point regarding coevolutionary dynamics in virus-host systems. The former set of models (Weitz et al. 2005; Beckett and Williams 2013) presumed that resistance was mediated by changes in surface receptors, so that adsorption and lysis were conflated. In models of immune-mediated defense, viruses inject their genomes into target host cells, and viral genomes are cleared from those target cells that possess immunity. Intracellular defense implies that the subset of targeted but immune hosts cause density-dependent mortality for viruses. This difference, as noted by Sieber and Gudelj (2014), has the potential to alter the nature of ecological and evolutionary dynamics, but this is certainly not the only possibility. Omitted from the discussion here are the importance of spatial heterogeneity (Forde et al. 2004; Heilmann et al. 2012) and trophic heterogeneity (Thingstad et al. 2014). To move in that direction requires steps toward connecting foundational models of virus-host interactions with specific environments.

Viruses have been isolated from the gut, the soil, vents, hot springs, lakes, and the ocean. Of these, the marine surface remains the focal environment for which, arguably, the most is known regarding the role of viruses in altering basic ecological and ecosystem processes. It is for that reason that ocean viruses—and their effect on ocean ecology—are the subject of the next two chapters.

5.5   SUMMARY

- Virus mutation and host mutations occur rapidly, so much so that new virus and host mutants are expected to arise on timescales similar to those for changes in total population.
- Host evolution to resistance can lead to virus extinction, but such extinction is not inevitable either in the lab or in the field.
- A single dominant host and virus type can, via a combination of mutation and selection, diversify into multiple virus and host types that coexist in one community.
- The interaction of multiple virus and host genotypes has fundamental effects on ecological dynamics.
- When multiple virus and host genotypes are present, their interactions can lead to novel ecological dynamics, such as apparently reversed predator-prey cycles.
- Moreover, the interactions of a single virus-host pair can lead to distinct modes of evolutionary dynamics, for example, fluctuating dynamics, arms races, and diversification.
- Arms races include the evolution of ever-increasing virulence among viruses and resistance among hosts.
- Fluctuating dynamics include locally well-adapted viruses, whose identities change as the host population evolves.
- Diversification events include an increase in the number of strains, even strains with distinct life history traits.
- Linking models of coevolutionary dynamics to specific virus-host systems is ongoing and a central challenge in quantitative viral ecology.

PART III

VIRAL ECOLOGY IN THE OCEANS:
A MODEL SYSTEM FOR MEASUREMENT
AND INFERENCE

CHAPTER SIX

Ocean Viruses: On Their Abundance, Diversity, and Target Hosts

6.1   WAYS OF SEEING

Imagine walking through a forest habitat, albeit with a purpose: to estimate how many birds there are, how many distinct species of birds are present, and, finally, the population-level growth and death rates of each type of bird. Easy, no? Certainly, such a census requires time, expertise, and repeated surveys. There will be challenges, for example, if the rare bird is neither seen nor heard during one or repeated surveys. Or, even if a bird is common, it may return to a survey site only seasonally. Finally, there may be common birds that are, nonetheless, hard to detect. But given sufficient resources, a survey could provide informative data on the diversity, distribution, and dynamics of bird species. In fact, academic and lay birders, routinely conduct such surveys for years. These surveys are so data rich that it is now possible to assess global-scale patterns in bird diversity, ecology, and aspects of evolutionary change (Jetz et al. 2007). Now, imagine a forest habitat in which all the robins, perhaps half of the thrush, and a sprinkling of warblers suddenly became invisible. Moreover, imagine that 90% of the trees and shrubs that the birds lived in and among also became invisible. It is true that such decreases, of varying scales, are taking place owing to habitat loss and other factors. However, the scenario you are being asked to imagine is one in which the birds, trees, and shrubs remain, but they cannot be detected. The situation sounds far-fetched, but in such a place what would you think you knew about the distribution, diversity, and population dynamics of the biota that turned out to be incomplete, misdirected, or perhaps just flat-out wrong?

This vision of a complex, diverse world that is nonetheless largely invisible to us is an apt reminder of the challenges we face in trying to grasp—to say nothing of understand—the complexity of the world of microbes and the viruses that infect them. It is this world, or at least this worldview, that predominated until the late 1980s. Bacteriophage and other viruses of microbes were used as the basis for spectacular advances in molecular biology and were themselves the subject of laboratory experiments in population and evolutionary biology. Yet, viruses remained a rather microscopic sidelight from an ecological or environmental science perspective. The state of the field was summarized by Bergh et al. (1989):

> The concentration of bacteriophages in natural unpolluted waters is in general believed to be low, and they have therefore been considered ecologically unimportant.

This belief was informed, in part, by the tools for measuring viruses available to that point. The plaque assay was, at the time, the standard method of counting the number of environmental viruses. As explained in Chapter 2, the plaque assay is a culture-based method. It remains the gold standard for counting virus particles, given a host-virus pair in culture. Given an initial environmental sample from which cells have been removed, the total number of plaques can be used to estimate the number of viruses able to infect, propagate, and lyse a target host or hosts. Because viruses are known to differ in the hosts they infact, if the wrong host is used or conditions are not ideal for virus propagation, then such methods may lead to significant errors in virus count estimates. Moreover, and perhaps more fundamentally, it is now widely recognized that the majority of hosts cannot yet be cultured. James Staley and Allan Konopka proposed the term "The Great Plate Count Anomaly" to consolidate various lines of evidence that the estimated number of bacteria from colony-forming unit assays were two orders of magnitude smaller than the inferred number of bacteria from production, division, and microscopy counts (Staley and Konopka 1985). It is hard to count viruses if one cannot even count their hosts.

To better estimate the population size of naturally occurring viruses, Bergh and colleagues Knut Børsheim, Gunnar Bratbak, and Mikal Heldal proposed a culture-independent counting method. This method involves centrifuging a sample; staining the water-free pellet with uranyl acetate, which binds to many proteins and lipids as well as to nucleic acids; and then viewing the sample at high magnification using transmission electron microscopy. The results can be seen in Figure 1 of Bergh et al. (1989). In retrospect, the micrograph hardly seems impressive. It is apparent that the sample contains many particles that are on the order of 1 μm in size and many smaller, circular particles on the order of 50 nm in size. The large particles are likely bacteria. What was truly innovative in this work was the hypothesis that the small particles were viruses, formally termed virus-like particles. The team observed virus-like particles with diameters ranging from less than 50 nm to upward of 100 nm.

They estimated total virus densities in excess of 2.5×10^8 / ml. In the authors' own words: these "viral counts are 10^3–10^7 times higher than previous reports on virus numbers in natural aquatic environments, which are based on counts of plaque-forming units of various bacteria." Subsequent studies supported these counts based on quantification with epifluorescence microscopy (Noble and Fuhrman 1998). In the words of Duncan Watts, "Everything is obvious, once you know the answer" (Watts 2012). The expression is rather apt in this case—the finding was discoverable for 50 years once the electron microscope was developed, if others had only thought to look!

This discovery facilitated rapid advances in the study of viral ecology. Knowing that viruses are abundant leads to many related questions: for example, how do viruses maintain such large total populations? Whom do they infect? How diverse are they? How important are they in modifying not only the fate of individual hosts but also the flux of nutrients and other small molecules in the environment? These questions are, in some sense, universal. They could be asked in the context of the surface, deep, or coastal oceans; in soils; or in microbiomes. Here, I will focus on aquatic ecosystems, and in particular, surface ocean waters, to address what theory and quantitative analysis can contribute to answering these questions. This chapter introduces the theoretical and modeling approaches necessary to estimate (i) viral abundance, (ii) viral diversity, and (iii) virus-host interactions. Then, Chapter 7 introduces a series of dynamic models for exploring the mechanisms underlying these emergent features of complex ecosystems.

6.2   COUNTING VIRUSES IN THE ENVIRONMENT

Modern methods for estimating the number of viruses in an environmental sample include electron microscopy, epifluorescence microscopy, and flow cytometry. The *Manual of Aquatic Virology* (MAVE) (Wilhelm et al. 2010) provides detailed procedures for implementing each of these methods, for example, in Ackermann and Heldal (2010) (for electron microscopy), Suttle and Fuhrman (2010) (for epifluorescence microscopy), and Brussaard et al. (2010) (for flow cytometry). All the methods share a common set of principles: the sample is physically and chemically prepared, then measured in the device, then labeled as a virus based on its physical features (e.g., estimated diameter), and/or interactions with the apparatus (e.g., degree of scattering). Despite the many protocols for estimating virus abundance, the classification step is usually researcher dependent; for example, "Viruses can generally be distinguished from cells by their staining characteristics. Viruses appear as bright pinpricks of light, whereas cells generally have discernable size." (Suttle and Fuhrman 2010). This leads to a challenge: how does one count small things with big, clumsy hands.[1]

The intuition and training of researchers has been essential to distinguishing virus-like particles from non-virus-like particles. Indeed, detailed comparisons of protocols with gold-standard approaches, including plaque assays and transmission electron microscopy, show that such classifications can lead to consistent estimates (Wilhelm et al. 2010), but standardization of the classification step is both warranted and feasible. Despite prior successes, the sample-to-abundance pipeline is hardly perfect. There are many new environments (e.g., the human microbiome (Minot et al. 2013; Barr et al. 2013)) in which the roles of the viruses of microbes are now being studied and for which new measuring protocols need be developed. Finally, it is no longer self-evident that virus should appear as pinpricks and that bacteria should appear with a discernible size. Giant viruses having discernible size infect amoeba and protists. Examples include the *Chlorella* virus, mimivirus, Mamavirus, and Leviathan, all of which exceed 300 nm in diameter, with the largest estimated virus reaching 400 nm in diameter (Xiao et al. 2005). Similarly, marine bacteria are often much smaller than routinely used lab strains. For example, individual cyanobacteria from the clades *Prochlorococcus* and *Synechococcus* are approximately 400 nm in diameter (Chisholm 1992). Likewise, the ubiquitous *Pelagibacter* clade includes individuals approximately 500 nm in diameter (Rappé and Giovannoni 2003). Furthermore, a ubiquitous actinobacter (representing 5% of the total surface microbe population in seasonal measurements) is estimated to be 300 nm in diameter (Ghai et al. 2013). Although viruses tend to be smaller than bacteria, this is not universally the case.

Improvements in sample preparation and imaging protocols also represent opportunities for theoreticians, and in particular imaging scientists, to improve sample measurement protocols. Appendix D.1 includes one such algorithmic protocol for standardizing the virus-particle estimation component of the sample-abundance pipeline as estimated via epifluorescence microscopy. It is not the only such protocol, but it is easily reproducible. The objective of the algorithm is to formalize the intuition of the researcher who must distinguish the signal due to virus-like particles from background noise. The noise includes background scattering and/or fluorescence associated with the sample preparation, and signal coming from other organisms in the sample, such as bacteria, diatoms, or zooplankton.

---

[1] Thanks to Steven Wilhelm for suggesting this turn of phrase.

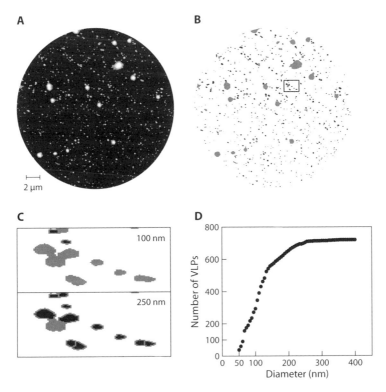

FIGURE 6.1. Estimating virus-like particle abundance from epiflourescence microscopy images. (A) Stained image of microbes and viruses, reprinted with permission of Jérôme Payet. Original sample collected on May 28, 2004, from the Amundsen Gulf, Station 200, and stained using Yo-Pro dye. (B) Digitally filtered image with each "particle" identified. Black particles denote virus-like particles, and gray particles denote particles not included, either because they are too large or because their shape is not "pinprick-like." See Appendix D.1 for details on the algorithm. (C) Change in the number of virus-like particles (VLPs) as a function of the size cutoff for the focal region of interest, designated with a rectangle in (B). In this case, two cutoffs are used: 100 nm and 250 nm, top and bottom panels, respectively. (D) The estimated number of VLPs increases with size cutoffs.

Figure 6.1 illustrates the results of this protocol. The original grayscale image was segmented, in this case with a threshold intensity value determined by Otsu's algorithm (Otsu 1979). The segmented image is further labeled according to whether the clusters appear "virus-like," in this case whether their effective diameter is smaller than 250 nm, and the eccentricity of the best-fitting ellipse of the cluster is less than 0.9. In this way, the computation of the number of virus-like particles (VLPs) can be completely automated given a sample image. Posterior estimates will depend on parameters, for example, whether a particle of size 50 nm or 150 nm is considered a virus-like particle. Selecting appropriate cutoffs is essential to improving estimates of the total abundances of both viruses and small bacteria. The last panel of Figure 6.1 illustrates the dramatic change in the number of virus-like particles with cutoff parameters. As is apparent, the distribution of sizes in environmental samples is not always bimodal. The variation in size cutoffs extends beyond what is typically considered for the identification of a VLP. Nonetheless, the point here is that a demarcation between pinpricks and cells of discernible size can be hard to establish. This area is underexplored, both from a methodological and environmental science perspective.

6.2.1   VARIATION IN THE ABUNDANCE OF MARINE VIRUSES

The study of marine viruses and their effects on microbial communities and ocean ecosystems has increased significantly since Bergh and colleagues' finding of elevated marine virus abundances in aquatic ecosystems (Bergh et al. 1989). As but one indication, the total number of papers whose topics include either "ocean viruses" or "marine viruses" (according to Web of Science) grew from 3 in 1989 to more than 300 in 2013. The number of citations to such papers grew from approximately 20 in 1990 to more than 12,000 in 2013. Many of these studies include the direct enumeration of total virus abundance. It is now evident that viruses are, in fact, one of the most abundant, if not the most abundant, biological entity in the surface ocean (Suttle 2005). A recent compilation of published and unpublished data from both surface ocean waters and marine sediments revealed significant variation in virus densities, from 10^4/ml to more than 10^8/ml (Figures 6.2A and B). What underlies the four orders of magnitude variation in marine virus abundance?

A comprehensive answer to such a question remains elusive. Indeed, any comprehensive answer must come with significant caveats. In particular, there are currently a few thousand direct measurements of virus abundances. Each measurement reflects significant planning, expertise, and (often) expense. The vast majority of these measurements were taken via oceanographic research vessels, of which virus ecology is usually a small component of a large suite of activities. It is an understatement to say that current measurements under-sample the 3.6×10^8 km^2 surface area of the global oceans, to say nothing of the ocean depths. Nonetheless, available measurements provide preliminary evidence for emergent patterns.

One evident pattern is that virus-like particle abundance is positively correlated with prokaryotic abundance (Figure 6.2). The reported fits (Danovaro

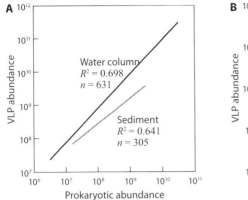
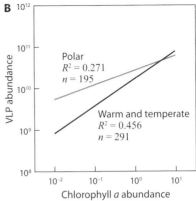

FIGURE 6.2. Marine virus-like particle abundances in the water column and in sediments. Lines denote best fit to individual measurements of virus abundance, prokaryotic abundance, and chlorphyll a in the water column and sediments, for both warm/temperate and polar systems. Abundances of bacteria and VLPs are per liter and per gram for the water column and sediment, respectively. Original analysis and methodology presented in Danovaro et al. (2011). (A) Correlations of virus abundances with prokaryotic abundances; (B) Correlations of virus abundances with chlorophyll a (units of μg/L) in the water column.

et al. 2011) are $\log_{10} V = 1.03 \log_{10} N + 0.660$ for the water column ($n = 631$, $R^2 = 0.698$, $P < 0.001$) and $\log_{10} V = 0.761 \log_{10} N + 2.35$ for the sediment ($n = 305$, $R^2 = 0.641$, $P < 0.001$), where V and N are virus and prokaryotic density, respectively. For those unfamiliar with virus-host models, such a correlation would appear to be expected, since the greater availability of bacteria and archaea would seem to imply more targets for viruses. Yet, viruses are thought to control, at least in part, the number of prokaryotes in a given community. In fact, the simplest model of virus-host dynamics leads to the conclusion that increases in resource supply lead to increases in virus, but not host, density. An alternative outcome may be that virus abundance is positively correlated with resource supply but uncorrelated with prokaryotic abundance. In fact, virus abundance is correlated with the amount of chlorophyll, a proxy for the productivity of the environment (Figure 6.2B). However, bacterial abundances are also correlated with chlorophyll and other indexes of ecosystem productivity (Williams and Follows 2011). These patterns are a key target for ecosystem-level studies and mechanistic models that include not only virus-host interactions but also the interactions between prokaryotes and other predators, like zooplankton. In explaining patterns, it may be necessary to consider the effects of specialization and generalism within populations, for example, as in analyses of zooplankton-phytoplankton biomass patterns along productivity gradients (Leibold 1989).

Working at global scales can be daunting. Instead, one may ask: what determines the site-specific variability in virus abundances? A few case studies will help illustrate the different types of regimes to be expected in the surface oceans. A major area of oceanographic research focuses on oligotrophic gyres. *Oligotrophic* is a term used to describe the majority of "blue ocean" environments, that is, with relatively low inorganic nutrients, for example, $<0.5\,\mu g/L$ of chlorophyll *a* in the surface. The term *gyre* refers to a rotating, circulating pattern in ocean currents, such as the North Pacific Ocean subtropical gyre. This type of oceanic region has been of interest for decades and for various reasons. For example, it is there that C. D. Keeling initiated the now-renowned Mauna Loa time series showing rises in atmospheric CO_2 over the past 50 years (Keeling et al. 1995; Williams and Follows 2011). A few researchers, including Grieg Steward, Alex Culley, and Jennifer Brum have all measured virus abundances in this gyre in the top 100 m of surface waters. They found total virus abundances on the order of 5×10^9 L (Culley and Welschmeyer 2002) and 10^{10} L (Brum 2005). Simultaneous measurements of bacterial densities were approximately one order of magnitude smaller.

All the ocean is not oligotrophic. Coastal waters receive significant organic and inorganic nutrient influx owing to river runoff, and upwelling events from the deep ocean is often associated with "bloom" events, in which the population abundances of plankton rise rapidly. Both coastal waters and bloom events are known to have elevated levels of viruses. For example, estimates of virus abundance exceeded $10^{11}/L$ in surface water measurements during a spring bloom in the Southern Pacific Ocean near New Zealand (Matteson et al. 2012; Strzepek et al. 2005). The total chlorophyll *a* concentration was $2.48\,\mu g/L$, nearly five times larger than at the Hawaii Ocean Time-series (HOT) site. And, unlike at HOT, bacteria abundances peaked at $1.5 \times 10^9/L$; that is, Wilhelm and collaborators observed virus to bacterium ratios exceeding 100:1 (Matteson et al. 2012). As is evident, distinct oceanic realms have significant differences in host and virus densities and in the emergent relationships between them.

These prior estimates also enable rough estimates of the total marine virus abundance in the oceans. Such estimates are admittedly coarse. For example, assuming virus abundances of $3 \times 10^9/L$ in 70% of the surface oceans and $1 \times 10^{10}/L$ in the remainder suggests an average surface ocean concentration of $5 \times 10^9/L$. If such averages are characteristic of the top 100 m of the ocean, then there should be 2×10^{29} marine viruses in the surface oceans. This estimate neglects viruses deeper in the water column and in marine sediments (Danovaro et al. 2011).

The abundance of viral genomes inside lysogens or infected cells is also poorly quantified. The fraction of lysogenic cells can be estimated in a culture-independent fashion. The general technique is to use an "inducing" agent, for example, a chemical like mitomycin C or direct UV exposure, which has been shown in many cultured systems to induce active virus propagation within lysogens (Paul and Weinbauer 2010). For example, in a study of marine surface waters in the Arctic Ocean, Jerome Payet and Curtis Suttle found that 5%–40% of microbial cells were inducible (Payet and Suttle 2013). Moreover, the fraction of inducible cells was found to be inversely correlated with free virus abundance, suggesting that total virus counts as measured using microscopy of free viruses will nonetheless underestimate the total number of viral genomes in the environment.

Finally, any discussion of virus abundances would be incomplete without mentioning the important fact that current staining methods that are pre-requisites for both epifluorescence microscopy and flow cytometry are not effective with RNA-based viruses (Steward et al. 2012) nor with ssDNA viruses (Holmfeldt et al. 2012). Grieg Steward and colleagues provocatively asked, "Are we missing half the viruses in the ocean?" (Steward et al. 2012). Perhaps. Then again, there are others, including Patrick Forterre, who caution that measurements of viruses-like particles do not necessarily guarantee that the particles are viruses (Forterre et al. 2013). For example, they could be gene-transfer agents (GTAs). GTAs are vesicle-bound particles containing nucleic acids, usually those of microbial hosts, that can be released and then taken up by other microbes. They are spherical and often on the order of 30–50 nm in diameter. In 2014, Penny Chisholm's group reported evidence of widespread vesicle production by cyanobacteria—these vesicles had an average diameter of ~75 nm (Biller et al. 2014). It is evident that the quantification of marine viruses, despite advances, remains a topic of active interest from both a methodological and scientific perspective.

6.2.2   ELEMENTAL RESERVOIRS IN MARINE VIRUSES

Chapter 2 presented a physical scaling model of the elemental content of virus particles (see Jover et al. (2014) for more details). As was explained, virus particles are nutrient rich; that is, they have high proportions of nitrogen and phosphorus compared with their carbon content relative to the stoichiometry of their marine microbial hosts. For example, viruses whose capsid diameters are approximately 60 nm are predicted to have a C:N:P ratio of 24:8:1. Contrast this with the Redfield ratio (106:16:1) for marine phytoplankton (Redfield et al. 1963). Alfred Redfield posited the existence of a ubiquitous elemental ratio in phytoplankton and marine detritus, and this ratio is a convenient departure point for studies even if exceptions are widespread (Martiny et al. 2013). The stoichiometry of individual microbial hosts varies by strain (e.g., heterotrophs are thought to be more nutrient rich than cyanobacteria (Suttle 2007)). Moreover, stoichiometry depends on physiology; for example, cyanobacteria grown in phosphorus-replete conditions have greater cellular proportions of phosphorus than do cells grown in phosphorus-depleted conditions (Bertilsson et al. 2003). The carbon requirements of cellular life reflect its increased need for structural support relative to virus particles. By way of contrast, Gram-positive bacteria have a cell wall that ranges from 20 to 50 nm in thickness and is characterized by a carbon-rich peptidoglycan (Schleifer and Kandler 1972). In summary, marine viruses, despite their small size and relatively low biomass, are nutrient rich. In a collaboration with Steven Wilhelm and Alison Buchan (Jover et al. 2014), we asked, how much carbon, nitrogen, and phosphorus is partitioned in marine virus particles at environmental concentrations? As a corollary, can viruses represent a substantial subcomponent of the dissolved carbon and nutrient content in marine waters?

The marine environment includes both organic and inorganic forms of carbon and nutrients not bound in organisms, that is, small molecules, molecular aggregates, marine snow, and other large pellets (Azam and Long 2001; Weinbauer et al. 2011). Focusing on the organic components, a rough classification of the types of organic matter can be achieved via size separation into small and large particles. Operationally, *dissolved organic matter* (DOM) is defined as any organic matter that passes through a filter with pore sizes in the range 0.22–0.7 μm. In contrast, *particulate organic matter* (POM) is defined as any organic matter that remains on the filter. Given a filter with pore size of 0.5 μm, DOM would include organic matter whose cross section was $\leq$0.5 μm. The operational definition of DOM includes nearly all viruses, and probably some bacteria and archaea. Once DOM is isolated from the rest of the sample, further chemical methods can be used to measure relevant atomic constituents, including organic carbon, nitrogen, and phosphorus. Specifically in the case of phosphorus, the chemical methods of accounting (Lomas et al. 2010) are likely to include the phosphorus bound in nucleic acids inside a protein shell of virus particles. This point was raised in Jover et al. (2014). In summary, virus particles are likely to contribute to current assessments of the total dissolved organic phosphorus (DOP) in marine environments, but the specific contribution of viruses to the DOP pool has not been systematically assessed. The total concentration of C, N, or P atoms bound within marine virus particles is a product of per-virus elemental content and virus density.

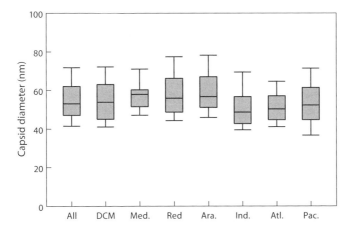

FIGURE 6.3. Variation in capsid diameter of marine viruses from distinct oceanic realms. The whisker-and-bar plot includes variation in diameter at the 10%, 25%, 50% (central bar), 75%, and 90% levels. The ocean realms include all surface data, all from the deep chlorophyll maximum (DCM), and then from distinct realms: Mediterranean Sea, Red Sea, Arabian Sea, Indian Ocean, Atlantic Ocean, and Pacific Ocean. Analysis is adapted from original data and presentation in the analysis of qTEM data in Brum et al. (2013).

Therefore, the DOP depends on virus size and on the number of viruses. Virus sizes range from diameters on the order of 30 nm or smaller, like the Leviviridae R17 and MS2 (De Paepe and Taddei 2006), to the large T-even phages with diameters approaching 100 nm (De Paepe and Taddei 2006), and even to "giant" viruses that can exceed 400 nm in diameter (Raoult et al. 2004; Fischer et al. 2010).

What is the typical size of virus capsids in marine surface waters? Jennifer Brum used quantitative analysis of variation in virus morphology and virus size via transmission electron microscopy (qTEM) to answer this question (Brum et al. 2013). In Brum's qTEM study, surface water samples were surveyed from the Mediterranean Sea, Red Sea, Arabian Sea, Indian Ocean, Atlantic Ocean, and Pacific Ocean. In all these surface water samples, the majority of viruses were nontailed, representing 65% to 80% of the total morphotypes. The remainder comprised myoviruses, podoviruses, and siphoviruses, in decreasing order of representation. Viruses, irrespective of morphotype, had capsid diameters that varied between approximately 20 nm and 200 nm, For tailed viruses, tail lengths of myoviruses were on the order of 150 nm; of siphoviruses, on the order of 210 nm; and of podoviruses, on the order of 15 nm. Despite this variation in type-specific sizes, the bulk of the variation found in marine viruses capsid sizes centered in the range of 50 to 70 nm (Figure 6.3).

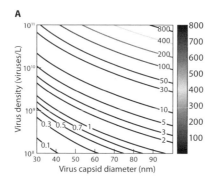

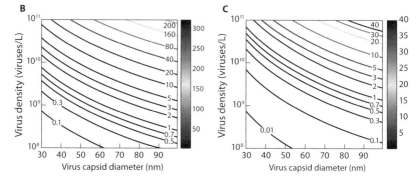

FIGURE 6.4. Total elemental content in marine virus particles as a function of virus abundance and average virus head size. The contour lines delineate combinations of diameter and density which yield the same predicted abundance in units of nmol/L for carbon (A), nitrogen (B), and phosphorus (C). Panel C is adapted from Jover et al. (2014) with permission.

The next component necessary to estimate elemental reservoirs in virus communities is virus densities. Figure 6.4 presents a range of possible concentrations of dissolved organic carbon (DOC), dissolved organic nitrogen (DON) and DOP bound in virus populations whose densities are $10^8 \leq V \leq 10^{11}$/L with average capsid diameters between 30 and 100 nm. The choice of virus densities reflects current estimates of natural variability in the marine water column (Figure 6.2). DOP bound in viruses is predicted to range from 0.1 nM to 20 nM if attention is restricted to virus sizes suggested by Brum et al. (2013). Recall that virus tails do not contain phosphorus; therefore, the estimates of DOP are robust to variation in virus morphology. Applying a similar series of calculations yields ranges of DON from 0.1 nM to 100 nM and DOC from 0.2 nM to 400 nM. These estimates are underestimates, as they do not include the contributions of virus tails, which contain carbon and nitrogen. A quantitative analysis of elemental contribution suggests that the contribution of tails is on the order of 10% that contained in virus heads (Jover et al. 2014).

How do these ranges of elemental reservoirs compare with the total dissolved elemental pools? Consider DOP first. Estimates of marine DOP in surface waters typically range from 50 to 250 nM (e.g., at oligotrophic sites in the Atlantic (Lomas et al. 2010), Pacific (Fujieki 2014), and Southern Oceans (Ruardij et al. 2005). Assuming that capsid diameters are 60 nm, then virus particles can represent greater than 5% of the DOP in the marine surface environment whenever the following condition is satisfied:

$$V > 3.5 \times 10^5 \ \text{DOP}, \tag{6.1}$$

where V is in units of particles/mL and DOP is in units of nM. Viruses are predicted to be a significant fraction of DOP if their total population exceeds 1.75×10^7/mL when DOP= 50 nM, or 3.5×10^7/mL when DOP= 100 nM, or 7×10^7/mL when DOP= 200 nM. A similar analysis reveals that viruses are unlikely to be a significant fraction of either DON or DOC in the marine surface environment.

This finding, at least for DOP, suggests a concomitant need to analyze the size of elemental reservoirs partitioned in viruses, and the role of viruses in modifying the flux of elements through and out of the marine surface. Virus particles may decay owing to UV-induced deactivation, aggregate on or part of marine snow, or adsorb to hosts. The relative rates of each of these processes in the marine surface warrants investigation. Another possibility is that viruses may be specifically targeted by grazers or nonspecifically taken up by filter-feeding organisms. This hypothesis was first raised by Gonzalez and Suttle (1993). If there is 0.01 fg of phosphorus per virion, then digestion of 100 viruses could provide the necessary phosphorus requirements for division of a nanoflagellate (Jover et al. 2014). The frequency and relevance of virus-targeting grazers remains an open question.

6.3   ESTIMATING VIRAL DIVERSITY

6.3.1   DIVERSITY METRICS

Diversity is an old, old concept. In the beginning, at least according to Genesis, Adam "gave names to all the cattle and to the birds of the sky and to all the wild beasts" (Genesis 2:20). This naming distinguished one type of living organism from another. Scientists have, for years, done the same. The names of old became the taxonomy of the near past, including kingdoms, orders, classes, families, and so on. In the study of microbes, and in particular of viruses, such demarcations are problematic. What distinguishes one bacterial species from the next? What distinguishes one virus species from the next? And do any such designations have biological meaning?

These questions remain highly debated and controversial, and the curious reader will readily find debates on such topics (May 1988; Ward 2002; Fraser et al. 2009; Mora et al. 2011). For now, it is worthwhile to step back and recognize that diversity is a concept meant to summarize the variation among distinguishable types of things. These things may be colors (of hats) or names (of people) or species (of organisms). When all the things in a group are of the same type, then diversity is low; when all of the things in a group are of different types, then diversity is high. How high and how low depend on the mathematical definitions. Viral diversity may be calculated with respect to genotype diversity, gene diversity, or even functional diversity. All such diversities require a metric, that is, a scale by which one arrangement of types can be compared with another.

Three of the most commonly estimated diversity metrics are Simpson diversity, Shannon diversity, and species richness (Schloss and Handelsman 2005). For a community with N individuals of S types, such that the number of individuals of each type is n_i, then the diversity of the sample is defined as:

$$\text{Simpson diversity}: \quad D_2 = 1 - \sum_{i=1}^{S} \left(\frac{n_i}{N}\right)^2,$$

$$\text{Shannon diversity}: \quad D_1 = -\sum_{i=1}^{S} \frac{n_i}{N} \log \frac{n_i}{N},$$

$$\text{Richness}: \quad D_0 = S.$$

These equations may appear cryptic, but they have (relatively) simple interpretations. *Simpson diversity* is equivalent to the probability that two randomly chosen individuals from the community are of different types. *Shannon diversity* is equivalent to the entropy of the community composition, which may be intuitive for physicists and computer scientists, and for others, may appear as intuitive as defining an Urdu word with a Finnish word.[2] In brief,

---

[2] Claude Shannon, after which the Shannon diversity is named, is considered the father of information theory. This field originated largely from his pioneering master's thesis. It is less well known that Shannon's PhD dissertation was on the subject of theoretical population genetics.

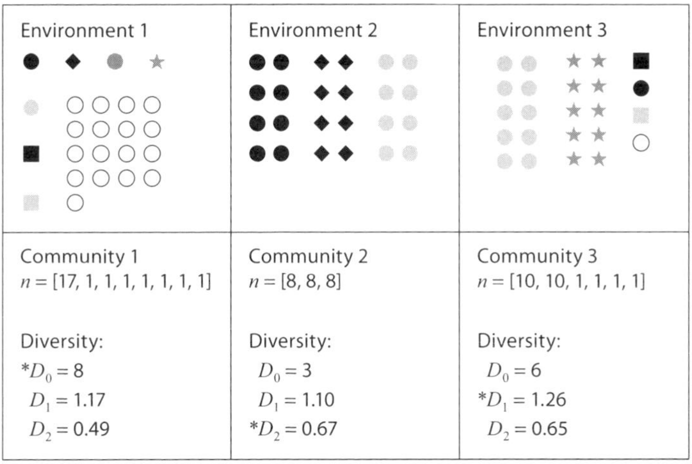

Environment 1	Environment 2	Environment 3
Community 1 $n = [17, 1, 1, 1, 1, 1, 1, 1]$	Community 2 $n = [8, 8, 8]$	Community 3 $n = [10, 10, 1, 1, 1, 1]$
Diversity: $*D_0 = 8$ $D_1 = 1.17$ $D_2 = 0.49$	Diversity: $D_0 = 3$ $D_1 = 1.10$ $*D_2 = 0.67$	Diversity: $D_0 = 6$ $*D_1 = 1.26$ $D_2 = 0.65$

FIGURE 6.5. Example of three different communities and associated diversity metrics. The three communities have the same total number of individuals ($n = 24$) but differ in their community composition. The three diversity metrics evaluated are richness (D_0), Shannon diversity (D_1), and Simpson diversity (D_2). Each community can be considered the most diverse, at least for one metric, denoted with an asterisk ($*$).

Shannon diversity is a proxy for the extent to which it is possible to compress a signal containing the equivalent information in the community composition $\mathbf{p} = (n_1/N, n_2/N, \ldots, n_S/N)$. When each type appears in equal abundance, then $n_i = N/S$, and the Shannon diversity, $D_1 = \log S$, is maximized, whereas when one type dominates, then $D_1 \to 0$. Finally, *richness* is, trivially, the number of different types in the sample. The use of subscripts 0, 1, and 2 for each of these diversity metrics is meant to convey another connection between a generalized family of diversity estimators called *Hill diversities* (Hill 1973; Haegeman et al. 2013; Chao et al. 2014).

As is apparent, the relative abundances n_i of each type have different weights in these diversity metrics. For example, two communities may have different numbers of types and still have the same Simpson diversity. Whereas, richness depends exclusively on the number of types, S, and so one community with one common type and many rare types will appear more "diverse" (according to the richness metric) than another community with a smaller number of types, all relatively uncommon. All too often it is assumed that one metric is as good as another; that is, they are equivalent, at least in terms of their relative ordering. This is not true. Figure 6.5 shows how three different community compositions, each containing the same number of organisms, can each be the most diverse, depending on whether diversity is measured in terms of richness, Shannon diversity, or Simpson diversity. Despite the use of the term "diversity" as a catch-all phrase, the appropriate diversity measure depends, ultimately, on the question. In the study of environmental viruses and microbes, many questions are comparative. For example, is the diversity at one location larger than at another, or does diversity vary with an environmental feature? The new challenge in asking such questions is that samples of viruses and microbes are nearly always a small portion of the community of interest. Therefore, methods are needed to infer the diversity of a community from observations of a sample. As it turns out, inferring the diversity of a large community from a small sample is not as straightforward as it may seem. This is the topic of the next section.

6.3.2   ROBUST ESTIMATION OF COMMUNITY DIVERSITY FROM A SAMPLE

The diversity of a sample is readily estimatable given a protocol for distinguishing between types of individuals within the sample. The same cannot be said of inferring the diversity of a community from a small sample. This problem can be understood by thinking about black swans. Most swans— as even the inattentive observer on a nature hike will affirm— are white.[3] A sequence of M measurements of the colors of swans in a sampled flock would, with near 100% certainty, be of the form W,W,W,...,W, where W denotes an observation of a white swan. The measured richness, $\hat{D}_0$, of the sample is 1, and the measured Simpson and Shannon diversity are $\hat{D}_1 = \hat{D}_2 = 0$. In other words, there is only one type of color and no variation in color within the sample. What then is the expected diversity of the *entire* population of swans? To estimate this diversity requires extrapolating beyond the current set of observations to the entire set of N_{tot} swans. Consider the scenario in which a single one of these N_{tot} swans is black. The true richness of the community is $D_0 = 2$, and the true Simpson diversity is (asymptotically) $D_2 = 2/N_{tot}$, whereas the true Shannon diversity is asymptotically $D_1 = \log(N_{tot})/N_{tot}$. The probability of observing this black swan in a (random) sample of N swans is $1 - (1 - 1/N_{tot})^N$, which can be approximated as N/N_{tot} so long as $N \ll N_{tot}$. As the black swan becomes ever rarer, it is harder to detect in a sample but also has a decreasing consequence on the difference between the measured diversity and the actual diversity—so long as attention is restricted to Simpson or Shannon diversity. In contrast, even as the black swan becomes rarer, the disparity between the measured species richness and

---

[3] Black swans, of the species *Cygnus atratus*, do exist and are endemic to Australia and New Zealand.

actual species richness remains constant. This reflects the fact that richness weights all species equally, irrespective of their relative abundance, including rare species, whereas rare species are the most difficult to observe. Formally, if S_{obs} is the number of observed species in the sample, then the estimated number of species in the community, $\hat{S}$, is

$$\hat{S} = S_{obs} + S_{rare}, \tag{6.2}$$

where S_{rare} is the expected number of unobserved species in the community. The subscript implies that those unobserved species are relatively rare, at least with respect to the sampling intensity.

It is apparent that the sample diversity and the true diversity can differ. It is less apparent whether the differences are likely to be small or, potentially, biased. Whatever the diversity metric, efforts to estimate the diversity of a community from a sample must involve some form of extrapolation to estimate the value of S_{rare}.

Consider a set of observations such that F_k is the number of species in a sample of size N such that each species is observed exactly k times. There are $S_{obs} \equiv \sum_{k=1}^{k_{max}} F_k$ species, such that $N \equiv \sum_{k=1}^{k_{max}} k F_k$. Classic probability theory would suggest that the most likely probability of each species would be equal to its observed frequency; for example, a species observed three times in a sample of 100 should have a community abundance of 3%. This estimate is biased, because rare species are less likely to be observed than common species. Common species appear, on average, to be slightly more common in a finite sample. Let $\hat{p}_k(N)$ be the true *community* frequency of a species which is observed k times in a sample of size N. In this way, it is convenient to introduce the notion of $\hat{p}_0(N)$, that is, the community frequency of an *unobserved* species. These species are considered *rare*, at least with respect to the sample. If there are S_{rare} such rare species, then their true relative abundance in the community is $S_{rare}\hat{p}_0$. Estimating the number of species in a community is then dependent on estimating $\hat{p}_0$ and by extension the number of unobserved species, S_{rare}.

Observations of the rarest species in the sample serve as an upper limit for the relative abundance of individuals from unobserved species. I. J. Good and Alan Turing derived an accurate estimator of the relative abundance of unobserved species (Good 1953; Gale and Sampson 1995). The Good-Turing theory presumes that unobserved species should have a maximal relative abundance equal to the observed total relative abundance of the rarest observed species in the sample, that is, the singletons. This leads to the estimate that $S_{rare}\hat{p}_0 = F_1/N$. The intuition behind this equivalence is that the prior probability that a newly examined individual is a new species, unobserved in the sample, is equal to the fraction of singletons in the sample. The number of unobserved species is therefore $S_{rare} = F_1/(\hat{p}_0 N)$. But what is $\hat{p}_0$? To move forward, consider a similar upper limit for the relative abundance of species observed only once. This relative abundance should be equal to the observed relative abundance of the second-rarest observed species in the sample, that is, the doubletons (those species observed twice in the sample). Via this logic, $F_1 \hat{p}_1 = 2F_2/N$, or $\hat{p}_1 = 2F_2/(N F_1)$. Finally, assume that $\hat{p}_0 \le \hat{p}_1$; that is, the true relative abundance of the rare species must be less than that of the singleton species. The expected number of rare species in the upper limit in which $\hat{p}_0 = \hat{p}_1$ is:

$$S_{rare} = \frac{F_1}{\hat{p}_1 N} = \frac{F_1^2}{2F_2}. \tag{6.3}$$

A lower limit on the number of species in the community is:

$$\hat{S}^- = S_{obs} + \frac{F_1^2}{2F_2}. \tag{6.4}$$

This is known as the *Chao-1 estimator* (Chao 1984; Colwell et al. 2004). The key point is that Chao-1 estimator is a *lower* limit on the expected number of unobserved species in the community. It is not the expected number of unobserved species in the community. The use of the Chao-1 estimator requires sufficient number of observations such that $F_1 > 0$ and $F_2 > 0$.

In practice, it is all too often forgotten that the estimator provides a lower limit. This might seem like a technicality, but it is not. Lower bounds and actual estimates need not be coincident. For example, consider two tourists wandering bewildered on their first trip to Manhattan, on a rainy and cloudy fall morning. Walking downtown they observe a skyscraper, taller than the cloud cover. The tourists estimate that skyscraper A is at least 400 feet tall assuming 10 feet per observed floor. The day improves and the tourists soon eye another skyscraper, this one at least 600 feet tall assuming, again, 10 feet per observed floor. That afternoon, the clouds and rain finally break, and the tourists, this time, better oriented and walking back uptown, observe the true heights and identities of the buildings: skyscraper A, the Empire State Building, is 1250 feet tall, and skyscraper B, the Chrysler building, is only 1050 feet tall. In other words, lower limits and actual values need not always coincide. Useful comparative statements require both lower and upper limits so as to provide guidance on the expected ranges of values to be compared. The same logic as in the example can also be used to provide an upper limit on species richness given a sample abundance distribution, $F_1, F_2, \ldots, F_{k_{max}}$.

The preceding logic assumes that unobserved species are, at most, as common as singleton species in the community; it is also possible that they are less common. In the limit of extreme rarity, all unobserved species will have a true community relative abundance of $1/N_{tot}$, where N_{tot} is the size of the community:

$$S_{rare} = \frac{F_1}{\hat{p}_0 N} = \frac{F_1 N_{tot}}{N},$$ (6.5)

such that an upper limit on the number of species in a community is

$$\hat{S}^+ = S_{obs} + \frac{F_1 N_{tot}}{N}.$$ (6.6)

Unlike the lower bound, the upper limit scales with the ratio between system and sample size. As mentioned before, this ratio often exceeds 10^{10} for under sampled viral communities. The gap between the lower limit and upper limit is very large. Therefore, lower bounds on viral species richness may, in fact, provide minimal useful information on the true, but unknown, viral species richness. It is difficult to compare the species richness of two communities without an accurate estimate of either. In summary, as in the analogy of estimating building heights by eye on cloudy days, quantitative methods used to infer species richness cannot see the part of the signal necessary to infer the property.

The consequences of this can be examined in a number of ways. For example, it is possible that two communities with vastly different true species richness values will have the same estimated lower limit on species richness according to Chao-1. As an illustrative example based on methods first presented in Haegeman et al. (2013), consider a microbial community sample from the surface oceans including 7068 distinct organisms that belong to 811 *operational taxonomic units* (OTUs) (Rusch et al. 2007). An OTU is a heuristic definition of a "species" in terms of sequence space, such that 16s rRNA sequences which have >97% similarity are clustered into a single OTU. An idealized community can be constructed from the sample data by assuming that the community includes additional rare species whose frequencies are each smaller than the most rare species in the sample; that is, they have true frequencies less than 1/7068. The cumulative relative abundance in the tail of the distribution should be F_1/N. The number of species in the tail can vary, so long as the relative abundance of all rare species, $\sum_i^{S_{rare}} p_i^{rare} = F_1/N$. In practice, different rare tails can be used. Here a power-law tail with exponent -2 as a function of the rank of the species was utilized. In turn, the relative

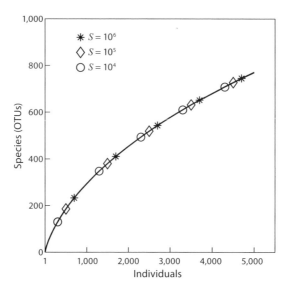

FIGURE 6.6. Predicting the number of microbial/viral species requires inferring the diversity in an entire community from a *relatively small* sample. Even if the true communities differ in diversity, the observed rarefaction curve (and associated species richness) may not reflect such differences. Methodology follows that described in the text; see similar analysis in Haegeman et al. (2013).

abundance in the community of each observed species in the sample must be corrected so that the total probability sums to unity. Applying a Good-Turing correction (Good 1953; Gale and Sampson 1995) leads to three in silico communities, each of which has 10^4, 10^5, and 10^6 distinct species, respectively, in the community.

A *rarefaction curve*, also known as a *collector's curve*, can be generated from each community, $s(n)$. The curve represents the expected number of unique species observed as a function of n collected individuals out of N in the sample. For a community with S_{true} species, each of which has a relative frequency of q_i,

$$s(n) = \sum_{i=1}^{S_{true}} 1 - (1 - q_i)^n. \qquad (6.7)$$

The rarefaction curve summarizes a community abundance distribution and contains the necessary information to extrapolate each of the diversity indices, D_0, D_1, and D_2. What is striking about the three in silico communities is that their rarefaction curves are indistinguishable, despite having dramatically different species richness (Figure 6.6). The true species richness measured in terms of OTUs could differ by multiple orders of magnitude, but this would not be apparent in the data. The Chao-1 estimator for this community is 4604. As was suggested earlier, the Chao-1 estimator is, in practice, an excellent lower bound on species richness, but it also not necessarily informative with respect to the actual species richness. For example, 4604 is much less than both 100,000 and 1 million species. In practice, the use of diversity metrics to estimate viral diversity of undersampled communities should focus on those estimators that measure diversity in a sufficiently weighted average. These estimators include Shannon diversity and Simpson diversity, though Simpson diversity is even more robust.

6.3.3   DIVERSITY OF VIRAL GENES AND GENOTYPES

The estimation of viral gene and genotype diversity is a new area of concern, made possible by technological advances in culture-independent sequencing of viral communities. This field of study is termed *viral metagenomics* (Edwards and Rohwer 2005). A hallmark 2002 study led by Mya Breitbart and Forest Rohwer analyzed ~1000 short viral DNA sequences from 200 L of seawater. Each extracted sequence was on the order of 1 kbp in length (Breitbart et al. 2002). Most of these viral sequences were unlike anything anyone had examined before. As was reported, "over 65% of the sequences were not significantly similar to previously reported sequences, suggesting that much of the diversity is previously uncharacterized" (Breitbart et al. 2002). In comparison, a single liter of seawater contains on the order of 10^{10} viruses, and if each has a genome approximately 50 kbp in length, then the total number of kbp sequences in a 200 L community should be on the order of 10^{14}. This is a factor of 10^{11} times greater than the sample.

The potential scope of a viral diversity study has increased nearly 1000-fold in terms of raw read length in the past decade, yet the problem of the unknown remains. A recent Pacific Ocean virome study presented a pipeline for organizing the sequence space into protein clusters (Hurwitz and Sullivan 2013). Prior knowledge with respect to the origins of sequences in the virome was described as follows: "the majority (87% photic zone, 91% aphotic zone) of the reads could not be classified based on sequence similarity to known taxa" (Hurwitz and Sullivan 2013). It is evident that any estimate of diversity will have to address questions that can be inferred given the limited amount of prior information available.

For other types of organisms, genome diversity is used as the basis for species definitions as well as phylogenomic classification. The OTU concept for bacterial species is made possible by the fact that all bacterial cells share universal markers. The 16S rRNA gene encodes for a functionally active RNA component of the bacterial ribosome. There is no analogous universal gene for viruses. The study of viral taxonomy, viral phylogenomics, and genome-based definitions of viral species is ongoing. One proposal is to utilize genetic markers that apply to a subset of the viral biosphere and that might be used to differentiate subsets of taxa (Rohwer and Edwards 2002). An alternative is to use contig analysis to circumvent the lack of universal marker genes for viruses by using the degree of overlap among sequences from a viral metagenome (Breitbart et al. 2002).

A *contig* is a concatenation of one or more overlapping sequences. Contig analysis assumes that if two short sequences in a metagenome overlap sufficiently, then this indicates that they were derived from the same virus type. These overlaps constitute a *contig spectrum*, whose elements, c_q, are defined as the number of contigs involving q sequences. For example, c_1 is the number of sequences that do not overlap any other sequence, whereas c_2 is the number of contigs involving a pair of sequences, and so on. Using this method, in silico communities were generated, each with different numbers of viral species. Contig spectra could then be derived from each community. The range of species richness values from synthetic communities whose contig spectra were similar to the measured contig spectra generated a posterior distribution from which to estimate viral diversity based on sample data. Alternatively, the Chao-1 estimator was also used to estimate lower bounds by interpreting the contig spectra as a form of species-frequency distribution. In brief, it was assumed that each contig represented a viral OTU and that the number of reads, q, in the contig denoted its abundance; hence, the values of c_i could be used within the Chao richness estimation formalism. This direct approach, known as PHACCS (Phage Communities from Contig Spectrum) (Angly et al. 2005), estimated the number of viral types in the 200 L seawater samples described earlier as between 374 and 7114 (Breitbart et al. 2002). Similar methods were used to conclude that viral diversity in marine sediment ranges between "10,000 and 1 million viral genotypes" (Edwards and Rohwer 2005). In both cases, estimates vary by at least one, if not two, orders of magnitude. The difficulty stems not only from the challenge of estimating diversity of a viral type when no universal marker is available but also from the fundamental fact that estimating species richness, whatever the context is incredibly hard, particularly when doing so from a small sample of a large community.

An alternative view of viral diversity can be gleaned from examining the diversity of viral protein sequences rather than viral genotypes. Putative protein sequences are inferred by identifying open reading frames (ORFs) in viral

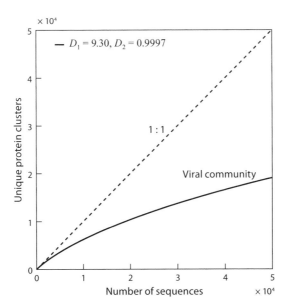

FIGURE 6.7. Rarefaction curve for protein clusters identified in a viral community metagenome. Samples were taken from marine surface waters in Monterey Bay, California. Original data from Deng et al. (2014). See the text for additional details on methodology.

metagenomes. The function of these ORFs remain largely uncharacterized. An increasingly popular method among empiricists is to group short sequences into protein clusters (Hurwitz and Sullivan 2013). The rationale is that doing so provides a lens into the diversity of distinct viral functional genes within a viral community. In some cases, such clustering may enable the generation of a hypothesis of a function for an ORF for which the function was not previously known. The drawback of such clustering is that given the large number of unknowns, the distinct ORFs in a cluster need not necessarily correspond to the same or similar function. Direct tests of virus protein functions in a community context remain elusive. Nonetheless, the same principle of estimating the diversity of species applies to protein clusters. Each cluster is considered a different type, and the number of sequences in a cluster denotes its abundance. The cluster-size distribution can be used to estimate the diversity of viral function of the sample and, by extension, that of the community.

An analysis of viral functional diversity is presented in Figure 6.7. The sample was taken from surface waters off Monterey Bay, where the community viral metagenome was sampled using procedures designed to maximize the yield of dsDNA marine viruses (John et al. 2011). Similar analyses of rarefaction curves for virus protein clusters have been used to identify environmental correlates with changes in viral functional diversity (Hurwitz et al. 2014). The relative frequency of distinct types of functions may, eventually, provide information on the ways that virus-host interactions vary in the oceans, particularly in the face of competition with other viruses and with zooplankton, and given limited nutrients.

6.4   VIRUS-MICROBE INFECTION NETWORKS

6.4.1   FROM SENSITIVITY RELATIONS TO INFECTION NETWORKS

The study of who infects whom has become a staple of virus-host research since Luria's early publication on "sensitivity relations" between bacteria and phage (Luria 1945). The target host of cross infection includes interactions among viruses and (i) bacteria (Weitz et al. 2013), (ii) archaea (Ceballos et al. 2012), and (iii) microeukaryotes (Allen et al. 2007). In the pregenomics era, exposure to viruses was a way to identify whether a newly isolated bacterial clone was, in fact, distinct from others in a culture collection. Phage susceptibility was often able to detect differences in bacteria where serological tests involving binding to antibodies failed. Even in the genomics era, cross-infection assays are often more sensitive to genetic variation than is marker-gene-based sequencing. Whether a virus can infect a set of given hosts appears to be weakly correlated with phylogenetic distance, for example, as in infectivity patterns of viruses that infect marine cyanobacteria from the *Synechococcus* or *Prochlorococcus* genera (Sullivan et al. 2003). The rationale for such discordance is that phylogenetic divergence is quantified on the basis of conserved rRNA genes. Differences in phage susceptibility are often driven by variation in surface receptors, whether proteins or carbohydrates, or other pathways that evolve faster than do marker genes. There is a disconnect between measures of bacterial taxonomy and of phage susceptibility.

To directly link the sequence of a host and a sequence of a virus with a predictive model of infectivity remains an unsolved problem. From an ecological perspective, a cross-infection assay provides direct information on the extent to which populations have diverged in terms of a functionally relevant trait. Cross-infection assays have become part of standard protocols—at least until recently, one of the general suite of microbiological methods to be used in the interest of thoroughness—despite the fact that it isn't readily apparent how the resulting data should be interpreted. The word *interpreted* is apt, as the data had been recalcitrant to quantitative analysis.

The generation of virus-host infection networks preceded the widespread adoption of methods to visualize, analyze, and interpret complex networks (Albert and Barabasi 2002; Newman 2003; Strogatz 2001). It may be relatively simple to assess whether two hosts or two viruses have different infection patterns, but finding patterns in larger networks is nontrivial. To understand why, consider a cross-infection assay with 10 hosts and 10 viruses. Such an experiment generates 100 data points, one for each of the elements of the 10×10 matrix, M. In matrix form, there are 10 rows, one for each host, and 10 columns, one for each virus, which corresponds to $10! = 10 \times 9 \times 8 \times \ldots \times 1$ combinations of distinct orderings to position one host per row, and a similar number for viruses. Thus, there are more than 1.3×10^{13} ways to visualize the network of interactions in a relatively small study. When there are 20 hosts and 20 viruses, the combinations rise to nearly 6×10^{36}. Visual inspection of a matrix is unlikely to reveal patterns without prior information on the ordering of hosts and viruses, such as a phylogenetic tree.

What patterns might there be in a virus-host infection network derived from cross infections among natural populations? The study of ecological relationships such as food webs and plant-pollinator interactions provides some guidance. In a food web (Cohen 1978; Cohen et al. 1990), an *edge* is present if a specific predator can consume a specific prey item. In a plant-pollinator network, an edge is present if there is an observed association between a specific pollinator and a specific target plant (Memmott 1999; Bascompte et al. 2003; Bascompte and Jordano 2013). In both cases, a *node* represents a population. A key hallmark of analyses of both systems is that nodes differ in their degree of interaction; for example, there are both specialists and generalist predators just as there as specialist and generalist pollinators. For example, given an interaction matrix M of size $m \times n$, one may define the connectance as $C = E/(mn)$, where E is the number of nonzero entries in the matrix. In network representation, E is the number of edges. How many edges are likely? Consider that all interactions are equally likely, with probability p, and that there are the same number of nodes in both sets in the bipartite network, $m = n$. A similar set of assumptions was invoked by the mathematicians Paul Erdős and Alfréd Rényi (Erdős and Rényi 1960), who argued that the probability that a node in a unipartite random network with n nodes would have k interactions should be

$$P(k) = p^k (1 - p)^{n-k} \binom{n}{k}.$$

(6.8)

The variation in degree of interaction in a random network follows a binomial distribution, in which each type should have, on average, the same number of interactions. When n is sufficiently large and $p \ll 1$, then the degree distribution within so-called Erdős-Rényi networks approaches that of a Poisson distribution with average degree $\hat{k} = pn$, such that $P(k) \approx e^{-\hat{k}} \hat{k}^k / k!$. In contrast, both food webs (Cohen 1978; Williams and Martinez 2000; Allesina et al. 2008) and plant-pollinator systems (Memmott 1999; Bascompte et al. 2003; Bascompte and Jordano 2007) have an unusually large number of nodes that have either many or few interactions. Indeed, the field of network science recognizes that the Erdős-Rényi network model is a useful point of departure for studying the actual structure of complex networks; but it also recognizes that this null model is a very poor model of actual structure, whether of social, biological, or physical interactions.

One way to consider deviations of actual networks from idealized networks is to compare the number of interactions, k_i, of node i with the expected number. The expected number of interactions for each node in a bipartite network is either Cm or Cn, depending on which population type is examined. The number of interactions can be thought of as samples from the marginal distribution of the interaction matrix M. In the present context of virus-host interactions, the comparison between observed and idealized degrees of interaction may be posed as follows: within a group of bacteria, are some variants relatively more or less susceptible to infection? Similarly, within a group of viruses, are some variants relatively more or less able to infect target hosts? Further, are there unexpected patterns of relationships among viruses and hosts? The answers to all these questions are yes and are the subjects of the next two sections.

6.4.2   TARGET PATTERNS WITHIN INFECTION NETWORKS

Consider a bipartite infection network $\mathbf{M}$ of size $m \times n$ representing the interaction among m host types and n phage types. Each element in the infection network represents the outcome of an infection assay, such as a spot or plaque assay. Infection assays are scored in terms of whether or not there is evidence of infection. In this Boolean scheme, $M_{ij} = 0$ denotes no infection, and $M_{ij} = 1$ denotes successful infection between host type i and phage type j. In some circumstances, the quantitative level of infection, for example, in terms of plaque-forming units, may be measured, in which case the value of M_{ij} may be a continuous response. The current discussion of infection network patterns focuses on analysis of Boolean infection networks, that is, those in which the entries in $\mathbf{M}$ are either 0 or 1. The extension of the present analysis

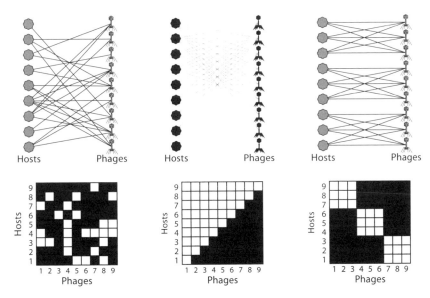

FIGURE 6.8. Cross-infection networks between phage and bacteria. The top panels are in a network layout, and the bottom panels are in a matrix layout. Both represent the same data. The leftmost set represents a random network. The middle set denotes a perfectly nested infection network. The rightmost set denotes a modular infection network. Reprinted with permission from Weitz et al. (2013).

to quantitative infection assays is ongoing. Given this notation, the number of links is $E = \sum_{ij} M_{ij}$. Further, $k_i = \sum_j M_{ij}$ and $d_j = \sum_i M_{ij}$ define the degree of hosts and viruses, respectively. The *degree* of a node is defined as the number of interactions it has with nodes of the other type. For sufficiently large networks, it is also possible to consider the distribution of the degrees, k_i and d_j, a distribution termed the *degree distribution* of rows and columns, respectively.

What are some of the possible structures that can arise in a virus-host infection network if M is not random? A recent series of papers introduced a network-theoretic approach to identifying patterns that could emerge in the cross infection of environmental microbial hosts and viruses (Flores et al. 2011; Poisot et al. 2011, 2012; Beckett and Williams 2013; Weitz et al. 2013). Two key patterns are modularity and nestedness (Weitz et al. 2013), for which idealized networks are shown in Figure 6.8. The *modularity* of a matrix refers to the extent to which the interactions in a bipartite network occur predominantly within mutually exclusive groups of viruses and bacteria. That is, a *module* denotes a set of hosts and viruses for which the viruses tend to infect hosts within the module and not outside it, and for which the hosts tend to be infected by viruses within the module and not outside it. For a matrix to be modular, the cross-infection structure should have a tendency toward an excess of in-group infections compared with a random matrix. The *nestedness* of a matrix refers to the extent to which the sets of interactions of hosts are embedded, one into another, and for which the sets of interactions of viruses are embedded, one into another. Such an embedding can occur, for example, if the host range of the most specialist virus is a subset of that of the second most specialist virus, whose host range is a subset of that of the third most specialist virus, and so on, until all host ranges are subsets of that of the most generalist virus. A similar embedding should occur for the *virus range* of hosts, that is, the set of viruses that can infect a given host. For a matrix to be nested, the cross-infection structure should have a tendency toward an excess of partial subsets within infection ranges compared with a random matrix.

Formally, the quantitative definition of the modularity of a bipartite network, as proposed in Barber (2007), is

$$Q = \frac{1}{E} \operatorname{Tr} \mathbf{R}^T \tilde{\mathbf{M}} \mathbf{T}. \tag{6.9}$$

Here, E is the number of interactions in the network, $\tilde{M}_{ij} = M_{ij} - k_i d_j / E$, $\mathbf{R} = [\mathbf{r}_1 | \mathbf{r}_2 | ... | \mathbf{r}_m]^T$ is an $m \times c$ index matrix, and $\mathbf{T} = [\mathbf{t}_1 | \mathbf{t}_2 | ... | \mathbf{t}_n]^T$ is an $n \times c$ index matrix, where c is the number of modules (Barber 2007). Here, $\mathbf{R}$ and $\mathbf{T}$ are Boolean matrices whose row sums must always be equal to 1. The column with the nonzero entry denotes the module associated with each host and virus (as indexed by the row). Thus, the modularity Q increases whenever there is an interaction between host i and virus j that have been grouped together in the same module. The lowest value of modularity is -1 and corresponds to the occurrence of interactions always between nodes from different modules in a very large network. In practice, such low values are rarely realized and usually result from a poor assignment of nodes into modules. The highest value of modularity is 1 and corresponds to the occurrence of interactions always between nodes of the same module in a very large network. Algorithms to quantify the modularity of a matrix must try to find the module assignments embedded in $\mathbf{R}$ and $\mathbf{T}$ that maximize Q. The combinations increase exponentially with the size of the network. Moreover, the number of modules is usually not known in advance. Multiple values of c must be tested, increasing the complexity of the task. In response, heuristic methods have been introduced to try and identify nearly optimal solutions. The widely used Bipartite, Recursively Induced Modules (BRIM) method is now available in a number of software packages (Barber 2007). The interested reader may consult the software and documentation associated with libraries for more details, for example, bipartite (an R-based library) (Dormann et al. 2008) or BiMAT (a MATLAB-based library) (Flores et al. 2014), for more information on implementation.

The quantitative definition of nestedness includes alternative metrics. Two of the most commonly used are the NTC (Nestedness Temperature Calculator) (Atmar and Patterson 1993) and NODF (nestedness metric based on overlap and decreasing fill) (Almeida-Neto et al. 2008). Despite these rather opaque names, both aim to quantify the degree to which cross-infection structure can be ordered in terms of nested sets. In NTC, the orders of the rows (bacteria) and viruses (columns) are shuffled so that many of the interactions appear as sets of each other. The convention here is that the most susceptible host is in the topmost row, and the most generalist virus is in the leftmost column (Figure 6.8). In practice, the heuristic algorithm of sorting rows and columns by their degrees, k_i and d_j, respectively, is nearly as effective as complex alternatives (Flores et al. 2011). Then, given an ordering of rows and columns, the value of nestedness depends on the extent to which the interactions differ from that of a perfectly nested matrix given the same size and number of interactions. The details of this method are presented in Weitz et al. (2013).

In NODF, the nestedness is defined as

$$N_{NODF} = \frac{\sum_{i<j} \mathcal{N}_{ij}^{\text{rows}} + \sum_{i<j} \mathcal{N}_{ij}^{\text{columns}}}{m(m-1)/2 + n(n-1)/2}, \tag{6.10}$$

where

$$\mathcal{N}_{ij}^{\text{rows}} = \frac{(\mathbf{r}_i \cdot \mathbf{r}_j)(1 - \delta(k_i, k_j))}{\min(k_i, k_j)}, \tag{6.11}$$

and

$$\mathcal{N}_{ij}^{\text{columns}} = \frac{(\mathbf{c}_i \cdot \mathbf{c}_j)(1 - \delta(d_i, d_j))}{\min(d_i, d_j)}. \tag{6.12}$$

In this definition, $\mathbf{r_i}$ and $\mathbf{c_i}$ are vectors that represents row i and column i, respectively, of the bipartite adjacency matrix, and $\delta = 1$ only when its two arguments are equal. By way of example, consider two rows of a 20×20 interaction matrix whose degrees differ (e.g., $k_i = 10$ and $k_j = 5$). Then, the potential contribution from such a pairwise combination should occur when the five interactions of row j are also present in row i. In that scenario,

$\mathbf{r}_i \cdot \mathbf{r}_j = 5$, and so $\mathcal{N}_{ij} = 1$. Alternatively, consider the case where none of the five interactions in row j is present in row i, in which case $\mathbf{r}_i \cdot \mathbf{r}_j = 0$, and so $\mathcal{N}_{ij} = 0$. Finally, consider the case where $k_i = 5 = k_j$. The value of $\mathcal{N}_{ij}$ must be zero in this case, because $1 - \delta(5, 5) = 0$; there cannot be decreasing filling between rows with the same number of interactions. In this way, NODF defines nestedness as the fraction of the $m(m - 1)/2$ row pairs and of the $n(n - 1)/2$ column pairs whose interaction ranges are subsets of one another. This method does not depend on ordering or labeling, though ordering of the matrix will help in visualizing the pattern. Multiple implementations are available for both the NTC and NODF metrics (Beckett et al. 2014; Flores et al. 2014).

The significance of any measured value of modularity or nestedness depends on a comparison with a null model. In other words, how different is the value of Q or N_{NTC} or N_{NODF} from what might be expected within an ensemble of random networks? The term *ensemble* refers to a collection of items that have similar, but not identical, properties. Indeed, the ensemble of random networks is meant to preserve some of the properties of the observed network. The two features that are invariably preserved in an ensemble are the size of the network and the average number of interactions, E. That is, every network in the ensemble will have size $m \times n$, and the average number of interactions in a random network in the ensemble should approach E. The Erdős-Rényi prescription for generating networks represents one class of null models. Another constraint can include the requirement that the degree of interactions in the observed network is preserved in the ensemble, either strictly or on average. Degree-constrained random networks represent a more restrictive class of null models. Whatever the choice, the significance of a pattern and the size of its effect can then be calculated by comparing the observed feature of the network with that of networks in the ensemble (Weitz et al. 2013).

It is important to keep in mind the complex relationship between degree distributions and network patterns, including nestedness and modularity. For example, consider the constraint that the degree distribution of a network must be strictly preserved. Further, assume that the network is perfectly nested, and the degree distribution is uniform. There is only one perfectly nested network with a given uniform degree distribution (if the node labels are fixed and specified). To understand why, consider the original network M. To preserve the degree of the original network requires identifying a pair of bacteria i and i' and a pair of viruses j and j' such that only j infects i, and j' infects i'. If that were the case, then one could swap the infections, thereby creating a new network that preserves the degree of all nodes (including that of i, i', j, and j'), as in the following schematic:

$$
\begin{array}{c|c|c|c|c|c|c|c}
 & 1 & \cdots & j & \cdots & j' & \cdots & n \\
\hline
1 & & & & & & & \\
\hline
\cdots & & & & & & & \\
\hline
i & & & 1 & & 0 & & \\
\hline
\cdots & & & & & & & \\
\hline
i' & & & 0 & & 1 & & \\
\hline
\cdots & & & & & & & \\
\hline
m & & & & & & &
\end{array}
\quad \xrightarrow{\text{swap}} \quad
\begin{array}{c|c|c|c|c|c|c|c}
 & 1 & \cdots & j & \cdots & j' & \cdots & n \\
\hline
1 & & & & & & & \\
\hline
\cdots & & & & & & & \\
\hline
i & & & 0 & & 1 & & \\
\hline
\cdots & & & & & & & \\
\hline
i' & & & 1 & & 0 & & \\
\hline
\cdots & & & & & & & \\
\hline
m & & & & & & &
\end{array}
\qquad (6.13)
$$

No such pairs of bacteria and viruses exist in a perfectly nested network. If virus type j is relatively more generalist than type j', it must, in a perfectly nested network, also infect i', because it infects all the hosts that any viruses below it in the ranking infect. Or, if type j is relatively more specialist than type j', it must, in a perfectly nested network, also infect i, because its host range includes all hosts infected by virus type j. Similar logic holds for the bacteria. Should one be surprised to observe nestedness given a particular set of degree distributions? No. But should one be surprised to observe that particular set of degrees? Yes, and even more so if one takes as a null model the Erdős-Rényi network, which tends to give a unimodal rather than a uniform degree distribution. The statistical consequences of any of these choices are the subjects of active debate (Fortunato and Barthélemy 2007; Ulrich et al. 2009; Fortuna et al. 2010). The next section considers random ensembles based on preserving the size and average connectivity of the network.

6.4.3   FEATURES OF INFECTION NETWORKS WITHIN THE MARINE ENVIRONMENT

The study of phage-bacteria infection networks in the marine environment owes a debt to the pioneering work of Dr. Karl-Heinz Moebus. Moebus was based at the Helgoland marine station, located on the island of Helgoland in the North Sea off the northwest coast of Germany. There, he and colleagues examined the relationship between viruses and microbial hosts both from the marine station and on cruises in the North Atlantic. The protocols Moebus refined (Moebus 1980; Moebus and Nattkemper 1981) have become, with some updating, standard practice in environmental virus isolation (Wilhelm et al. 2010). In brief, Moebus isolated environmental bacteria using growth media that favored fast-growing heterotrophic microbes, such as members of the *Vibrio* clade. Then, he exposed populations of bacteria grown from a single culture to bacteria-free seawater that contained marine viruses and evaluated them for lysis. When clearing took place, the contents of the flask presumably held many of the viral progeny that resulted from infection and lysis of the target host. Moebus then plated these viruses back on a lawn of the target host population, picked individual plaques, and then grew them again on the target host to yield a virus isolate. With a growing number of microbial hosts and viruses in the Helgoland collections, it became possible to begin to evaluate a virus for the potential to infect other hosts.

The most striking example of Moebus's cross-infection work was published in 1981 in Helgoland's own research journal: *Helgoländer Meeresunter-suchungen* (Moebus and Nattkemper 1981). In it, "phage-host cross-reaction tests were performed with 774 bacterial strains and 298 bacteriophages." Individual strains were collected at a series of 48 stations in the Atlantic Ocean. The total scope of the experiments represents more than 230,000 pairwise infection assays. Many of the hosts were resistant to all isolated phage. Further, the only way to distinguish a host strain in this assay was via its susceptibility profile. As a consequence, only a reduced subset of the interaction data was reported in a foldout table in the original journal publication. The computer-assisted digitized result focusing on the interactions between 286 host types and 215 phage types is depicted in Figure 6.9 (Flores et al. 2013). The hosts and viruses were, predominantly ordered by station. There is evidently blockiness to the pattern of infections, perhaps associated with geography given the route of the expedition from east to west across the Atlantic, but any visual analysis is based on but one of the very large number of permutations. The authors did notice that "sensitivity marks are distributed unevenly" and that there appeared to be clusters of interacting types. They attributed these clusters to differences in infectivity arising from geographic separation.

The advantages of network-based analyses is evident when the same dataset is visualized, albeit in a layout highlighting the modularity inherent within the data (Figure 6.10). Here, the rows and columns have been shifted so that those phage that tend to infect the same hosts are grouped together. In addition, blocks have been drawn for each cluster corresponding to the modules identified by the BRIM algorithm (Flores et al. 2013). There are 49 blocks in which 94% of all interactions occurred within modules. The measured modularity of $Q = 0.795$ was significant ($p < 10^{-5}$) and of large effect. Although the composition of modules does exhibit a station effect, many hosts and viruses from different stations are included inside modules. Statistical analysis revealed that the number of stations represented inside a module was less than that expected by chance; hence, phage and host isolated at a subset of stations tend to constitute modules. These stations have a geographic

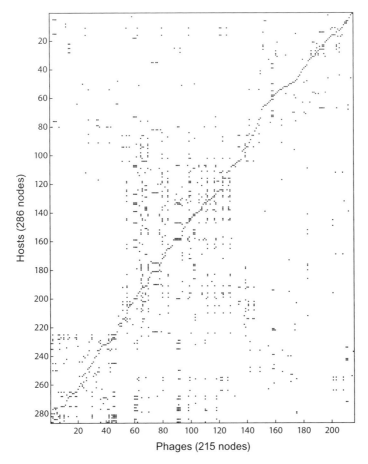

FIGURE 6.9. Phage-bacteria infection network as collected by Karl Moebus and colleagues (Moebus and Nattkemper 1981). Image reprinted with permission from Flores et al. (2013). Black squares denote infection.

signal, but it is not so clear as to demarcate geographically colocated stations. Individual viruses were found to infect hosts located thousands of kilometers away. Similar analyses of phage-host interactions from soil communities, for example, those of *Pseudomonas aeruginosa* isolates and associate phage, have found that viruses are better adapted to co-occurring bacteria than to bacteria isolated from other habitats (Koskella et al. 2011; Buckling and Brockhurst 2012). Extending such logic to the oceans requires consideration of the physical flow of water masses as well as environmental factors that shape the composition of hosts and viruses. For example, does coevolution among viruses and hosts drive divergence in infectivity, for example, owing to

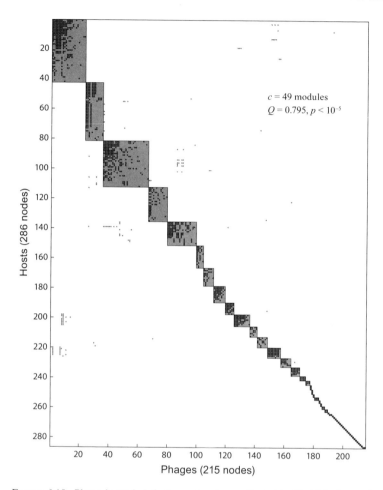

FIGURE 6.10. Phage-bacteria infection network as collected by Karl Moebus and colleagues (Moebus and Nattkemper 1981). Image reprinted with permission from Flores et al. (2013). Data were resorted to highlight modularity in the cross-infection network.

specialization on those hosts in the community? Or is allopatric separation of hosts followed by specialization of viruses on locally adapted hosts the norm? Answers to these questions cannot be found in Moebus's study, as it predated the standard use of genetic analysis of microbes and their viruses. Perhaps this is why this early work is often ignored, but it should not be.

Indeed, the scope of the Moebus study contrasts with the typical size of cross-infection assays. A meta-analysis of 38 distinct virus-host infection networks found that the typical size of such assays included 19 host isolates

A

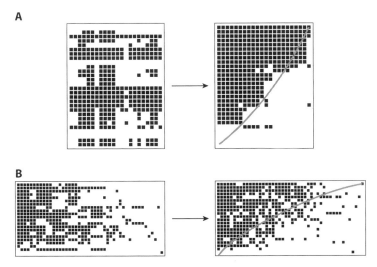

B

FIGURE 6.11. Phage-bacteria infection networks from the marine environment. Reprinted with permission from Weitz et al. (2013). (A) Infections between phage (columns) and *Flavobacterium* (rows), a fish pathogen. Data from Stenholm et al. (2008). (B) Infections between phage (columns) and a marine heterotroph (rows). Data from Holmfeldt et al. (2007). In both (A) and (B), the left matrix represents the original formatting, whereas the right matrix has been resorted to highlight nestedness. The line in the right matrices denotes the isocline of perfect nestedness (i.e., all interactions would be above the isocline).

and 11 viral isolates, of which 65 of the nearly 200 possible interactions were positive (Flores et al. 2011). The studies largely focused on a set of closely related bacteria; for example, given available information, the taxonomic classification of bacteria in any given study differed, at most, at the genus level. Given that scale, a significantly nested pattern was observed in 27 of the 38 studies. Two examples are included in Figure 6.11, both taken from studies that focused on marine systems—one involving *Flavobacterium psychrophilum*, a fish pathogen, and the other involving *Cellulophaga baltica*, a marine heterotroph. As shown in the previous chapter, nestedness may be the result of a coevolutionary process in which hosts increase their resistance in response to phage targeting and in which phage expand their host range to target these newly evolved hosts. Investigation of the internal structure of modules in the Moebus study reveals that 7/13 of the largest modules are significantly nested (Flores et al. 2013). Different processes involving both specialization—associated with enhanced modularity—and coevolutionary range expansions—associated with enhanced nestedness—may be operating concurrently. Or, as theoretical work has suggested, a single coevolutionary mechanism could be responsible for both patterns, with information on quantitative levels of infection seemingly crucial to providing further insight (Beckett and Williams 2013).

In addition to foundational questions regarding coevolution and ecology, the study of complex infection networks in natural systems raises a number of issues of practical concern: for example, which hosts do viruses infect, and can the target host of a virus be predicted from sequence alone? It is evident that viruses are, relatively speaking, specialists when compared with other sources of microbial mortality, such as grazing. Like many topics in phage biology and ecology, it would seem that Salvador Luria anticipated many of the challenges and limits ahead. Luria (1945) speculated on the potential for widespread cross infection:

> It is interesting, from the standpoint of bacterial taxonomy, that while bacterial viruses may be active on species belonging to different genera, chiefly within the family *Enterobacteriaceae*, no virus has ever been found to be active on members of different families.

Recently, Hyman and Abedon (Hyman and Abedon 2010) and Meaden and Koskella (Meaden and Koskella 2013) compiled examples of published studies in which a single phage isolate was shown to infect and lyse on multiple genera. Three such examples are (i) *Pseudomonas* and *Erwinia* (Koskella and Meaden 2013); (ii) *Salmonella*, *Klebsiella*, and *Escherichia* (Bielke et al. 2007); and (iii) *Pseudomonas*, *Sphaerotilus*, and *Escherichia* (Jensen et al. 1998). To this list must be added individual cyanophage isolates that can infect individual cyanobacteria from either *Prochlorococcus* or *Synechococcus* (Sullivan et al. 2003). In this last case, phylogeny provided some information on the likelihood of cross infectivity and, perhaps, even the basis for host-range variation in the isolated phage. For example, the morphotype of isolated viruses was correlated with the ecological niche in which the host is found; for example, podoviruses were isolated from high-light-adapted *Prochlorococcus*, and myoviruses were predominantly isolated from *Synechococcus* strains (Sullivan et al. 2003). The myoviruses had a much broader host range than did podoviruses; for example, myoviruses were able to infect isolates from both genera, whereas the podoviruses were able to infect at most one additional host isolate beyond that on which they were isolated. It remains an open question whether similar principles relating morphology to cross-infectivity exist in other marine host-virus systems.

To move forward, new methods are needed to probe the statistical limits to cross-infectivity, keeping in mind the following principles: (i) even a single base pair change can modify the degree of permissiveness to viral infection from highly permissive to completely resistant; (ii) there are almost no reported instances of cross infection of two hosts by a single virus isolate if those hosts differ taxonomically at the family level or above. The study of Jensen et al. (1998) is notable for being perhaps the only published claim of a phage whose host range includes bacterial isolates from distinct families. Culture-independent methods are likely to be essential to probing the network of cross infection in marine environments and elsewhere. In this same vein, the study of Moebus and Nattkemper (1981) quantified two types of infection levels, possibly attributed to lytic and lysogenic events. A recent study of *Cellulophaga baltica*, a ubiquitous ocean heterotroph, found that differences in infection strength may be the results of distinct environments (Holmfeldt et al. 2014). In nutrient-enriched environments, lysis was favored, whereas in nutrient-depleted environments, lysogeny was favored. This finding is consistent with large-scale environmental assays of the frequency of lysogeny (Laybourn-Parry et al. 2007; Payet and Suttle 2013; Hurwitz et al. 2014). A sharper focus on quantitative rates of lysis and alternative cells fate will be required to better understand cross infection.

Finally, the study of any complex network requires some consideration of the effect of sampling. In the study of food webs and mutualistic networks, the issue of sampling arises because not all links are measurable (Bersier et al. 1999; Martinez et al. 1999; Poisot et al. 2012). For example, a predator may not eat a particular prey while under observation, or the prey may not be identifiable from gut content analysis. In the case of virus-host networks, another type of sampling issue arises: the majority of viruses and hosts in the surface ocean cannot yet be cultured. This implies a sampling challenge of nodes and associated links in the network. Methods to directly assess virus-host interactions, including viral tagging (Deng et al. 2014) and sequencing of polonies will be part of any effort to quantify infection networks among both culturable and nonculturable viruses and hosts.

6.5   SUMMARY

- Viruses are extremely abundant in the oceans, with estimates of virus-like particle densities ranging from approximately 10^4 to 10^8/ml.
- Total virus abundance correlates with prokaryotic abundance and other abiotic proxies, including chlorophyll.
- Virus abundance is estimated to be at its highest in coastal environments, during blooms, and in sediments.
- Viral diversity remains elusive. Those features of viral diversity that are estimable include Shannon and Simpson diversity and should be utilized instead of attempting to estimate the total number

of virus "species" in the community based on measurements of a
small subsample.

- Viral diversity includes genotypic, genetic, and functional diversity.
- It is evident that individual viruses infect more than one host type, and individual hosts are infected by more than one virus.
- The cross-infection networks in natural systems include evidence of specialization, as measured by modularity, and hierarchical order, as measured by nestedness.
- The basis for emergent infection networks remains unresolved.
- A full accounting of viral diversity, whether genetic or functional, requires improvements in culture-independent methods to probe virus-host interactions.

CHAPTER SEVEN

Virus-Host Dynamics in a Complex Milieu

7.1 ROSENBLUETH AND WIENER'S CAT

More than 25 years ago it was discovered that aquatic viruses can reach densities exceeding 10^{11}/L (Bergh et al. 1989). And more than a decade has passed since viral metagenomics revealed the incredible potential diversity of uncultured virus communities (Breitbart et al. 2002). Greater than 75% of viral sequences in these original metagenomic studies had no closely related hits in curated sequence databases. This situation remains largely unchanged to-day (Hurwitz and Sullivan 2013). High abundances and diversities may appear impressive, but are viruses, in fact, ecologically relevant in the environment? In this chapter, I examine how models of virus-host interactions can be used to help infer viral effects on microbial communities and complex ecosystems, by focusing on dynamics taking place in marine surface waters.

One way to reason through the potential significance of marine viruses is to recognize that viruses must infect and eventually lyse their hosts to compensate for their own decay in the environment (Wommack and Colwell 2000; Weinbauer 2004). Environmental residence times of viruses in marine surface waters are estimated to be on the order of 1 day (Suttle and Chen 1992). Given virus densities of 10^7/ml, then approximately that many viruses per milliliter must be produced per day. If burst sizes are approximately 50, then 2×10^5 microbial hosts per milliliter per day must be lysed to maintain a relatively constant virus population, whereas if burst sizes are approximately 200, then 5×10^4 microbial hosts per ml per day must be lysed. Microbial densities vary in the environment, though total densities on the order of 10^6/ml are typical (Williams and Follows 2011). This back-of-the-envelope calculation suggests viruses could lyse 5%–20% of the standing stock of microbes every day! Then again, the same back-of-the envelope calculation could suggest viruses are responsible for 1%–4% daily clearance given virus densities are on the order of 2×10^6/ml. Similarly, viruses could be responsible for $\sim$50% daily clearance if virus densities are on the order of 2.5×10^7/ml. Moreover, these estimates presume that each bacterial type has the same risk of being killed by a virus. The actual risks are likely to be heterogeneous. How much of relative mortality is, in fact, under the control of viruses, and how else do viruses affect microbial communities?

Understanding the ecological role of viruses requires considering multiple factors beyond those already introduced in the models of Chapters 3–5. First, virus-host interactions are not strictly pairwise (see Chapter 6). That is, viruses infect more than one host, and hosts are infected by more than one virus. Viruses may infect two strains of the same host species or even two strains of different species, and in some circumstances, different genera. Second, unlike in chemostat models, resources necessary for cell maintenance, growth, and division are not drawn only from the "outside." They may be fixed via autotrophic processes, whether light- or chemically induced. In many natural environments, including oligotrophic regions of marine surface waters, resources are sparse, limiting, and hotly contested. Available resources are often regenerated or recycled; that is, they were most recently partitioned inside a living organism before being released as waste, exudate, or even cell debris. Third, viruses are not the only source of mortality—Far from it. Microbes are subject to other forms of mortality, including death by predation. The predators of microbes in the marine environment are largely eukaryotic zooplankton.

How does one model infection networks, carbon and nutrient cycles, and the interplay between parasitism and predation, in practice? Thankfully, there is precedent for developing models that include such realism, or complexity, depending on whether you view a glass half-full or half-empty. Mathematical models of microbial population biology owe much to Bruce Levin's seminal work with Frank Stewart and Lin Chao (Levin et al. 1977), which itself owes a debt to Alan Campbell's model of virus reproduction on a single host (Campbell 1961). These models are highly influential in efforts to determine how viruses affect and coexist with microbial populations in the environment. For example, the core assumptions of Campbell, Levin, Stewart, and Chao re-appear in models of environmental virus-host dynamics as proposed by Heldal, Bratbak, and colleagues (Bergh et al. 1989), Middelboe (Middelboe 2000) and T. Frede Thingstad, Takeshi Miki, and colleagues (Thingstad and Lignell 1997; Thingstad 2000; Rodriguez-Valera et al. 2009; Miki et al. 2008). Many features of these models were introduced specifically to address challenges arising from environmental microbiology, including the consideration of infection networks, zooplankton grazing, and nutrient recycling.

But interactions do not tell the whole story in a nonlinear dynamical system. The parameters and rates governing interactions are also essential.

In that sense, the efforts of Alexander Murray and George Jackson are equally important to understanding viral effects on complex microbial communities (Murray and Jackson 1992). Together, Murray and Jackson adapted Howard Berg and Edward Purcell's famous study on the physics of chemoreception (Berg and Purcell 1977) to virus-host interactions in the marine environment. In doing so, they provided a basis for quantitative estimates of the crucial life history traits, such as contact rates between individual viruses and hosts, that underlie any effort to model the ecological dynamics of microbial and viral populations.

In summary, conclusions drawn from ecological models of virus-host interactions should be reevaluated in a community context. That, in essence, is the aim of the current chapter. In taking a step toward complexity, it is worth recalling the caution imbued in this quote from Arturo Rosenblueth and Norbert Wiener: "[T]he best material model for a cat is another, or preferably the same cat" (Rosenblueth and Wiener 1945). Hence, whatever nods to realism introduced here will be partial, intentionally so. They reflect the intuition of scientists, this another among them, who believe that such features are crucial to driving changes in natural systems. They also reflect a tendency of theorists to try to match the scope of complexity to the types of data being measured. In practice, environmental microbiologists studying the marine environment measure cross-infections, relative rates of grazing, and multiple processes involved in nutrient cycling. Whether these are the right preoccupations will, like all science, be revealed ... eventually.

7.2   MANY VIRUSES AND MANY HOSTS

7.2.1   DYNAMIC MODEL OF COMPLEX VIRUS-HOST COMMUNITIES

How do many different virus types coexist with each other and their hosts? There is no one correct answer to this question. The natural world is sufficiently diverse that there may be different mechanisms operating in different environments and at different moments in time. The mechanisms may include specific types of virus-defense systems operating within hosts, or counterdefense systems operating within viruses. The mechanisms may also be ecological, arising owing to changes in the structure of the microbial community even if the mode of resistance and counterresistance is unchanged. Coexistence may also be enhanced by generation of novel types in the community and/or the influx of novel types from without. In practice, ecologists often break down a problem into components, progressively increasing in complexity, to gauge the

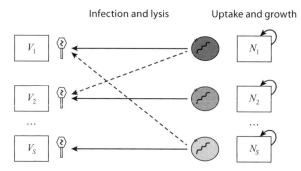

Infection and lysis   Uptake and growth

FIGURE 7.1. Schematic of a dynamic model including complex cross infection between viruses and hosts. In this example, there are S virus types and S host types. Each host takes up resources (modeled implicitly) and is subject to virus infection and lysis. Viruses increase in number with each lysis event and decrease in number owing to decay and other events (not depicted). Lines between host and virus nodes denote the ability of that virus type to infect and lyse the host type. The type of the line indicates that the efficiency of infection and lysis can vary between virus-host pairs.

effect of that component on the overall system behavior. Yet, complex feedback loops and nonlinearities often pose a problem to such a piecewise diagnosis.

One way to simplify the diversity of a natural system is to focus on interactions at one scale. For example, the first models introduced in this book focused on population dynamics arising from interactions between a single virus-host pair (Chapter 3). Later chapters considered variations in which two or more strains of the same species interacted. These strains were meant to represent closely related strains of the same viral type, presumably differing in a very small number of genomic sites, that infect host strains from within the same bacterial species. Some viruses can infect multiple bacterial strains in a species, differing substantially in their genome sequences. Other viruses can infect bacteria from more than one species, and even from more than one genus (see Chapter 6). How are community dynamics expected to change given realistic cross-infection networks?

To begin, consider a system of S host strains and S viral strains. Hosts compete for implicitly defined resources, while viruses infect different subsets of the host community. In this model, N_i is the density of host i and V_j is the density of virus j. Unlike in Chapter 3, here cross infection is represented by a network of interactions, $\mathbf{M}$ (Figure 7.1). $\mathbf{M}$ is a *bipartite* network consisting of virus and host nodes, and interactions between them. The network can be represented as a matrix of 1s and 0s. An edge, $M_{ij} = 1$, between host i and virus j is present whenever viruses of population j can infect and lyse individual bacteria from population i. The absence of an edge, $M_{ij} = 0$, represents the fact that viruses within a given population j cannot infect individual bacteria from population i (see Chapter 6 for an extended introduction to virus-host infection networks). The ecological dynamics of a set of S host and S viral populations given a cross-infection network $\mathbf{M}$, can be written as

$$\frac{dN_i}{dt} = r_i N_i \left(1 - \frac{\sum\limits_{i'=1}^{S} N_{i'}}{K} \right) - \sum_{j=1}^{S} \overbrace{M_{ij} \phi_j N_i V_j}^{\text{lysis of host } i \text{ by phage } j} ,$$

$$\tag{7.1}$$

$$\frac{dV_j}{dt} = \sum_{i=1}^{S} \overbrace{M_{ij} \phi_j \beta_j N_i V_j}^{\text{lysis of host } i \text{ by phage } j} - m_j V_j.$$

This model structure was introduced in Jover et al. (2013). Adsorption rates and burst sizes are assumed to depend only on the virus type; that is, $\phi_{ij} = \phi_j$, and $\beta_{ij} = \beta_j$. Further, resistance is extracellular, so that viruses adsorb only to hosts they can infect and lyse.

At steady state, the sum of new virions released in lysis events must balance per-virion mortality. Similarly, per capita growth of hosts must balance the sum of lysis events by all virus strains. The rate of lysis varies with each strain. The assumption that virus-host interactions are governed by a type I functional response enables the following analytical solution of equilibrium densities (Jover et al. 2013):

$$\mathbf{M}^{\mathrm{T}} \vec{N}^* = \vec{h}, \qquad \mathbf{M} \vec{V}^* = \frac{\vec{r}}{\vec{\phi}} \left(1 - \frac{\sum\limits_{i'=1}^{S} N_{i'}^*}{K} \right), \tag{7.2}$$

where, $\mathbf{M}^{\mathrm{T}}$ denotes the transpose of the interaction matrix, and $\vec{N}^*$ and $\vec{V}^*$ are vectors of the equilibrium densities. Further, $\vec{h}$ is a vector whose elements are $m_j/\phi_j \beta_j$, $\vec{r}$ is a vector whose elements are r_i, and $\vec{\phi}$ is a vector whose elements are ϕ_j. A fixed-point equilibrium for which both host and virus populations coexist with positive densities is possible only if $\mathbf{M}$ is invertible and if $K > \sum_{i=1}^{S} N_i^*$. Together these conditions imply that the network and parameters matter. Some cross-infection networks may not enable coexistence while others do. Combinations of viruses and hosts may have the same cross-infection network but different infection and/or growth rates, leading

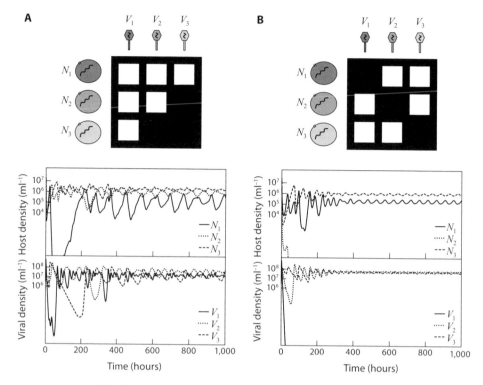

FIGURE 7.2. Network structure changes the fate of interacting virus and host populations. The dynamical system corresponds to Eq. 7.1 given interaction networks **M** in each column. Both networks have the same size and number of interactions. Parameters for both networks are $\vec{r} = (0.53, 0.31, 0.17)$, $K = 10^7$, $\vec{\phi} = (0.650.260.60) \times 10^{-8}$, $\vec{\beta} = (45, 75, 69)$, $\vec{m} = (0.91, 0.23, 0.84)$. Densities for viruses and hosts are in units of ml^{-1}.

to alternative outcomes. Figure 7.2 illustrates this first point, by showing how distinct network structures can lead to distinct outcomes. In one case (Figure 7.2A), all three host types and three virus types persist, in the other (Figure 7.2B), only two host and two virus types persist. That who infects whom should affect coexistence seems intuitive. The challenge is to predict who persists, and why, given overlapping ranges of coinfection and different life history traits.

7.2.2   COEXISTENCE GIVEN OVERLAPPING INFECTION RANGES

Chapter 6 provided evidence that viruses can infect many hosts, that hosts can be infected by many viruses, and that the cross-infection networks between viruses and hosts have recurring structures. These structures include nested

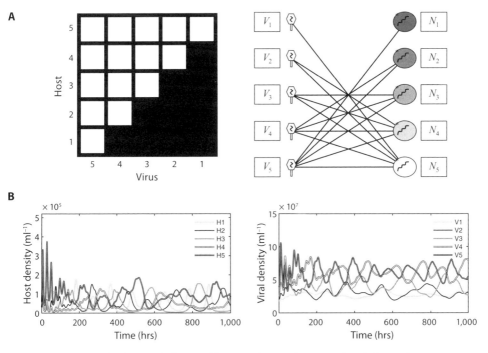

FIGURE 7.3. Hosts and viruses with a nested network can coexist via fluctuating dynamics. (A) Example of a perfectly nested virus-host infection network, – visualized in both matrix and graph form. Host 1 is the most resistant, and host 5 is the most susceptible to infection, quantified in terms of the number of viral types that can infect a given host. Similarly, virus 1 has the narrowest host range, and virus 5 has the broadest host range. (B) Dynamics of host and viral coexistence in the model. Adapted from Jover et al. (2013). Parameters for the model include the life history–associated traits of hosts: $K = 10^7$ cells/ml, and $\vec{r} = (0.3162, 0.6325, 0.8367, 1.1, 1.14)$ hr^{-1} (for hosts 1 to 5); a common infection rate for all virus-host interactions: 5×10^{-9} ml/(cells $\times$ hrs); and life history–associated traits of the viruses: $\vec{m} = (0.009, 0.027, 0.0548, 0.0922, 0.1392)$, and $\vec{\beta} = (20, 40, 60, 80, 100)$ (for viruses 1 to 5).

and modular patterns of cross infection. Nested infection networks pose an interesting dilemma, namely, how do hosts and viruses coexist given this particular type of overlapping range of cross infection? Consider the idealized nested cross-infection network in Figure 7.3. Host susceptibility varies from well defended, like host 1, which is susceptible to only one virus, to permissive, like host 5, which is susceptible to all five viruses. Likewise, the host range of viruses varies from narrow, like virus 1, which infects a single host, to broad, like virus 5, which infects all five hosts. Dynamics of hosts and viruses with this infection network coexist (Figure 7.3B), but this coexistence appears problematic. For example, why isn't virus 1 outcompeted by the other viral types? It infects a single host type, yet all other viral types also infect that same host type. Similarly, why doesn't host 1 outcompete all other hosts? It is infected only by virus 5, but so are all the other hosts.

One answer to these questions centers on the fact that network structure, taken alone, does not determine the fate of a population. Life history traits matter. This is evident to biologists, but not always in the field of network science, where network structure has been used as a bellwether for system behavior. To borrow an old adage, network structure has a vote, but not a veto. The equilibrium solutions of host and viral densities (Eqs. 7.2) can be adapted to the specific structure of a nested infection network $\mathbf{M}$ with S hosts and S viruses. As was explained in Chapter 3, viruses can limit the densities of hosts. The production of new viruses must be balanced by virion loss. In the example, where $S = 5$, the specialist virus type 1 infects host 5. The density of host 5 is set by the life history traits of virus 1: $N_5^* = m_1/\phi_1\beta_1$. Then, at equilibrium the birth and death of type 2 viruses must satisfy

$$\left(\beta_2\phi_2 N_4^* + \beta_2\phi_2 N_5^*\right) V_2 = m_2 V_2. \tag{7.3}$$

Substitution of the density of N_5^* yields

$$N_4^* = \frac{m_2}{\beta_2\phi_2} - \frac{m_1}{\beta_1\phi_1}. \tag{7.4}$$

For this host population to have a positive equilibrium density, the combined effects of the life history traits of virus 2 must be worse than those associated with virus 1. The combination of life history traits of relevance here is the ratio $h_i \equiv m_i/(\beta_i\phi_i)$. Given a single virus-host pair, this ratio is equivalent to the inverse of the basic reproductive number of a virus, $\mathcal{R}_0$. In other words, the most specialist and the second-most specialist virus can control two host populations without driving them to extinction if the most specialist virus would, if alone, increase more rapidly on its single target host. An alternative explanation is that there must be a trade-off between host range, the number of hosts a virus can infect, and host-use efficiency, the extent to which a virus exploits the hosts it can affect. The logic of how two viruses can draw down two host populations can be extended, such that the equilibrium population densities of all S host populations can be written, generally, as

$$N_S^* = h_1, \quad N_{S-1}^* = h_2 - h_1, \quad \dots, \quad N_1^* = h_S - h_{S-1}. \tag{7.5}$$

Again, the first condition for coexistence of all hosts, as limited by a virus community, is that there must be a strict trade-off between host range and host-use efficiency. How strict the trade-off must be when the network is close to, but not perfectly, nested remains an open question.

A similar approach can be used to predict the densities of viruses. Given the example of a community with 5 hosts and 5 viruses, the equilibrium condition for the host targeted by the fewest types of viruses (host 1) is:

$$r_1(1 - N_{\text{tot}}/K) = \phi_5 V_5^*. \tag{7.6}$$

The total density of host populations,

$$N_{\text{tot}} = \sum_{i=1}^{5} N_i^* = \frac{m_5}{\beta_5 \phi_5}, \tag{7.7}$$

is set by the life history traits of the most generalist virus. The generalist virus has an equilibrium density of

$$V_5^* = \frac{r_1}{\phi_5} \left(1 - \frac{h_5}{K} \right). \tag{7.8}$$

Similarly, the equilibrium condition for the host targeted by the second fewest types of viruses (host 2) is

$$r_2(1 - N_{\text{tot}}/K) = \phi_4 V_4^* + \phi_5 V_5^*, \tag{7.9}$$

such that

$$V_4^* = \frac{(r_2 - r_1)}{\phi_4} \left(1 - \frac{h_5}{K} \right). \tag{7.10}$$

For both virus types 4 and 5 to coexist, there must be a trade-off between host growth rate and susceptibility to lysis. The host most difficult to infect must grow less efficiently on available resources than the host next most difficult to infect. The same logic applies to the entire set of hosts, given that the general equilibrium densities of viruses in a community characterized by a perfectly nested infection network is (Jover et al. 2013)

$$V_S^* = \frac{r_1}{\phi_S} \left(1 - \frac{h_S}{K} \right), \quad V_{S-1}^* = \frac{(r_2 - r_1)}{\phi_{S-1}} \left(1 - \frac{h_S}{K} \right), \quad \cdots$$

$$V_1^* = \frac{(r_S - r_{S-1})}{\phi_1} \left(1 - \frac{h_S}{K} \right).$$

In summary, there must be a strict trade-off between permissiveness of hosts to infection and host growth rate for all viruses to coexist.

Now that a condition for coexistence has been established, the relevant question becomes, what is the predicted distribution of densities among viruses and hosts given a nested coinfection network? Previous theories of growth rate versus defense trade-offs suggested that the most abundant host types are those that are the most well defended (Thingstad 2000). In the marine environment, some of the strongest evidence in support of this hypothesis derived from observations of high densities of the ubiquitous pelagibacter SAR11, of which individuals classified as a single OTU can reach abundances of 10^5/ml. SAR11 was not known to be infected by any viruses—that is, until 2013, when Stephen Giovonnani and collaborators announced they had identified viruses infecting SAR11—the pelagiphage (Zhao et al. 2013). Not only were pelagiphage of SAR11 discovered, they are now thought to be among the most abundant virus types in the entire ocean, with recent estimates derived from metagenomic read analysis suggesting that they represent 10%–25% of the entire ocean surface virus population (Kang et al. 2013). It would seem possible for a bacterial strain to reach relatively large abundances even while being subject to extensive viral infection.

What does the current trade-off theory predict? Consider the preceding example and the differences between hosts 2 and 3:

$$\Delta N = N_2^* - N_3^* = (h_4 - h_3) - (h_3 - h_2) = h_4 - 2h_3 + h_2. \qquad (7.11)$$

This difference is equal to the finite approximation to the curvature of the virus efficiency. The general formula for curvature of any discrete function, x_i, is $x_{i+1} - 2x_i + x_{i-1}$. Although the trade-off condition requires that successive values of the virus efficiency, measured in terms of h_i, increase, there was no such precondition on the curvature of the virus efficiency. Hosts that are more permissive to virus infection can be the most abundant if costs of resistance are sufficiently strong. The strength of the trade-off can be measured in terms of the curvature. This result can be generalized for arbitrary values of S, with the exception of the most permissive host, whose density is set not by the difference in viral efficiency but, rather, by the viral efficiency of the most specialist virus type 1. A similar analysis can be applied to virus densities, such that the difference in virus densities is proportional to the growth-rate curvature. A specialist virus can be the most abundant if it infects a host that grows much faster than the host most similar to it in terms of susceptibility to infections, that is, the difference in the traits of host S and $S-1$.

Trade-offs between infectivity and efficiency of viruses and between defense and growth rate for hosts are found in model systems, but not invariably so (Chapter 4). The extent to which such trade-offs can be generalized to marine microbial hosts and associated viruses remains unclear (Breitbart 2012). The extent to which the most abundant or rare types of hosts and viruses correlate with their absolute degree of infectivity or the differences between their infectivity and that of their closest competitors also remains unclear. Whatever the experimental protocol, a key take-away point from the present analysis is that understanding the coexistence of hosts and viruses in natural systems requires considering trade-offs in a community, rather than a pairwise, context. Given that such in situ studies are nontrivial, it is important to ask, is the predicted trade-off and subsequent coexistence a feature of the particular assumptions of the model, or can it be generalized? The present analysis began with a resource-implicit description of virus-host interactions. Such dynamics permit neutral coexistence of all host types even in the absence of viruses. One might ask, can a nested cross-infection network promote greater diversity than existed when viruses were not present? And if so, are coexistence and resulting densities mediated by trade-offs and strength of tradeoffs? Answering these questions requires a model framework in which only one host persists in the absence of viruses, such as, that of chemostat dynamics.

7.2.3 MODEL EXTENSIONS AND GENERALIZATIONS

The model introduced in Eq. 7.1 can be extended from a resource-implicit framework to a resource-explicit framework (Korytowski and Smith 2014). Utilizing the chemostat dynamics introduced in Chapter 3, this model becomes

$$\frac{dR}{dt} = -\omega(R - J_0) - \sum_{i=1}^{S} \gamma_i R N_i$$

$$\frac{dN_i}{dt} = \epsilon_i \gamma_i N_i R - \sum_{j=1}^{S} M_{ij} \phi_j N_i V_j - \omega N_i, \qquad (7.12)$$

$$\frac{dV_j}{dt} = \sum_{i=1}^{S} M_{ij} \phi_j \beta_j N_i V_j - \omega V_j,$$

where a Type I functional response has been used for the uptake of resources. Here, the uptake rates γ_i and the conversion efficiencies ϵ_i may vary between strains. A well-developed theory of chemostat dynamics has already shown that only one host type can persist in the absence of viruses, so long as there is a unique maximum value of $\epsilon_i \gamma_i$ (Smith and Waltman 1995). The rationale is that each of the hosts competes for the same resource, and only the host that, when alone, would draw down resources to the lowest level will dominate.

This theory is known as R^* *competition theory* and recurs in studies of ecological competition (Tilman 1985, 1994). It was introduced in Chapter 3. For host competition in a chemostat, the equilibrium density of resources in the absence of viruses is

$$R^* = \frac{\omega}{\max\{\epsilon_1 \gamma_1, \epsilon_2 \gamma_2, \ldots, \epsilon_S \gamma_S\}} = \frac{\omega}{\epsilon_c \gamma_c}, \qquad (7.13)$$

and the equilibrium density of the winning host is

$$N^* = \frac{\omega (J_0 - R^*)}{\gamma_c R^*} = \epsilon_c \left(J_0 - R^* \right), \qquad (7.14)$$

where the index of the best competitor is defined as $i = c$. In this system, hosts must be able to persist on a resource concentration below that of the inflow; that is, $R^* < J_0$. With these values as a baseline, what are the possible equilibrium solutions given the inclusion of viruses that infect hosts via a nested cross-infection network $\mathbf{M}$?

Given a nested network $\mathbf{M}$, the densities of hosts are controlled, successively, by viruses:

$$N_S^* = \tilde{h}_1, \quad N_{S-1}^* = \tilde{h}_2 - \tilde{h}_1, \ldots, \quad N_1^* = \tilde{h}_S - \tilde{h}_{S-1}, \qquad (7.15)$$

where $\tilde{h}_j = \omega/(\beta_j \phi_j)$. This model also predicts that there must be a trade-off between viral infectivity and fitness for hosts to coexist. Given these viral-limited host densities, the equilibrium resource density is

$$R^* = \frac{J_0}{1 + \frac{1}{\omega} \sum_i^S \gamma_i N_i^*}, \qquad (7.16)$$

which must be less than J_0. The growth rate of hosts (at equilibrium) can be written as $r_i \equiv \epsilon_i \gamma_i R^*$. Finally, the strain-specific virus densities are

$$V_S^* = \frac{r_1 - \omega}{\phi_S} \quad V_{S-1}^* = \frac{(r_2 - r_1)}{\phi_{S-1}} \quad \cdots \quad V_1^* = \frac{(r_S - r_{S-1})}{\phi_1}.$$

Virus coexistence is conditioned upon the existence of a trade-off between host growth rate and defense and that between viral infectivity and fitness. The coexistence is often characterized by aperiodic, fluctuating dynamics.

This series of results was developed and formalized in Korytowski and Smith (2014). First, when hosts and viruses coexist, their dynamics are persistent; that is, dynamics continue indefinitely with positive densities for all hosts and viruses. The value of such an analysis is that it shows the inherent robustness of the dynamics with respect to perturbations or fluctuations, even if nothing in biology continues forever. This result generalizes prior results showing persistent dynamics of single virus-host pairs given chemostat dynamics (Smith and Thieme 2012). Second, a nested virus-host network with $S - 1$ types can be successively invaded by an additional virus-host type if the nested network structure is preserved. Such invasion provides an additional mechanism where by viruses can stimulate the diversification of communities. The Sneppen group proposed a similar result while examining a unified framework for evaluating the effect of diagonal and nested infection networks on coexistence (Haerter et al. 2014),

In summary, nested infection networks are both consistent with, and may even stimulate, diversity of viruses and hosts. From a theoretical perspective, an outstanding challenge remains: is there a robust statistical relationship between network structure and diversity? This question is of practical relevance given that networks in natural systems are rarely perfectly nested (or perfectly anything). Whatever the outcome of such analysis, its relevance will be heightened only if coexistence is considered in light of other pressures. Two such pressures—that of competition for limited nutrients and grazing by predators—are treated next.

7.3   NUTRIENTS AND THE VIRAL "SHUNT"

7.3.1   NUTRIENTS RELEASED UPON LYSIS

In 1999, two key papers, one by Fuhrman (1999) and one by Wilhelm and Suttle (1999), heralded an emerging recognition of the dual role of viruses in marine systems. Although it had long been evident that viruses are deleterious to the microbes they infect and lyse, these two papers synthesized growing evidence that viruses could stimulate the productivity of a microbial community. To quote Wilhelm and Suttle:

> Viruses divert the flow of carbon and nutrients ... by destroying host cells and releasing the contents of these cells into the pool of DOM in the ocean.

This process was termed the *viral shunt* (Wilhelm and Suttle 1999). The operational definition of DOM in aquatic ecology is any organic matter that passes through a filter with a pore size range of 0.2–0.7 μm. Case studies, both prior and subsequent to 1999, have found that viruses can stimulate both autotrophic and heterotrophic production, presumably by increasing the bioavailability of dissolved organic carbon (Middelboe et al. 2003), nitrogen (Shelford et al. 2012), and iron (Poorvin et al. 2004) or a combination thereof (Gobler et al. 1997) induced by viral lysis. A key feature of all these studies is the use of relatively complex communities, so that the release of organic matter into the environment has the potential to stimulate productivity, at least by those cells not targeted by viruses.

The viral shunt was estimated, by Fuhrman (1999), to be responsible for increasing the organic carbon uptake of bacteria by 33%, and by Wilhelm and Suttle (1999), to increase bacterial production and carbon mineralization rates by 27%. These estimates were derived by balancing fluxes between distinct microbial and organic matter pools and so cannot be used diagnostically to evaluate what drives variation in the strength and consequences of the viral flux. Moreover, in both approaches, the net effect of viruses was to shift the relative proportions of the flux through an aquatic food web rather than, potentially, to affect the total magnitude of both the fluxes and reservoirs. Although it is clear that some amount of such elements will be released upon lysis, it is not clear how much will be released and how such fluxes compare with other sources of input and removal.

Quantifying the amount of carbon and nutrients released upon lysis requires some information on the relative carbon and nutrient content of viruses and their microbial hosts. Such content varies with taxa, environmental conditions, physiological state, and other factors. The most important difference is size—larger organisms, whether cyanobacteria, heterotrophic bacteria, or eukaryotic microbes, will have more carbon and nutrients, but how much so depends on their elemental stoichiometry. A baseline for estimates of the stoichiometry of microbes is the hypothesis, now more than 50 years old, that there exists a universal molar ratio of carbon to nitrogen to phosphorus among marine detritus materials and marine microbes themselves. This *Redfield ratio* of 106:16:1 (C:N:P) was proposed by Albert Redfield in 1962 (Redfield et al. 1963). There are many departures from this ratio (Martiny et al. 2013); for example, heterotrophic bacteria are thought to have relatively lower ratios of C:N and C:P than do cyanobacteria (Suttle 2005). Nonetheless, the Redfield ratio provides a starting point for assessing the potential of the viral shunt for virus-host interactions across scales.

Three scenarios will help illustrate this point, with parameter values drawn from virus-host systems commonly found in marine surface waters: (i) pelagibacteria, (ii) cyanobacteria, and (iii) eukaryotic autotrophs and their respective viruses. Pelagibacteria are small, ubiquitous heterotrophs whose total densities of individual microbes within specific clades, such as SAR-11 and SAR-116, can exceed 10% of the total microbial community in the marine surface. Cyanobacteria are small autotrophs that are both

TABLE 7.1. The relative sizes and elemental contents of three host-phage systems. Parameter estimates are derived from compilations (Brown et al. 2006; Bertilsson et al. 2003; Benner et al. 2013; Kang et al. 2013). The elemental content for all hosts is assumed to be strictly Redfield (i.e., with molar ratios of 106:16:1). The carbon content of the bacteria is consistent with a volume-based estimate using a wet-mass density of 1 g/ml, of which 20% is carbon. The carbon content of *E. huxleyi* is based on an estimate of ≈8 pg/cell (Benner et al. 2013). The mass ratios are not 106:16:1 because of the differences in atomic masses of carbon, nitrogen, and phosphorus. The elemental content of viruses is size dependent, as predicted in Jover et al. (2014), with elemental ratios C:N:P of approximately 20:6:1 for viruses of genome length 100 kbp. Pelagibacter sizes (Yooseph et al. 2010; Rappé and Giovannoni 2003), pelagivirus sizes (Zhao et al. 2013), cyanobacteria sizes (Bertilsson et al. 2003), cyanophage sizes (Jover et al. 2014), and eukaryotic host and virus sizes (Brown et al. 2006) are all approximate.

Organism	Volume (μm^3)	C (fg)	N (fg)	P (fg)
Pelagibacter (e.g., HTCC1062)	10^{-1}	20	3.5	0.49
Pelagivirus (e.g., HTVC010P)	6.5×10^{-5}	0.02	0.0078	0.0025
Synechococcus (e.g., WH8103)	1	200	35	4.9
Cyanophage (e.g., P60)	1.1×10^{-4}	0.033	0.013	0.0045
Microeukaryote (e.g., *E. huxleyi*)	125	8000	1400	190
Eukaryotic virus (e.g., EhV)	1.8×10^{-3}	0.42	0.18	0.082

ubiquitous in the marine surface and highly diverse; that is, many clades and ecotypes have been identified whose distribution varies in space and time, and with depth (Follows et al. 2007). Cyanobacteria, predominantly of the genera *Prochlorococcus* and *Synechococcus*, reach abundances exceeding 10^5/ml. Finally, there are plentiful eukaryotic autotrophs, including ubiquitous coccolithophores and diatoms, that inhabit the surface waters. Of these, coccolithophores, characterized by their sometimes calcium-rich coats, are often associated with blooms, exceeding 10^5/ml, despite being 100 to 1000-fold larger on a per-cell basis than cyanobacteria. Individual microbes of each of these host types are commonly infected by viruses. Pelagiviruses, cyanophage, and phycodnaviruses that infect pelagibacteria, cyanobacteria, and *E. huxleyi*, respectively, are model systems in marine virology. Table 7.1 provides approximate sizes and elemental content of cells and examples of viruses known to infect them.

There is a large size disparity between viruses and hosts. Of relevance here is the difference in elemental composition, which is the basis, conceptually at least, for the viral shunt. Quantifying how much of the elements inside a host are returned, upon lysis, to the environment requires another bit of information: where do the elements in a virus derive from? That is, do viruses acquire all the necessary carbon, nitrogen, and phosphorus for their genomes and capsid from outside the host cell, postinfection? Or do viruses acquire all that they need from carbon and nutrients already present in the host upon infection, for example, by catabolic degradation of cellular components and, then, anabolic synthesis of viral components?

The notion that viruses acquire elements postinfection from outside the cell has its origins in early phage biology. Seymour Cohen and colleagues used the radioactive phosphorus isotope ^{32}P to trace the origins of the phosphorus in the packaged genome in virus particles (Cohen 1951; Weed and Cohen 1951). They showed that approximately 80% of phosphorus was derived from outside the host cell (i.e., acquired de novo postinfection), for infections of *E. coli* with select T-even phages. In this limit, the amount of element released into the environment as DOM is approximately that contained in the host:

$$q_{\text{lysate}} \approx q_{\text{host}}, \qquad (7.17)$$

where q denotes the mass of the total elemental content of interest. The story for marine phage infections is different. A similar ^{32}P study was conducted on marine phage populations from surface seawater, led by Farooq Azam (Wikner et al. 1993). They concluded:

> [M]arine phages may derive the majority of their nucleotides from the nucleic acids of the bacterial host. This metabolic strategy may be common in the marine phage community, and contrast the origin of nucleotides in coli-phages, suggested to be mainly derived by de novo synthesis.

In this limit, the amount of element released into the environment is approximately equal to the difference between the original host content and that contained in the virus burst; that is,

$$q_{\text{lysate}} \approx q_{\text{host}} - \beta q_{\text{virus}}, \qquad (7.18)$$

where β is the burst size of viruses (Figure 7.4).

This latter limit can be further leveraged for analysis of virus-host systems in the marine environment. Given a virus-host pair, each with fixed elemental content, the amount of each element released as DOM and available for subsequent uptake and assimilation can be estimated. The predicted fraction of the host element released as cellular debris, rather than as virus particles, decreases with increasing burst size. Hypothetical limits on burst sizes can be obtained, again assuming that all elements must be repurposed from the host cell. There are obvious issues with this second approximation; for example, use of elements from the cell wall and membrane would make it impossible for the cell to continue to function, even with virus-derived metabolic genes.

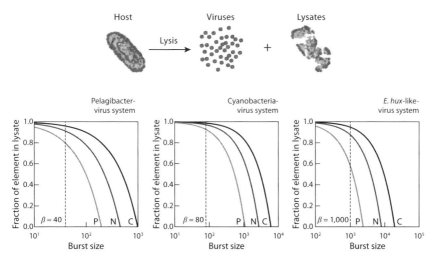

FIGURE 7.4. Virus-induced recycling of organic carbon and nutrients. (Top) An illustration of the process by which the contents of a host cell are released, upon lysis, as either cell debris or virus particles. (Bottom) The predicted fraction of the elemental content (C, N, and P) of the host that is released as debris as a function of burst size, where (approximate) observed burst sizes are marked with a vertical dashed line. In each panel, the burst size on the log-scale x-axis is plotted against the fraction of the host content not bound in virus particles, or $1 - \beta q_{\text{virus}}/q_{\text{host}}$, on the y-axis. The values of q_{virus} and q_{host} for each system are listed in Table 7.1.

Nonetheless, such a comparison may provide a test of potential upper limits to burst size given distinct types of limitation. Figure 7.4 gives the results of these numerical experiments for the three virus-host systems in Table 7.1. Because of the stoichiometric mismatch between virus and host, phosphorus return to DOM decreases faster than does nitrogen or carbon return. As a consequence, observed burst sizes are closer to putative upper limits for P availability than for either C or N.

The actual utilization of host resources lies between these two limits. For example, some viruses encode metabolic genes that can modify the cell's metabolism during infection. The resulting "virocell" may have enhanced photosynthesis, nutrient uptake and use—including that of N and P—and sphingolipid metabolism (Bidle and Vardi 2011). Even as viruses reuse host cell components, they also acquire new elements to be synthesized directly into viral products, whether RNA, DNA, protein, or even lipds. Empirical studies are needed to provide more information on the quantitative levels of partitioning of elements. For now, Eq. 7.18 will be used as the starting point for integrating virus-induced recycling into ecological models, as in the next section.

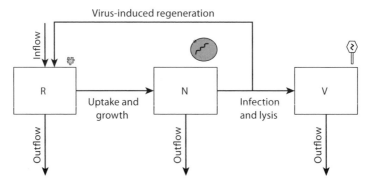

FIGURE 7.5. Schematic of an explicit resource, host, and virus model in which a portion of the elemental content of hosts is recycled into the bioavailable pool.

7.3.2   DYNAMIC MODEL OF VIRUS-INDUCED REGENERATION OF NUTRIENTS

The effects of virus generation of nutrients can be examined in a simplified model. Consider the following resource-explicit model in which virus lysis regenerates a fraction p of the difference in nutrients contained in the host microbe versus that in the virus burst (see Figure 7.5). This model extends the chemostat dynamics introduced as Eq. (3.1):

$$\frac{dR}{dt} = -\omega(R - J_0) - \gamma RN + \overbrace{p(q_N - \beta q_V)\phi NV}^{\text{nutrient regeneration}},$$

$$\frac{dN}{dt} = \epsilon \gamma RN - \phi NV - \omega N, \tag{7.19}$$

$$\frac{dV}{dt} = \beta \phi NV - \phi NV - \omega V.$$

The equilibrium concentrations, in the form (R^*, N^*, V^*), arising from this model can be compared with those of a similar model (introduced in Chapter 3) with explicit resource uptake but without virus-induced regeneration:

Without viruses: $\left(\frac{\omega}{\epsilon\gamma}, \epsilon J_0 - \omega/\gamma, 0\right)$,

With viruses but without

virus-induced regeneration: $\left(\frac{J_0}{1 + \frac{\gamma}{\bar{\beta}\phi}}, \frac{\omega}{\bar{\beta}\phi}, \frac{\epsilon\gamma R^* - \omega}{\phi}\right)$,

With viruses and virus-induced regeneration: $\left(\frac{J_0 - \sigma\omega/(\bar{\beta}\phi)}{1 + \frac{\gamma}{\bar{\beta}\phi}(1 - \sigma\epsilon)}, \frac{\omega}{\bar{\beta}\phi}, \frac{\epsilon\gamma R^* - \omega}{\phi}\right)$, where $\sigma \equiv p(q_N - \beta q_V)$. Both models predict virus control of the hosts to the same density. The reason is that in models of virus-host interactions without zooplankton, the addition of resources, whether by influx or via regeneration, is redirected to the controlling population rather than to the target population. This idea of trophic transfer can depend on functional responses; however, the two models differ in their prediction of both resource concentrations and viral concentrations.

Intuition would suggest that resources should increase if viruses, via lysis, regenerate nutrients. The relative change in R^* can be quantified by comparing $\left(\frac{J_0 - \sigma \omega / (\tilde{\beta} \phi)}{1 + \frac{\gamma}{\beta \phi}(1 - \sigma \epsilon)} \right)$ with the predicted R^* without regeneration: $\left(\frac{J_0}{1 + \frac{\gamma}{\beta \phi}} \right)$. The equilibrium with regeneration can be written in the form $(a + \sigma c_1)/(b + \sigma c_2)$, where the equilibrium without regeneration is written in the form a/b, where $a = J_0$, $b = 1 + \sigma/(\tilde{\beta}\phi)$, $c_1 = \omega/(\tilde{\beta}\phi)$, and $c_2 = \epsilon c_1$. Then, in the limit of small σ, the condition for virus-induced regeneration to increase total nutrient concentration is

$$\frac{\epsilon \gamma J_0}{1 + \frac{\gamma}{\beta \phi}} > \omega.$$

This condition is precisely the requirement for coexistence of hosts and viruses given chemostat dynamics and a Type I functional response. When regeneration is small, nutrient concentration increases. Given virus-induced regeneration of nutrients, then virus concentration increases as well. Although the host population is not expected to increase, its productivity should, because there are more resources available for each host cell to acquire and utilize for growth. Whether the principle of virus-induced enhancement of resource pools and of host productivity extends to a more general scenario—specifically, an aquatic food web with grazers—is the subject of the next section.

7.4   VIRUSES AND GRAZERS

7.4.1   DYNAMIC MODEL OF PARASITES AND PREDATORS

Zooplankton are ubiquitous eukaryotes that consume smaller microbes. They are obligate heterotrophs, deriving all their organic carbon from other organisms. Zooplankton range in size from a few microns in body length, like the flagellated or ciliated nanozooplankton, to hundreds of microns in body length, like the megazooplankton. Zooplankton are active swimmers and, in some cases, can follow chemical trails to track their prey. Once in proximity to a potential prey item, they utilize a number of different strategies to consume target organisms, including filter feeding and other primitive forms of "eating," for example, by using cilia to draw in bacterial cells into their gut, where they are digested. Zooplankton are often size selective in their feeding and can consume other zooplankton in addition to bacterial and eukaryotic phytoplankton. A review of the size dependence of many zooplankton life history traits is available in Hansen et al. (1997).

There is a long history of modeling zooplankton interactions with phytoplankton in aquatic environments (Klausmeier et al. 2004; Follows and Dutkiewicz 2011; Stock et al. 2014). The most widely used are nutrient-phytoplankton-zooplankton models, so-called NPZ models (Franks 2002). The core principle of NPZ model variants is that phytoplankton depend on dissolved nutrients to reproduce and that zooplankton depend on the availability of phytoplankton. Combining models of zooplankton predation of phytoplankton and microbes with models of virus-host dynamics can leverage many of these established principles. The seemingly most straightforward way to consider the dynamics of a food web that includes parasites is with a single target host that can be eaten by zooplankton or infected and lysed by a virus. Given a chemostat-like environment with washout rate ω, the population dynamics of hosts, viruses, and zooplankton can be written as

$$\frac{\mathrm{d}N}{\mathrm{d}t} = rN(1 - N/K) - \phi NV - \overbrace{\psi NZ}^{\text{grazing}} - \omega N,$$

$$\frac{\mathrm{d}V}{\mathrm{d}t} = \beta \phi NV - \phi NV - \omega V, \tag{7.20}$$

$$\frac{\mathrm{d}Z}{\mathrm{d}t} = \overbrace{\theta \psi NZ}^{\text{grazing}} - \overbrace{mZ}^{\text{basal loss}} - \omega Z,$$

where the new zooplankton-associated terms are labeled (see Figure 7.6 for a schematic). In particular ψ is the grazing rate, m is the basal loss term due to respiration and exudation by the zooplankton, and θ is the conversion efficiency of grazing. Here, the focus is necessarily on a subset of zooplankton, for exampe, an aggregated nanomicro-zooplankton pool that take up hosts. Additional loss terms due to the predation of zooplankton by higher predators are not considered.

The interesting feature of such models is that they fail, miserably, because they predict that viruses and zooplankton cannot coexist. The benefit of such failure is that such models also have the potential to shed light on conditions required for coexistence. Viruses act as top-down predators in this model

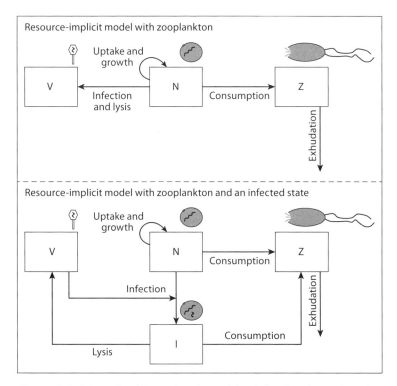

FIGURE 7.6. Schematic of interactions in models of virus-host interactions that include a zooplankton population. (Top) Resource-implicit model in which zooplankton graze upon host cells. (Bottom) Resource-implicit model in which zooplankton graze upon both uninfected and infected cells.

without zooplankton. If only viruses are present, then host density should approach $N^* = \omega/((\beta - 1)\phi)$, whereas if only zooplankton are present, then host density should approach $N^* = (m + \omega)/(\theta\psi)$. But when both are present, these two densities are different, and one "predator" outcompetes the other. The winning predator is the one that draws down the microbial host to a lower level. For the mathematicians among the readers, it is worth noting that there is a highly constrained set of parameters under which these two densities are equal, called *knife-edge coexistence*, because any infinitesimal change in parameters undoes the conditions. There is no generic equilibrium coexistence in this model of competition between a parasite and a predator for a single host.

It would seem that this first attempt at combining viruses and zooplankton was too simplistic, because viruses and zooplankton do coexist in natural systems. The issue may have been the assumptions built into the model. For example, the functional form of zooplankton consumption of microbial cells is often more complicated. The rationale is that zooplankton require time to handle and consume each prey item. Models of handling time therefore posit that when the resource is at very high densities, uptake of prey will approach a saturating level, corresponding to the limitations of handling rather than contacting a prey item. Given saturation, consumption should be be modeled as $aNZ/(Q_N + N)$, where a is the maximal rate of zooplankton digestion of host cells when they are plentiful, and Q_N denotes a host density at which zooplankton grazing reaches half its maximal rate. This new model of virus-host-zooplankton dynamics can be written as

$$
\frac{dN}{dt} = rN(1 - N/K) - \phi NV - \overbrace{\frac{aN}{Q_N + N}}^{\text{Type-II grazing}} Z - \omega N,
$$

$$
\frac{dV}{dt} = \beta \phi NV - \phi NV - \omega V, \tag{7.21}
$$

$$
\frac{dZ}{dt} = \overbrace{\frac{\theta aN}{Q_N + N}}^{\text{Type II grazing}} Z - mZ - \omega Z.
$$

Despite this added realism, the problematic nature of the model has not changed. Now, a zooplankton-only environment would drive down host cell densities to $N^* = Q_N(m + \omega)/(a\theta - m - \omega)$, while a virus-only environment would drive down host cell densities to $N^* = \omega/((\beta - 1)\phi)$. Only one "predator" can win. Inclusion of a Type II functional response increases the predicted density of host cells when controlled by zooplankton. To the extent that there is a qualitative change in model prediction, it is possible for viruses to more frequently dominate and outcompete zooplankton in a model with handling time than in a model without handling time.

   The objective here is to use simple models to reconcile the coexistence of microbial hosts, viruses, and zooplankton in the environment. Although the first two attempts failed, they suggest that perhaps the mechanisms of coexistence reflect key differences between the way viruses interact with hosts and the way zooplankton interact with their prey. Viruses do not immediately convert host cells into new progeny. Two models of latency were introduced in Chapter 3, one including a fixed, finite time delay between infection and lysis and the other including an exponentially distributed delay. An infected host cell can be removed from the chemostat after infection but before lysis. Although in the models in Chapter 3 the removal was due to washout, here the removal may be due to predation. Zooplankton can consume infected hosts, thereby short-circuiting the viral infection process, which directly inhibits virus populations rather than indirectly via competition for a shared host resource. A dynamical model of uninfected hosts, infected hosts, viruses, and zooplankton can be written as:

$$\frac{dN}{dt} = rN(1 - N/K) - \phi NV - \overbrace{\psi NZ}^{\text{grazing of uninfected hosts}} - \omega N,$$

$$\frac{dI}{dt} = \phi NV - \eta I - \overbrace{\psi IZ}^{\text{grazing of infected hosts}} - \omega I,$$

$$\frac{dV}{dt} = \beta \eta I - \phi NV - \omega V,$$

$$\frac{dZ}{dt} = \overbrace{\theta \psi (N + I)Z}^{\text{total grazing}} - mZ - \omega Z.$$

(7.22)

Here, zooplankton consume both uninfected and infected cells. Thus, a virus infection can end in washout, cell lysis, or the termination of infection via zooplankton consumption. As a consequence, zooplankton consume both hosts and viruses, altering the nature of the interaction. Does such a change lead to a qualitative change in outcomes?

The equilibrium solutions of this coupled system of ordinary differential equations are not particularly informative. Instead, consider a set of parameters that describe the entire system: $(r, K, \phi, \psi, \omega, \eta, \beta, \theta, m)$ and that permit coexistence of two kinds:

(i) With viruses only: $(\frac{\omega}{\phi \tilde{\beta}}, \frac{\phi N^* V^*}{\eta + \omega}, \frac{r(1 - N^*/K) - \omega}{\phi}, 0)$, (7.23)

(ii) With zooplankton only: $(\frac{m + \omega}{\theta \psi}, 0, 0, \frac{r(1 - N^*/K) - \omega}{\psi})$, (7.24)

where the coexistence is in the form of (N^*, I^*, V^*, Z^*), and $\tilde{\beta} = \beta \eta/(\eta + \omega) - 1$. In the former case, is it possible for a small zooplankton population to invade the resident virus population? Similarly, in the latter case, is it possible for a small virus population to invade the resident zooplankton population? If the answer to only one of these questions is yes, this implies that one kind can invade and likely replace the other (Geritz et al. 2002). However, if the answer to both of these questions is yes, this implies that it will be difficult for either viruses or zooplankton to exclude the other. The reason is that if the system was to approach either a virus-free or a zooplankton-free equilibrium, then a small population of viruses or zooplankton would increase in number, respectively.

The potential for mutual invasibility in this system can be examined numerically. The invasion fitness of a zooplankton population when rare, given a virus-induced equilibrium, is

$$r_{Z|V} = \theta \psi (N_V^* + I_V^*) - m - \omega, \tag{7.25}$$

where N_V^* and I_V^* are the equilibrium concentrations of uninfected and infected hosts, respectively, given top-down limitation by viruses. Similarly, the invasion fitness of a virus population when rare, $r_{V|Z}$, given a zooplankton-induced equilibrium, is determined by the largest eigenvalue of the (linearized) subsystem

$$\frac{dI}{dt} = \phi N_Z^* V - \eta I - \psi Z^* I - \omega I, \tag{7.26}$$

$$\frac{dV}{dt} = \beta \eta I - \phi N_Z^* V - \omega V, \tag{7.27}$$

where N_Z^* and Z^* are the equilibrium concentrations of uninfected hosts and zooplankton, respectively, given top-down limitation by viruses. The subsystem dynamics tracks the total viruses in the system, that is, the sum of free virus particles and those inside infected cells. Figure 7.7 reveals that even when more realistic infection dynamics are included, viruses and zooplankton are still unlikely to coexist when competing for a single host. The analysis includes the evaluation of two sets of 10^5 different parameter combinations of which nearly 16,000 led to potential coexistence. The two sets differed only in the baseline intensity of zooplankton grazing (Figure 7.7). In either set, there was not a single instance of mutual invasibility—providing further support for fundamental barriers to coexistence among viruses and zooplankton when competing for a single host resource, as per competitive exclusion theory (Armstrong and McGehee 1980).

Nonetheless, numerical sampling does not rule out the possibility of coexistence, particularly coexistence in a fluctuating steady state. Such coexistence may arise because zooplankton uptake and loss includes additional nonlinearities and/or because nonlinearities arise from alternative models of virus-host interactions, such as fixed delays between infection and lysis. Without evaluating yet another model, the same numerical procedure just described can be used to identify potential coexistence regimes. In the preceding trial, $>99\%$ of cases led to invasion by only one population when rare. A small number of trials corresponded to conditions under which invasion was not possible

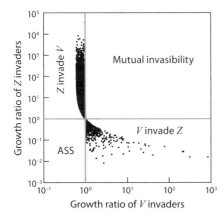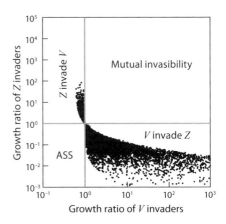

FIGURE 7.7. Numerical examination of invasion dynamics of viruses given a zooplankton-induced equilibrium, and vice versa. Resident equilibria were examined given a Latin Hypercube Sampling of parameter space centered on the parameter values $r = 0.16$, $K = 2.2 \times 10^7$, $\phi = 10^{-9}$, $\eta = 0.5$, $\beta = 50$, $\omega = 0.05$, $m = 0.01$, $\theta = 0.005$, and either $\psi = 10^{-4}$ (left) or $\psi = 10^{-6}$ (right). In all cases, parameters were allowed to vary 10-fold above and below. For each parameter combination, putative steady states were identified with viruses and with zooplankton (but not both). Then, the invasion fitness of each when rare was calculated (as described in the text). Each point represents a single pair of growth ratios (of rare viruses invading a zooplankton equilibrium, on the x-axis, and of rare zooplankton invading a virus equilibrium, on the y-axis). Note that for plotting purposes, the growth ratio, rather than the invasion fitness is used. The growth ratio of zooplankton is the ratio of its production rate to its loss rate. The growth ratio of viruses is the exponential of its fitness. Invasion is possible when invasion fitness is positive or, equivalently, when the growth ratio exceeds 1. ASS = alternative stable state. *Key point: For the majority of parameters, viruses could invade zooplankton, or zooplankton could invade viruses, but both could not occur. This implies that one type would outcompete the other for host resources.*

by either population, when rare. Such conditions are often associated with alternative steady states and, potentially, stable coexistence equilibrium among all component populations. Investigation of the long-term dynamics of the model at parameter combinations associated with putatively alternative stable-state conditions helped identify instances of long-term, fluctuating coexistence among infected and uninfected hosts, viruses, and zooplankton (Figure 7.8). This finding highlights the potential consequences of oft-ignored details of the virus-host infection process on conclusions derived from complex community models. Further investigation of these interactions alone, or as part of complex food webs, is warranted.

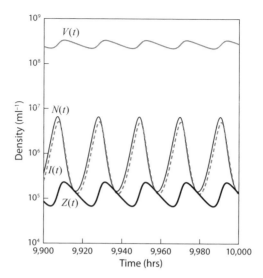

FIGURE 7.8. Long-term oscillations among viruses (V), hosts (uninfected, N, and infected, I), and zooplankton (Z). Parameters for this simulation were: $r = 1.55$, $K = 9.56 \times 10^7$, $\phi = 2.68 \times 10^{-9}$, $\eta = 0.093$, $\beta = 101$, $\omega = 0.042$, $m = 0.080$, $\theta = 0.007$, and $\psi = 5.65 \times 10^{-6}$ (where time is in hours, and volumes are in milliliters). The resulting dynamics focused on the time period 9900–10,000 hours, long after transient dynamics had dissipated. The parameter set was identified via a Latin Hypercube Sampling as having the hallmarks of an ASS, with either a virus-free or a zooplankton-free equilibrium.

7.4.2 ZOOPLANKTON, VIRUSES, AND THE CONTROL OF COMPLEX MICROBIAL COMMUNITIES

The interaction between viruses and zooplankton is not restricted to competition for a single host. In fact, a key difference between viruses and zooplankton is that viruses tend to be relatively more specific in their target host cells than are zooplankton. Virus infection is often strain specific; that is, small differences in chemical properties of receptors and/or intracellular regulation may block virus adsorption or render infection ineffective. Modifications to extracellular receptor structure or intracellular defense mechanisms need not affect host palatability or digestibility by zooplankton. This is not to say that zooplankton eat any microbe. Zooplankton are known to be size selective in their feeding, with a common principle being that consumption is often targeted to microbes 10% in size relative to predator body dimensions (Wirtz 2012).

These ideas informed Frede Thingstad's proposed alternative explanation for the control of microbial communities, namely, viruses may control the relative success of individual microbes even if it is grazing by zooplankton that determines the overall size of the microbial community (Thingstad and Lignell 1997; Thingstad 2000). The core idea is similar in spirit to that of models introduced by Daniel Janzen and Joseph Connell (Janzen 1970; Connell 1970). in which specific pathogens promote diversity among their target species. The "Janzen-Connell" hypothesis was developed to explain tropical forest diversity. A similar model, as applied to marine microbiology, is known as the "kill the winner" (KTW) model, introduced in the late 1990s. The KTW model makes the following assumptions: (i) all microbes compete for a common pool of resources; (ii) all microbes, except for one population, are susceptible to virus infection; (iii) all microbes are subject to zooplankton grazing; (iv) viruses infect only a single type of microbe. Different variants of the model have been presented, both by Thingstad (Thingstad and Lignell 1997; Thingstad 2000) and collaborators (Rodriguez-Valera et al. 2009; Winter et al. 2010). A common thread to these models is that they include high potential host and viral diversity while enforcing the relative specificity of virus infection compared with that of zooplankton grazing.

To evaluate the consequences of this commonality, consider a microbial community including S bacterial strains, $S - 1$ viral strains, and one zooplankton population, whose dynamics can be written as

$$\frac{dN_i}{dt} = r_i N_i \left(1 - \frac{\sum_{i'=1}^{S} N_{i'}}{K} \right) - \phi_i N_i V_i - \psi N_i Z - \omega N_i,$$

$$\frac{dV_j}{dt} = \tilde{\beta}_j \phi_j N_j V_j - \omega V_j,$$

(7.28)

$$\frac{dN_S}{dt} = r_S N_S \left(1 - \frac{\sum_{i'=1}^{S} N_{i'}}{K} \right) - \psi N_S Z - \omega N_S,$$

$$\frac{dZ}{dt} = \sum_{i=1}^{S} \theta \psi N_i Z - m Z - \omega Z,$$

where the indices i and j in the first two equations run from 1 to $S - 1$, and where $\tilde{\beta}_j \equiv \beta_j - 1$. In this model, host strains $1 \ldots S - 1$ are subject to virus

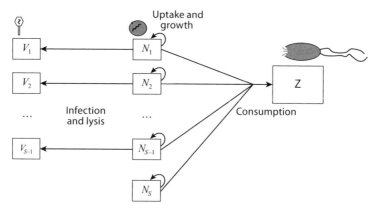

FIGURE 7.9. Schematic of interactions among $S-1$ virus strains, S host strains, and a generalist zooplankton. Loss terms, including washout and basal loss, are not depicted.

infection and grazing, whereas strain S is subject only to grazing (Figure 7.9). Viruses are specialists in this model, while the zooplankton are generalists. All interactions between two populations, either hosts and viruses or hosts and zooplankton, are modeled via Type I functional responses. Moreover, the generalist grazer is assumed to consume all hosts at the same rate, ψ.

At equilibrium, the total size of the microbial community is set by zooplankton grazing:

$$\sum_{i=1}^{S} N_i = \frac{m+\omega}{\theta\psi}. \tag{7.29}$$

In this sense, the zooplankton "control" the microbial reservoir. When this total microbial density is less than K—the total carrying capacity in the absence of zooplankton or viruses—then the number of zooplankton is determined by the life history traits of the most well-defended microbe. Host strain S cannot be infected by viruses, but it can be grazed upon, such that

$$Z^* = \frac{r_S\left(1 - \frac{m+\omega}{\theta\psi K}\right) - \omega}{\psi}. \tag{7.30}$$

Because viruses infect and lyse only a single host, then each of these host densities is set by its specific virus "predator":

$$N_i^* = \frac{\omega}{\tilde{\beta}_i\phi_i}, \tag{7.31}$$

for $i = 1 \ldots S - 1$. The density of the resistant microbial population must be the difference between the virus-controlled and zooplankton-controlled totals; that is,

$$N_S^* = \frac{m + \omega}{\theta \psi} - \sum_{i=1}^{S-1} \frac{\omega}{\tilde{\beta}_i \phi_i}. \tag{7.32}$$

Finally, the density of each virus population is determined by the total zooplankton grazing:

$$V_i^* = \frac{r_i - r_S}{\phi_i} \left(1 - \frac{m + \omega}{K \theta \psi} \right). \tag{7.33}$$

This model, despite its complexity, reveals a number of principles. It predicts that there must be trade-off between resistance and growth rate. The particular form arising in this model is that the growth rate of the resistant host, r_S, must be less than that of all other hosts for virus-host pairs to emerge (i.e., $r_i - r_S > 0$). It also predicts that virus densities increase linearly with the growth rate of their target microbial hosts. Both these results coincide with those predicted by the KTW model (Thingstad 2000), albeit the KTW model used the term *nutrient affinity* with a constant efficiency to describe uptake, rather than a fixed, strain-specific growth rate, as is used here. Given a hierarchy of growth rates, where r_S is the lowest, then there is a broad range of parameters that permit coexistence. To illustrate this point, Figure 7.10 shows the ensuing dynamics in a case where $S = 10$. The coexistence is mediated by the fact that one of the hosts is not subject to virus infection. In practice, this condition might arise because the density of the most resistant host is below a sustainable level, that is, arising ecologically rather than owing to evolution to complete resistance. As modeled, the viruses and zooplankton have different niches, *sensu* G. Evelyn Hutchinson (1957). They can coexist given a *set* of hosts rather than a single target host.

Beyond the issue of coexistence, this model also begins to address the question of the relative importance of viruses versus zooplankton in the mortality of microbial hosts. With the exception of host strain S, the fraction of deaths of host strain i due to viral lysis is

$$f_i^{(V)} = \frac{\phi_i V_i^*}{\phi_i V_i^* + \psi Z^* + \omega}. \tag{7.34}$$

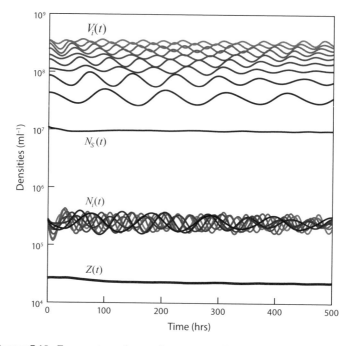

FIGURE 7.10. Emergent coexistence in a system with S host types, $S-1$ virus types, and one zooplankton type. Parameters are the same as in Figure 7.7, with the exception of $\psi = 10^{-6}$ and r, which ranges linearly from 1.57 to 0.157 for host types 1 to S. The lowest value of r corresponds to the resistant host type S.

At steady state, this equation can be rewritten (after some algebra) as

$$f_i^{(V)} = 1 - r_S/r_i. \qquad (7.35)$$

The model predicts that at equilibrium, viruses become an increasingly important source of strain-specific microbial mortality for those microbes that grow faster (in which case, $r_i \gg r_S$, and so $f_i^{(V)} \to 1$). Once again, a core prediction of these models is that life history traits matter to microbial community structure and ecosystem functions. Improved quantification of process rates are required to better quantify the environmental impacts of viruses.

7.4.3 PUTTING ALL OF IT, OR AT LEAST SOME OF IT, TOGETHER

Assessing viruses in a community context would seem to require integrating aquatic food webs with virus-host interactions, yet an explicit virus pool has not yet been integrated into Earth system models nor into the mainstream pipeline of predictive microbial oceanographic models (Follows et al. 2007; Stock et al. 2014). Instead, higher-order loss terms are included in tracking the population dynamics of microbes, whether cyanobacteria or heterotrophs. This higher-order loss term is proportional to the population density squared (Stock et al. 2014). The rationale for this choice is that viruses track host populations and therefore will become increasingly important at high host population levels. On one level, such an approximation is obviously problematic, as it implies that virus and host densities are coincident. Such coincidence need not be the case, given that predicted phase relations usually involve a quarter phase lag (Chapter 3) but can be decoupled, as in cryptic oscillations (Chapter 4) or have inverted lags (Chapter 5). Including implicit higher-order terms also makes it difficult to evaluate how much of mortality is likely due to viruses versus other causes.

Integrating viruses into biogeochemically relevant models requires evidence that such inclusion could help resolve mismatches between predictions of Earth system models and observations. Current large-scale Earth system models already capture significant spatiotemporal variability in carbon and nutrient concentrations and fluxes. If viruses are, in fact, ecologically relevant, then the relevant question is, do models with viruses recapitulate known patterns, and do they predict novel features beyond those of virus-free models? In a few cases viruses have been integrated into aquatic food web models. One example is the KTW model described earlier, though it does not include explicit consideration of organic nutrient pools. Another example is Miki's work on organic matter regeneration in freshwater systems (Miki et al. 2008), which does not include explicit consideration of inorganic nutrient pools. Whatever the form of virus-host interactions, it is likely that including them in current models would lead to results discordant with simulations. In other words, the models would stop working. Why is this the case? In practice, by not explicitly including viruses in models ascribes their effect to other components, such as modifying grazing rates of zooplankton. Including viruses would necessitate systematic reexamination of parameters, pertaining not only to virus-host interactions but also to those relevant to other processes in the system.

An extended multitrophic model including viruses was recently proposed by a consortium of authors who participated in a National Institute of Mathematical and Biological Synthesis working group on "Ocean Viral Dynamics."[1]

---

[1] Steven Wilhelm, and I were the co-organizers of this working group, and I am the lead author of the first modeling paper arising from this collaboration (Weitz et al. 2015).

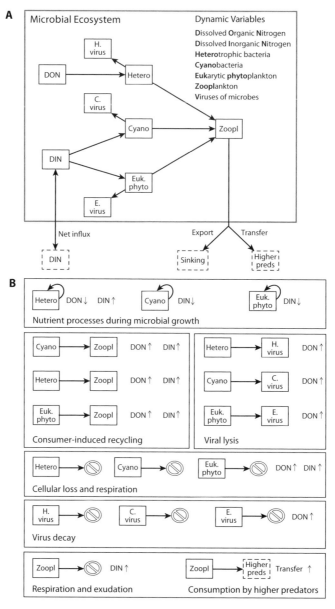

FIGURE 7.11. Schematic of the NPHZ-V model. Reprinted with permission from Weitz et al. (2015). The original caption reads: "(A) Interactions between populations and nutrients where arrows denote the flow of materials in the system. Note that decay processes are not depicted and the secondary effects on nutrients are depicted in panel B. (B) Nutrient cycling in the system. Each of the different processes affect the levels of dissolved organic nitrogen (DON) and dissolved inorganic nitrogen (DIN), where the symbols ↑ or a ↓ denote whether a given process increases or decreases that particular pool, respectively."

The model includes the interactions among organic and inorganic nutrients, phytoplankton (both cyanobacteria and eukaryotic autotrophs), heterotrophic bacteria, zooplankton, and viruses (Figure 7.11). This NPHZ-V model includes a relatively simple NPZ component with Type I functional responses and a generalist grazer that can predate upon all three microbial components. Viruses in the model have the potential to be both deleterious to their target hosts and stimulatory to nontargeted hosts. Virus lysis is modeled via a Type I interaction term, that is, proportional to the product of microbial host and virus densities. Each lysis event releases DOM into the environment. The magnitude of the virus-induced regeneration of organic matter is proportional to the lysis rate multiplied by the difference in nutrient content between hosts and that contained in the virus burst. For systems to reach steady state, the order of the nonlinearity of the inputs must differ from that of the outputs. For example, in a birth-and-death model, if births and deaths are both linear in population size, then the population will either grow or decline exponentially with the exception of the knife-edge steady state that occurs when birth rates match death rates. In contrast, when there is a nonlinear feedback of population size on birth or death, then a steady state can be reached. In the NPHZ-V model, the closure of the model was intentionally simplified to model a highly retentive microbial food web. Inorganic nitrogen is assumed to enter the system at a rate proportional to the difference between nitrogen content above and below the nutricline. Further losses from the system are due to export of pellets, that is, waste products of zooplankton, and transfer of zooplankton biomass to higher trophic levels, that is, via an implicit and constant nonlinear predation term. The full model is presented in Weitz et al. (2015) and in Appendix E.1.

Notably, the equilibrium population densities and resource concentrations can be solved analytically. This solution enables large-scale evaluation of the sensitivity of model outcomes to parameter variation. Some of these parameters generate "realistic" model output. That is, the predicted equilibrium solutions are similar to those reported from productive marine surface environments, for example, bacterial densities of 10^8/L, virus densities exceeding 10^9/L, zooplankton on the order of 10^4/L, organic nitrogen of approximately 5 μmol/L, and inorganic nitrogen on the order of 0.1 μmol/L. The model steady state is also consistent with documented fluxes within surface marine microbial ecosystems; for example, simulated NPP is in the range of 860–1400 mg C/m^2/day[1], in line with realistic biological variation in the surface oceans.

With these core features in hand, one can ask, how do ecosystems differ in the presence or absence of viruses? The key take-home messages are twofold. First, total microbial biomass tends to increase with viruses. This increase occurs despite the direct negative effects of viruses on target hosts. The lysing

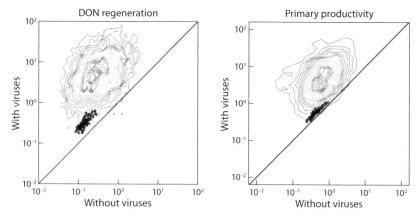

FIGURE 7.12. Stimulatory effects of viruses in the context of complex marine surface food web models. The contour lines represent the probability distribution of pairwise comparisons for the 4,366 model replicates (parameter distributions are in Weitz et al. (2015), where the solid line denotes the 1:1 line. The fact that contour lines lie above the 1:1 line implies that simulated ecosystems with viruses have higher DON regeneration (left) and primary productivity (right) than those without viruses. All axes are in units of μmol/(L$\times$ days). Images adapted from Weitz et al. (2015) with permission.

of hosts by viruses increases the availability of dissolved organic nitrogen in the microbial loop (Figure 7.12, left). The increase depends on the inclusion of multiple host types, so that lysis of autotrophs enhances heterotrophic productivity, which increases the release of remineralized nutrients, which, in turn, causes an increase in net primary productivity and phytoplankton pool size (Figure 7.12, right). The increase in remineralization, largely by heterotrophs, is a direct consequence of the increased uptake of organic nitrogen. Viruses increase the productivity of both heterotrophs and autotrophs, so long as both are present. Overall, the indirect stimulatory effects of the total virus community compensate for the direct negative effects on the cells they infect. These core results may appear paradoxical, yet the mechanisms underlying both results are consistent with identified interactions between viruses and their hosts, and the cascade of nonlinear feedback in the food web. In essence, viruses function on a continuum between top-down and bottom-up control.

This result is not the endpoint of analysis. The NPHZ-V model does not include much of the complex size structure in the ocean surface, nor does it include the differences in stoichiometry between populations or the possibility of flexible stoichiometry. Much more research can, and should, be done. Nonetheless, the model presents a synthesis of effects that have been identified but not yet integrated into surface ocean models. This effort provides further support for studies of the role of viruses in modifying ocean ecology. Presumably, similar integrated efforts will need to be developed in other environments like the human gut (Minot et al. 2013; Dutilh et al. 2014) and wherever else viruses are both abundant and persistent.

7.5   SUMMARY

- Models of complex food webs including viral parasites reflect innovations in the study of basic interactions as well as the quantitative estimation of life history traits.
- These models can be used to investigate the variation in the effect of viruses on ecological processes, given three key features: (i) complex multistrain and multispecies communities, (ii) nutrient regeneration via viral lysis of hosts, and (iii) competition between zooplankton and viruses for hosts.
- Life history trade-offs are predicted to be essential for the coexistence of many viruses with many hosts, given overlapping networks of infection.
- These hypothesized trade-offs include (i) trade-off between host growth rate and viral defense and (ii) trade-off between viral fitness and host range of viruses.
- Trade-offs in a community context remain an important target for *empirical* studies.
- Models predict that when viruses regenerate nutrients, both resources and viruses will increase in concentration.
- The strength of the feedback between nutrients and virus production requires further evaluation.
- Coexistence between zooplankton and viruses is predicted to be difficult when they are competing for a single host.
- Coexistence between zooplankton and viruses is facilitated in a community context, where niche differentiation between a generalist grazer and specialists viruses is possible.
- When viruses are included in ecosystem models they are found to stimulate ecosystem-level functioning despite having negative effects on target hosts.
- Few models combine viruses-host interactions with a consideration of complex nutrient and trophic dynamics; therefore, a synthesis remains incomplete and should be the subject of ongoing work.

CHAPTER EIGHT

The Future of Quantitative Viral Ecology

8.1   CURRENT CHALLENGES, IN THEORY

My favorite poem is the one that starts "Thirty days hath September" because it actually tells you something.
                                                          — *Groucho Marx*

One way to discuss the future of quantitative viral ecology is to consider its past. Given that this monograph has done just that—for what else is a synthesis but a synthesis of past knowledge—I turn to the present instead. By present, I mean the current challenges that seem to be in the forefront of research seminar topics, breakout poster sessions, and conversations at cafes and at bars (at least those in the cafes/bars near scientific meetings). These challenges are idiosyncratic, as they must be. I will make them even more so by focusing on theoretical challenges of two kinds: "forward" problems and "inverse" problems. I present these knowing that theory, perhaps like poetry, can appear inscrutable and, to some, not strictly necessary. Nonetheless, many a project would have benefited from having involved theory and theoreticians at an earlier stage.

What is the difference between a forward and an inverse problem? The difference is illustrated by the following example. Consider two very large prime numbers: 1231 and 8233. Then, multiply these two numbers together. Can you do it? Of course you can, at least with a calculator or any simple spreadsheet program. This is an example of a forward problem, in which you perform a calculation given some known condition. Now, consider that someone else has already performed the calculation and has given you the resulting product: 10,134,823. They then ask you to figure out which two prime numbers they multiplied. Can you do it? Unlikely, at least not without a computer and a specialized program. The fact that the forward problem is so easy and the inverse problem is so hard is the basis for the widely used RSA encryption method of modern cryptography. In that method, the public key is based on the product of two large prime numbers, but encrypted messages can be decrypted only if one knows the values of the two large prime numbers.

The study of quantitative viral ecology is replete with forward and inverse problems. Multiple examples of forward problems have been presented in this monograph. For example: simulate the expected dynamics of a set of virus populations interacting with a set of host populations when environmental conditions and life history traits are known. This forward problem, and scenarios like it, are relatively easy to evaluate, at least from a computational perspective. An example of an inverse problem might be: given a set of time-series data of virus and host populations can one infer who infects whom? This scenario is a very difficult problem. It is difficult, because without preknowledge of interactions, nonlinearities and latent population variation can lead to many similar time series with very distinct infection networks and many divergent time series with very similar infection networks. Therefore, confidence in predictions of the dynamics of actual biological populations of viruses and hosts remains limited because of uncertainty in the suitability of model approximations of the underlying biological processes and interactions.

As in this example, significant challenges in quantitative viral ecology center on the problem of determining how mechanisms at one scale emerge from and feed back-to affect processes occurring at other scales. Integrating processes and patterns across scales requires advances in solving both forward and inverse problems. Throughout this text, I have largely treated viruses as killers, with some mention of their affects on cellular- level processes. In fact, the mechanisms by which viruses modify host cell metabolism after infection, but before lysis, may have dramatic effects on ecosystem functioning. This modification has led some, including Patrick Forterre, to suggest a new organizing concept for the study of environmental viruses: the virocell (Forterre 2013). The idea is that the infected cell should be the object of study, less so the virion. This suggestion seems particularly apt in the study of giant viruses that dramatically modify the cellular state. This modification is highlighted by the appearance of an alternative pseudonucleus called the *viral factory*, where new giant viruses are formed. The modification is so substantial that there are even viruses of viruses, the so-called virophages, that require giant virus infection of host cells to propagate (La Scola et al. 2008; Fischer et al. 2010; Taylor et al. 2014). The idea that viruses modify of cells also has gained currency in studies of the physiology of infected cyanobacteria (Lindell et al. 2005), roseobacter (Ankrah et al. 2014), and coccolithophores (Vardi et al. 2012). Obviously, empirical studies are essential here, but so is theory, to quantify and better characterize the modification of host state by viruses. Viruses contain many auxiliary metabolic genes that can modify the uptake and utilization of nutrients and even of light, postinfection. In environments where infection is common, such modifications could lead to changes in the net flux of nutrients within and from the microbial loop. In turn, the outcome of virus infection is likely to be modified by the nutrient state of the community. A key complexity to keep in mind is that the there are *many* forms of nutrients with highly variable degrees of accessibility by microbes. In summary: cellular- and ecosystem-scale processes are linked by virus infections as part of a nonlinear feedback loop.

Connecting viruses with nutrient dynamics and other ecosystem interactions is nontrivial and would seem to require a synthesis of dynamics and interactions occurring at vastly different scales (Brussaard et al. 2008). Recall the ever-useful quote by P. W. Anderson: "More is different" (Anderson 1972). It is evident that virus-host interactions include changing abundances of strains but also changes in the identities of genotypes. Evolutionary dynamics, to say nothing of coevolutionary dynamics, have only begun to be incorporated into complex models of virus ecology, whether in the surface oceans or elsewhere. This is a positive development, as coevolutionary dynamics are certainly taking place in the environment. Signatures of such dynamics have largely been ignored, whether in virus-host systems or in consumer-resource dynamics more generally (Hiltunen et al. 2014). But it is not yet evident which aspects of ecosystem functioning are affected by coevolution (Lennon and Martiny 2008). Therefore, it is also not yet evident how many of the details of the coevolutionary process are necessary to understanding virus effects on ecosystem functioning.

Constructing a synthesis of virus effects from cells to ecosystems also requires consideration of the very different spatial scales inherent to viral ecology—from nanometers to kilometers. In 1992, Simon Levin raised the problem of "pattern and scale in ecology" (Levin 1992). This landmark paper demonstrated how and why features of a complex ecological system can differ when viewed at distinct spatial scales. In the present monograph, I have considered multiple temporal scales as well as different scales of complexity, from cells to ecosystems, but I have almost completely neglected the question of viral *spatial* ecology. The reasons for this are many, not least of which is that empirical information on environmental virus ecology has emphasized the study of diversity over many disparate locations rather than deep sampling in nearby spatial domains. The primary focus of viral spatial ecology has been on the expansion of fronts in viral plaques (Fort and Méndez 2002; Abedon and Yin 2006; Abedon and Culler 2007). The situation is changing (Gómez and Buckling 2011; Koskella et al. 2011), and with good reason—existing theory shows that the spatial connectivity among hosts and viruses can alter the nature of coevolutionary dynamics (Kerr et al. 2002; Forde et al. 2004; Kerr et al. 2006; Heilmann et al. 2010, 2012; Haerter and Sneppen 2012; Sieber et al. 2014).

Realistically, there is a long way to go to such a synthesis of virus effects on ecology and evolution across scales. An anecdote: I presented a visiting seminar on virus ecology a few years ago, where I concluded with a mention of a working group I co-led with Steven Wilhelm on ocean viral dynamics. One of our objectives was to develop an integrated marine surface food web model with parasitism that could reproduce known variation in community structure, nutrient concentrations, and fluxes (see Chapter 7). A colleague asked, skeptically, in a one-on-one chat after the seminar: "Do you really think you can ever do that?" Although I can't recall my exact response, I do remember its gist, which was that the process of trying will almost certainly lead to new discoveries in biology, ecology, and perhaps even physics that would not have been possible without attempting to integrate viruses into complex food webs in the first place. In fact, along the way, a subset of us (Steven Wilhelm, Alison Buchan, and I, along with two students) realized that a key parameter in any such model was the elemental content of virus particles. Surprisingly, only a few estimates of carbon per virion were available. In response, we collaboratively developed a first-principles model of the elemental content of viruses (Jover et al. 2014). This model revealed new mechanisms by which viruses may play a larger-than-appreciated role as a reservoir of certain elements (like phosphorus) and even transform, rather than merely shunt, nutrients back into the microbial loop. The specification of elemental content therein also helped establish a multitrophic model of virus effects on an idealized surface marine ecosystem (Weitz et al. 2015).

Yet, in neither of these projects did we close the loop on virus effects at global scales or on the feedback of climate change on viruses. Virus stability and interactions with hosts will be affected by changes in water temperature, salinity, and acidification. Small changes unfolding on the global scale can have large effects. In the words of Brussaard and colleagues, the oceans are characterized by "global-scale processes with a nanoscale drive" (Brussaard et al. 2008). The scaling-up requires careful thought, particularly in light of a changing planet.

8.2   ON THE FUTURE OF QUANTITATIVE VIRAL ECOLOGY

> "There is nothing new under the sun".
>
> —*Ecclesiastes, 1:9*

The titular question of this section has been in the back of my mind while writing this book. Similar questions must be in the minds of many others, because to care about a field (and the science therein) also means to care about its future. In the present case, I have tried to consider the future by having a particular date in mind: 10 years from now. This choice of timeline was inspired by discussions I had with colleagues who authored science books. Two timescales emerged for science book completion from those conversations: 1 year and 10 years. The former seem to have ended up as monographs and the latter, as textbooks. The present book you are holding, and nearly finished reading if you have read from front to back, is of the former variety. As I started writing, and now, as I near the end, I became increasingly aware that it is simply not the right time for a textbook on "quantitative viral ecology." Instead of writing the book for the future, I focused on the book for the present. At this moment in time, I hope a monograph on quantitative viral ecology influences, rather than predicts, the future of the field. In viral ecology, as in the rest of science, there is a natural cycle to all ideas, including revision and, ultimately, replacement. If the future looks unlike the present, so much so that the needs of the community demand a far different text 10 years from now, then I think this monograph will have succeeded.

Indeed, in writing this book, I was keenly aware of the limitations of what I could say with certainty on the state of quantitative viral ecology. As should be apparent from reading this monograph, a disconnect remains between what we know about a few model systems and the state of virus-host dynamics and their consequences in the environment, in environmentally relevant conditions. That disconnect is partially a limitation of technology. It is just more than 25 years since the advent of increasingly reliable estimates of environmental viral densities, 10 years or so since the first viral metagenomes were analyzed, and only in a few years ago that partner organisms in virus-host interactions could be quantified without requiring that both viruses and host be available in culture. Hence, my focus throughout this book has largely been on developing a body of theory, models, and theory-driven insights on model virus-host systems for which some prior knowledge regarding interactions was available. This included knowledge of fundamental interactions modes and, in some cases, even the rates of such interactions. Will the present set of insights be able to be generalized?

I believe a similar text 10 years hence will have much more to say on the natural (ideally, unbiased) variation of virus numbers, diversity, interactions, spatial distributions, and effects. The current insights from models and model systems are all likely to apply; for example, that viruses can draw down host abundances, drive oscillations, modify nutrient processes and feedback, and be part of complex coevolutionary dynamics, But these processes require more contextualized examples and direct quantification. To this end, I would hope that the lessons learned from colleagues in phytoplankton and plant ecology would also take root in viral ecology: that functional traits matter, a lot, not "just" diversity (Violle et al. 2007). I have tried to emphasize this point throughout the present monograph, for example, by showing how changes in life history parameters determine the relative success or failure of hosts and viruses, whether in pairs or in a community context. But estimating viral traits is not easy—it's extremely hard. And more data will require better statistical tools. Therefore, the field of viral ecology will likely see increasing contributions from those scientists capable of leveraging large-scale data in a way that provides meaningful information about ecological processes.

Perhaps such data will eliminate the need for the types of relatively simple models presented here. It may eventually be possible to develop predictive statistical models of virus-host co-occurrence and cross-correlation. The input to such models may be the spatiotemporal abundances of viruses, hosts, and zooplankton at many locations. But the cautionary component of big-data science is that more facts do not necessarily translate into more knowledge. Even with the increasing availability of viral-associated data, I believe there will still be a need for "small viral ecology" (to adapt the term "small systems biology" favored by Kim Sneppen (Sneppen 2014)). I believe this not just because of my interest in the area but, rather, because large, predictive models are often right for the wrong reasons. Ryan Gutenkunst, Jim Sethna, and colleagues have developed a theory of parameter dependence of complex models that shows many parameter sensitivities are far more "sloppy" than had been expected (Gutenkunst et al. 2007). Being right for the wrong reasons may seem innocuous; for example, who cares why a prediction works, so long as it does? However, it becomes precarious when conditions change, and what used to be a good predictive model suddenly becomes a bad one, but no one is quite sure why.

Will the future of viral ecology also include new discoveries that could not have been anticipated by the current paradigms? I believe the answer is, again, yes. Ideally, a future text on advances in quantitative viral ecology will not just be a recapitulation of the same interactions, processes, and players. To the contrary, all indications are that the study of viral ecology is still on track to reveal fundamentally novel modes of interactions occurring in natural environments that are truly new to science. Examples from the past few years alone include (i) the discovery of giant viruses whose structure (both physical and genomic) bring into question the origins of cellular (and noncellular) life and (ii) the growing body of evidence for metabolic repurposing of host cells during infection. We still don't know the answer to what would seem like foundational questions: What determines who infects whom in the environment? How much of ecosystem functioning is under the control of viruses? How much do viruses govern the natural evolution of microbes? The answers to such questions may vary among focal environments given different virus-host systems.

Where will such new systems come from? In the final part of this monograph, I focused on marine surface waters to link small and relatively simple models with large and relatively complex environments. But viruses of microbes are present in the human gut, in waste reactors, in terrestrial soils and in deep sediments, in the air, in built environments—in other words: everywhere science (and scientists) have looked. I also anticipate that progress in the study of viral ecology will help scientists in other domains of expertise think about viral effects in other systems. These effects may have been ascribed to other components or as a source of unexplainable noise, but may also have viral origins.

All this leads me to one final thought about the future of quantitative viral ecology. I began my career in physics and received a PhD in physics, despite having decided early on in my graduate training that whatever "physicist" I would be, it would certainly not be conventional—for many reasons. Physicists who have turned toward biology, chemistry, and even sociology have had their reasons as well. One insight that I did glean is that a healthy scientific field requires a balance of theory and empiricism. In my view, viral ecology presently has too little theory and too few theorists. And, in my view, this is a problem that the field must face if it is to become quantitative viral ecology. The same view has been espoused in efforts to integrate more theory into microbial ecology (Prosser et al. 2007). This transition from viral ecology to quantitative viral ecology leads me to a closing set of thoughts.

I have intentionally left the adjective *Quantitative* in this title, despite knowing that it will not sit well with all readers. Nearly all measurements are quantitative, and therefore all science, involving the measurement of the natural world—including viral ecology—must be quantitative. Yet, there are two senses in which this term does not quite fit with viral ecology as it has been practiced. First, for purely technological reasons, many measurements in viral ecology have not been comparable, in a quantitative sense. This may come as a surprise to outsiders but is a fact well known to insiders. For example, any culture-based or even PCR-based method for measuring the abundance of two strains, whether host or viral, has a strong potential bias. Even if strain 1 is measured to be more abundant than strain 2, the converse may be true. This is a major reason why a number of experimental groups have invested significant time and resources into improving the sample-to-X pipeline, where X may be a sequence, trait, abundance, or other feature of a virus or virus community. Second, there have been relatively few attempts to integrate what virus ecologists have measured into models of microbial communities. Whatever integration has occurred has focused almost exclusively on lab-based population and evolutionary dynamics. Crucially, there is a pressing need to understand the ecological role of viruses of microbes, which requires an increasing number of theorists—whether of the big data, small systems, or other varieties—to investigate the interactions between microbes and the viruses that infect them. My final hope in this respect is that if the current text has made it easier and more accessible for theorists to do so and, concomitantly, for empiricists to engage with theorists who want to do so, then the book has succeeded. I don't imagine I will ever have a tally, but one can dream.

TECHNICAL APPENDIXES

Viral Life History Traits

A.1 MEASURING VIRAL LIFE HISTORY TRAITS: A QUANTITATIVE PERSPECTIVE

This appendix explains a set of core techniques in the measurement and quantification of viral life history traits. These techniques are not new. To the contrary, they were developed by the founding generation of phage biologists as part of efforts to understand foundational principles in molecular biology, cell biology, and virology. The utility of this appendix, like that of all the appendixes in this book, depends on the background of the reader. A physicist, engineer, mathematician, or computer scientist should find the explanations approachable, as they are nearly a faithful representation of the principles underlying efforts to quantify viral life history traits. They are also approachable precisely because they do not recapitulate all the details in actual laboratory protocols. These principles can be adapted for use with the viruses of archaea and microeukaryotes, and indeed for the viruses of target cells inside "macro"eukaryotes. Moreover, with the increasing interest in the study of viruses of microbes, many research groups are working with phage or other viruses of microbes for the first time and for which lab protocols and/or rationales for estimating traits may not yet have been established. A biologist in such a group will likely learn something new about the nature of the inference process that is all too often taken for granted.

This appendix is not meant as an alternative to a detailed lab protocol but, rather, as a complement to such a protocol. Detailed lab protocols can be found in modern virology textbooks (Dimmock et al. 2007), as stand-alone guides (Clokie and Kropinski 2009a, 2009b), and in the online *Manual of Aquatic Viral Ecology* (Wilhelm et al. 2010). Nonetheless, neither lab protocols nor this appendix should be applied as an inferential tool without careful consideration of the nature of the problem at hand. The protocols described here are intended for use in the examination of interactions of a population grown from a single viral isolate with a population of hosts grown from a single host isolate. This may sound restrictive, but the interaction of pairs of viruses and hosts are the building block of complex interactions. Moreover, this type of procedure is in common use. These techniques are also applicable to cases where the viruses and/or hosts are themselves mixtures of many populations.

Finally, the last, and perhaps most important, aim of this appendix is to lay out the experimental steps for viral life history trait estimation alongside the core mathematical principles, so that both theorists and experimentalists can have a better common ground for communicating regarding what is being measured and why.

A.2   A CORE TECHNIQUE: THE PLAQUE ASSAY

Viruses, like bacteria and microeukaryotes, can be isolated. *Isolation* means to be separated from the rest of the community so that only a single population is present in a sample. The isolation of bacteria, for example, requires the use of selective media in which the bacterium of interest is likely to grow, picking a "colony," and then growing a population from that colony. This process is repeated multiple times so that the bacteria that remain in the flask or on the Petri dish are exclusively the recent descendants of a single bacterium or, technically, of a single *colony-forming unit* (CFU). By analogy, the isolation of viruses also requires the use of selective media in which the virus of interest is likely to grow; the selective media for viruses is a target host. In practice, a population of target hosts is grown to high density, then a candidate sample of viruses is added; some of these viruses infect and lyse cells, thereby releasing new viruses. The resulting sample has many viruses, cell debris and, possibly, living cells. Additional chemical treatments are often added to eliminate cells so that the sample contains only viruses and debris. The viruses are then diluted successively and added to agar plates that have a layer of media and a layer of target hosts (a "top" agar). Viruses that infect and lyse cells will release new viruses that will subsequently infect nearby host cells, leading to a clearing in the top agar. These clearings are called *plaques* and are analogous to colonies. Each plaque is founded by a single virus, or a *plaque-forming unit* (PFU), just as a colony is assumed to have been founded by a single bacterium or colony-forming unit (CFU). Viruses selected from a single plaque can be added back to a population of hosts and the process repeated to isolate a very large population derived from a single viral type.

The plaque assay is considered the gold standard for characterizing the number of viruses in a sample, analogous to counting CFUs for bacteria. It can be used to initially assess the titer of viruses in an isolated population or as part of kinetic assays to estimate life history traits. Other assays used to measure viral densities, that is, other techniques for "counting" virus particles,

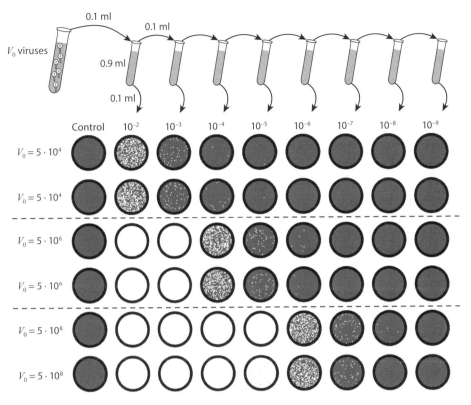

FIGURE A.1. Simulating a plaque assay for different viral concentrations. (Top) Protocol for the simulations. An initial concentration V_0 of viruses is successively diluted by factors of 10 and then diluted once more before being plated onto an agar plate with a top layer of susceptible hosts. (Bottom) Synthetic results of plaque assay "experiments." The control (gray) is a top agar with only bacteria. Each white circle represents a plaque. There are two replicate pairs per value of V_0, where different sets of pairs are separated by a dashed line. Each row is the result of an in silico numerical simulation of a dilution series, followed by visualization in which the ratio of plaque radius:plate radius is 1:50. The listed value of V_0 is the initial concentration of viruses for the synthetic plaque assay.

include higher-throughput techniques such as flow cytometry (Brussaard et al. 2010), epifluorescence microscopy (Suttle and Fuhrman 2010), and electron microscopy (Ackermann and Heldal 2010).

Consider a solution of pure viruses; that is, it contains viruses at density V_0 as well as (possibly) cell debris but no living cells. How does one estimate the density V_0 using the plaque assay? The top panel of Figure A.1 provides a schematic of the assay. Initially, a flask contains a pure virus culture at a density of V_0 viruses/ml, where V_0 is the value to be determined. Each of the eight, smaller test tubes to the right contains 0.9 ml of sterile media or buffer. Initially, 0.1 ml of the original virus solution is added to the first tube, which is then mixed. Thus, if V_0 was the initial density of viruses, then the density of viruses in the first tube is now approximately $V_0 \frac{0.1}{0.1+0.9} = 0.1 V_0$. Then, 0.1 ml of the solution in the first tube is transferred to the second tube, so that it contains viruses at an (approximate) density of $0.01 V_0$. The same series of steps is continued so that the tubes contain pure virus populations at densities ranging from $0.1 V_0$ to $10^{-k} V_0$, where k is the number of test tubes, in this case $k = 8$. Previously, a pure bacterial population that is known to be susceptible to the virus should have been grown, usually to mid-log phase (i.e., to a density in which the bacterial population is doubling at its maximum rate), then poured on top of a growth media and allowed to regrow. The Petri dish has a first layer of growth media (the "bottom" agar) and a second layer of bacteria (the "top" agar). Given the series of diluted virus populations, a small inoculum of viruses (in this example, 0.1 ml per tube) is poured on top of the agar plates. So *begins* the plaque assay.

What happens next depends on the initial (unknown) concentration of viruses, V_0. Consider the first dilution in the series, in which $0.01 \times V_0$ viruses have been added to the agar plate (as the concentration is $10^{-1} V_0$, and 1/10 of the sample is plated). If V_0 is, in fact, 5×10^8 viruses/ml, then approximately 5×10^6 viruses will be added. A single virus is sufficient to yield a plaque. This equivalence, by no means trivial, was first established in the late 1930s (Ellis and Delbrück 1939). Although plaque sizes vary based on host, virus, and media, a typical plaque radius is 1 mm (Gallet et al. 2011). If each virus creates one plaque, then the maximum total clearing area will be $5 \times 10^6 \times (\pi 1^2) \approx 1.6 \times 10^7$ mm^2. Agar plates are typically 100 mm across, such that their total area is approximately 7.85×10^3 mm^2, which is much smaller than the expected total area of plaques. As a consequence, the entire top layer of bacterial cells should be lysed when exposed to viruses at sufficiently high titer, even when taking into account repeated infections of the same host cell given high viral titer. This lysis will be visually apparent, such that the top agar appears clear. In contrast, an agar plate not exposed to viruses should appear opaque, the result of increased light scattering off the dense concentration of bacterial cells grown in a selective media.

The bottom panels of Figure A.1 illustrate these concepts via an in silico plaque assay that incorporates the stochastic nature of the dilution and plating steps. Examine the last two sets of rows in which an in silico plaque assay experiment has been conducted. The control plate appears opaque, whereas there are so many viruses in the first plate (again, 5×10^6 on average) that the agar plate appears clear. The same result applies even when the initial virus density is $V_0 = 5 \times 10^6$, but not when it was $V_0 = 5 \times 10^4$. Consider again the naive expectation that when $V_0 = 5 \times 10^4$ the total area of plaques should be 1.6×10^3 mm^2, which is approximately 20% of the total agar plate. As should be apparent, once the applied virus density becomes sufficiently low, then individual plaques may be visible, counted, and used to estimate the true, but unknown, viral concentration V_0. It should also be apparent that if the virus solution is diluted too far, then there may not even be a single virus left, so that there is no difference between the control and the treatment plates. Such example can be seen in the rightmost column in the bottom panel of Figure A.1. It is for this reason that the virus solution must be *successively* diluted so that the number of plaques v can be counted at a dilution level k and then this number, $v(k)$, used to infer V_0.

Plaques are often counted by hand, although automated methods exist. To focus on the core principles of the plaque assay, the assumption here is that it is possible to reliably estimate the number of plaques given at least some fixed dilution level k. Choosing the right dilution level k must balance countability versus estimatibility. If k is too small (i.e., the viral solution is not sufficiently diluted), then plaques will overlap, and estimates of p will be poor. If k is too large (i.e., the viral solution is diluted too far), then there will be very few viruses and an increased stochasticity due to the dilution process can introduce large posterior uncertainty in V_0. The best choice of n depends on the size of plaques and other features and can be determined by comparing the inference of V_0 given two alternative dilution levels. What are the consequences of estimating the (true, but unknown) value of $V_0 = 5 \times 10^6$ at dilutions corresponding to 10^{-4} or 10^{-5}. In the former case, there should be 500 individual virus particles that interact with bacteria on the agar plate, whereas there should be 50 individual virus particles in the latter case. The dilution introduces stochasticity in the number of viruses remaining after each transfer. To a first approximation, the dilution of 10^{-k} represents the independent probability that any given virus is added to a given top agar plate. Technically, there are correlations in fluctuations between tubes that are introduced via the successive dilution mechanism, but that is a second-order effect. The probability that v viruses of an initial population of V_0 viruses will be present given a dilution level k is

$$P(v|V_0, k) = p_k^v (1 - p_k)^{V_0 - v} \binom{V_0}{v}, \qquad (A.1)$$

where $p_k = 10^{-k}$, and $\binom{V_0}{v} \equiv \frac{V_0!}{v!(V_0 - v)!}$. This is the binomial probability distribution and enumerates the probability that exactly v events, each with probability 10^{-k}, will take place out of V_0 trials. When p_k is small and V_0 is very large, then the binomial is well approximated by a Poisson distribution, such that the probability, $\mathcal{L}$, of v viruses being present at a dilution level k can be written as

$$\mathcal{L}(v|V_0, k) = \frac{\lambda^v e^{-\lambda}}{v!}, \tag{A.2}$$

where $\lambda = 10^{-k} V_0$. The expected number of viruses is just

$$\bar{v} = \lambda = 10^{-k} V_0, \tag{A.3}$$

and the variance in the number of viruses is also λ. The standard deviation, σ, in the number of viruses is equal to the square root of the mean, and, as a consequence, the coefficient of variation (i.e., $\sigma(v)/\bar{v}$) scales as $\bar{v}^{-1/2}$. This would imply that counting more plaques by diluting less is inevitably better. Although this is true (in theory), in practice, counting more plaques may introduce technical barriers, and, eventually, diluting less leads to plaque overlap and potential biases in estimation of V_0. Nonetheless, this analysis suggests that measuring a value of v implies an inherent posterior variation in estimating the true value of V_0.

The Poisson distribution approximates the probability of observing v plaques at a dilution level k given an initial population of V_0 viruses. Yet, an empiricist is interested in the related question, what is the probability that there are V_0 viruses in the population given an observation of v plaques? This question can be posed formally in terms of $P(V_0|v)$. Here, we can apply Bayes's rule (noting that $p(a|b)p(b) = p(b|a)p(a)$ and write

$$P(V_0|v)p(v) = P(v|V_0)p(V_0), \tag{A.4}$$

and then rearrange terms so that

$$P(V_0|v) = \frac{P(v|V_0)p(V_0)}{p(v)}. \tag{A.5}$$

The value of $P(v|V_0)$ is the Poisson likelihood function derived earlier $\mathcal{L}(v|V_0, k)$. Further, $p(V_0)$ is the prior distribution of the virus density. This could be any positive value in a fixed range, given an uninformative prior. Finally, $p(v)$ is the probability that the data count is actually what is measured. Given a count value v, the probability that the true value of density is either V_0

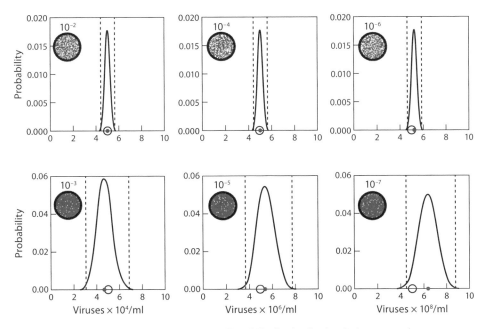

FIGURE A.2. Inferred maximum likelihood distribution for the viral concentration V_0/ml. The three columns denote experiments in which $V_0 = 5 \times 10^4$, 5×10^6, and 5×10^8 (from left to right). The top panel corresponds to inferences based on plates with approximately 500 plaques; the bottom panel corresponds to inferences based on plates with approximately 50 plaques. The solid line denotes the maximum likelihood distribution, and the vertical dashed lines denote the 95% confidence interval for V_0. The open circle on the x-axis denotes the true V_0, and the closed diamond on the x-axis denotes the maximum likelihood estimate of V_0.

or V_0' can be compared:

$$\frac{P(V_0|v)}{P(V_0'|v)} = \frac{\mathcal{L}(v|V_0)}{\mathcal{L}(v|V_0')} \frac{p(V_0)}{p(V_0')} \tag{A.6}$$

$$= \frac{\mathcal{L}(v|V_0)}{\mathcal{L}(v|V_0')}, \tag{A.7}$$

which is just equivalent to the ratio of likelihoods! This has been a long way to say something rather straightforward: that the most likely value of V_0 is one that maximizes the likelihood of having observed v plaques. For the Poisson distribution this is $\hat{V}_0 = 10^k v$. The advantage of the likelihood approach is that it also yields a 95% confidence interval. Figure A.2 illustrates this 95% interval for the cases of $V_0 = 5 \times 10^4$, 5×10^6 and 5×10^8, respectively, given two alternative dilution levels. The solid curves are the likelihood that there are

V_0 viruses given an observation of v plaques (shown in the inset of each panel). The vertical dashed lines demarcate 95% confidence interval for the values of V_0. These points, $V_0^{0.025}$ and $V_0^{0.975}$, can be found via the cumulative of the Poisson distribution. The key point is that the inferred value, $\hat{V}_0$, and the true value, V_0, are not the same, yet $\hat{V}_0$ remains the best estimate the experimentalist can make given the observations. The second key point is that the number of measurements matters, particularly in defining the posterior uncertainty in V_0 given the estimate $\hat{V}_0$.

A.3   PROTOCOLS FOR LIFE HISTORY TRAIT ESTIMATION

Mortality rate (m): Viral *mortality rate* or *decay rate* represents the decline of infectious virus particles in a population over time. The decay of virus particles is often modeled as a first-order kinetic process with a decay rate m, such that there is probability m per unit time that a virus particle will transition from infectious to noninfectious. This process can be modeled at the population level as

$$\frac{dV}{dt} = -mV, \tag{A.8}$$

such that the number of infectious virus particles should decline exponentially as

$$V(t) = V_0 e^{-mt}. \tag{A.9}$$

The decay of a viral population can be observed experimentally by isolating viruses in a cell-free flask or tube and then successively plating. Then, the estimated mortality rate, $\hat{m}$, is equal to the slope of the best-fit line between the logarithmically transformed estimates of $V(0), V(1), \ldots, V(n)$ and the times at which those estimates were taken $t_0, t_1, \ldots, t_n$. Some care should be taken in spacing the observations. In practice, observations usually occur at fixed intervals. The 95% confidence intervals for the point estimate of a slope are available in most standard statistical packages.

Adsorption rate (φ): *Adsorption rate* is defined as the irreversible loss of free virus from an environment due to binding to host cells. Adsorption is generally modeled as a first-order kinetic process in which viruses and hosts come into contact through diffusive processes

$$\frac{dV}{dt} = -\phi NV. \tag{A.10}$$

Anticipated values of the maximal value of ϕ based on diffusion limitation alone are discussed in Chapter 2. However, in practice, the realized adsorption rate is smaller owing to factors such as sparse receptor availability, hetero-geneous placement of receptors on cell surfaces, and secondary processes preceding adsorption such as phage diffusion on cell surfaces (Rothenberg et al. 2011). From an experimental point of view, the suite of methods for estimating adsorption rate generally proceed as follows. First, viruses and hosts of known titers are combined. Then, at fixed time intervals, a subsample of the mixture is taken, which is then manipulated (either physically or chemically or sometimes both) such that only free viruses are left. The concentrations of viruses in these samples are then measured by plaque assays. If the assays are conducted over time periods during which hosts are not actively multiplying or dying (either by lysis or via other mechanisms), then $N(t) \approx N_0$, and the free virus population is expected to decay as

$$V(t) = V_0 e^{-\phi N_0 t}. \tag{A.11}$$

The key assumptions are twofold: (i) the viral decay rate, m, is much lower than the viral adsorption rate, ϕN_0; and (ii) host density does not change during the measurement process. As before, the estimate of ϕ involves fitting a line to the logarithmically transformed estimates of $V(t)$ at times $t_0, t_1, \ldots, t_n$ to yield a regression slope of α, Then, the adsorption rate can be estimated by

$$\phi = \frac{\alpha}{N_0}, \tag{A.12}$$

where N_0 is the estimated host density. A recommended set of units for ϕ is ml/(cells $\times$ hr), but no matter what the units, it is important to realize that the slope of the decay of free viruses is not equal to the adsorption rate. The slope of the decay is equal to the adsorption rate multiplied by cell density. To standardize comparisons, it is strongly recommended to divide the measured decay constant by initial cell density, as noted. Here, the total uncertainty in ϕ should include contributions from the uncertainty in α, the measured slope, and in N_0, the density of bacteria.

The one-step growth curve: Estimating burst size (β) and latent period (τ or η): Use of a one-step growth curve is an experimental procedure whose aim is to simultaneously estimate the burst size and latent period of a virus given a target host (Hyman and Abedon 2009). The procedure is designed to study the lytic lifestyle of viruses and presumes that viruses will generate a burst of β viruses at a time approximately τ after adsorption to hosts. The procedure is generally implemented as follows. A viral population at concentration V_0 is mixed with a host population at concentration N_0 for a period of time δ. In some cases, this mixing takes place at low temperatures, lower than the physiological optimum of the host, so that adsorption can take place but host and virus replication is slowed, if not temporarily halted altogether. Remaining free viruses can be removed by a combination of physical and chemical treatments, such as, pelleting cells and removing the supernate. What is left are infected cells and uninfected cells. Plating a sample with a top agar of bacteria leads to an estimate of the number of "infective centers" (Ellis and Delbrück 1939). If the sample derives from the start of the experiment, then the number of infective centers is equal to the number of infected cells. If the sample derives from later in the experiment, then cells may already have been lysed, releasing many phage progeny. This increase in phage progeny leads to an increase in the number of infective centers. Care must be taken to choose this second time point after the expected latent period but not so long as to include multiple infection cycles. The ratio between the final number of infective centers and the initial estimate is the burst size. Further, the midpoint of the rise is an initial estimate of the latent period. Other subtleties may emerge. For example, viruses are assembled prior to lysis. If one is interested in the eclipse period as well as the latent period, then artificially liberating viruses from infected cells prior to performing a plaque assay can provide additional information. Some methods for simultaneously infering the latent period, eclipse period, and burst size have been proposed (Rabinovitch et al. 1999).

Probability of lysogeny (p): The *probability of lysogeny* is defined at the cellular level as the likelihood that a virus infecting a cell will integrate its genome with that of the host and persist stably during successive divisions. These outcomes are in contrast to alternatives, such as, initiating the lytic pathway or being a failed infection. The experimental procedure for measuring the probability of lysogeny can be classified into (i) single-cell tracking and (ii) population-level assays. The details of single-cell tracking are described in Zeng et al. (2010). They depend on incorporating genetic markers into the phage and on microscopy techniques that can be used to visually track the fate of infected cells. Population-level assays are more broadly applied. The problem confronting any effort to estimate the population-level fraction of lysogens is how to distinguish lysogens from uninfected cells. In practice, two approaches are taken. In the event that the phage can be genetically modified to include an antibiotic-resistance cassette, then application of said antibiotic can discriminate between total cells and lysogens (see an example with phage λ (Zeng et al. 2010)). When such integration is not possible, an alternative is to induce lysogens to enter the lytic mode via the application of UV light or other inducing agent. Then, the number of lysogens can be measured using a plaque assay method.

In both approaches, a known density of viruses, V_0, is mixed with a known density of host cells, N_0. The mixing is often done initially at low temperatures, e.g., $4°$ C, to enable adsorption of nearly all free viruses. The use of a low-temperature mixing step is meant to inhibit both infection and the postinfection regulation of host cells by the viral genomes, thereby synchronizing the infections. Then, the mixed population is placed in a normal-temperature environment to enable integration of adsorbed viruses and expression of the antibiotic resistance gene carried by the viruses. At this point, the experimental flask contains free viruses as well as cells that are either uninfected, lysogenized, or en route to lysis. A mixture of physical and chemical techniques can be used to remove free viruses and cells fated for lysis. At this point, the procedures differ. In the case of genetically modified phage with antibiotic-resistance cassettes, aliquots of cell-containing media are added to two types of plates—one containing an antibiotic and one without. In this way, the concentration of lysogens, N_l, and the concentration of uninfected cells, N_-, can be estimated (Zeng et al. 2010). When such genetic modification is not possible, then aliquots of cell-containing media are applied as part of a plaque assay, and a lysogen-inducing agent is included in the media (e.g., UV, a chemical stressor, on starvation). Thus, the concentration of infective centers (corresponding to lysogens), N_l, and the concentration of uninfected cells, N_-, can be estimated (Kourilsky 1973). The probability of lysogeny, $p = N_l/N_0$, is often expressed as a function of the population *multiplicity of infection* (MOI = V_0/N_0).

Population Dynamics of Viruses and Microbes

B.1 HOST-ASSOCIATED LIFE HISTORY TRAITS

Models that describe virus-host interactions require the specification of host-associated life history traits. Recall that resources taken up on a per-host basis can be modeled via a Type II functional response, $f(R) = \gamma R/(Q + R)$, such that the per-host division rate as a function of resource availability is approximated as $\epsilon f(R)$. In this formulation, three key host-associated traits for growth of a host population on a single limiting nutrient include ϵ, the yield of host cells given uptake of resources (cells/μg); γ, the maximum uptake rate of resources (μg/(cells $\times$ hr)); and Q, the half-saturation constant associated with resource uptake (μg/ml). The specific values of such traits will vary with the host type and limiting resource under consideration. A complete analysis is beyond the scope of this appendix. Instead, an illustrative example provides both specific parameter values and, to some extent, a procedure for estimating these parameter values for other host-resource combinations.

In their initial study of phage-bacteria population dynamics, Levin and colleagues utilized a glucose-limited chemostat (Chao et al. 1977). For reference, the molecular formula of glucose is $C_6H_{12}O_6$, such that there is approximately 0.4 μg of C for every 1 μg of glucose. An *E. coli* cell growing in mid-log phase contains approximately 200 fg of carbon (Milo et al. 2010). For every 1 μg of glucose taken up, there is enough carbon to synthesize $0.4\,\mu g\,200\,fg = 2 \times 10^6$ cells. Yet, resource conversion is not 100% efficient. If glucose is converted with 10% efficiency into cell biomass, then $\epsilon = 0.1 \times 2 \times 10^6 = 2 \times 10^5$ cells/μg glucose. This estimate can be integrated with the maximum division rate to estimate the maximum uptake rate. The maximum division rate of *E. coli* B in saturating-glucose conditions is on the order of 1/hr. Therefore, $1 \approx \epsilon\gamma$, or $\gamma = 5 \times 10^{-6}$ μg/(cells $\times$ hr). These estimates can be compared with the inferences of these parameters for *E. coli* B (Levin et al. 1977): $\epsilon \approx 3.8 \times 10^5$ cells/μg, and $\gamma \approx 1.92 \times 10^{-6}$ μg/(cells $\times$ hr).

Finally, the half-saturation constant of glucose uptake can be estimated by fitting a saturating response function to growth rates under different resource concentrations. The estimate of $Q \approx 3.92$ μg/ml is due to Jacques Monod (Monod 1949), and this value was carried forward by Levin et al. (1977).

The key point is that minimal information on cell size and growth rate can be used to estimate relevant life history parameters in a population dynamics model. For example, given measurements of both cell size q (in fg) and maximum growth rate r_{max}, the values of ϵ and γ can be estimated by assuming a 10% yield of resources taken up into biomass. First, $\epsilon \approx 10^8/q$ cells/μg, where q is in femtograms. The value of ϵ may approach or exceed 10^7 for smaller cells, on the order of 10–30 fg, typical of ocean heterotrophs (Fagerbakke et al. 1996). Second, the maximum uptake rate can be estimated as: $\gamma = r_{max}/\epsilon$. Faster-growing organisms must take up nutrients at a faster rate, given identical cell sizes and yields. The estimates of ϵ, γ, and Q are integrated into virus-host population dynamics models in the text.

B.2   LINEAR STABILITY ANALYSIS OF A NONLINEAR DYNAMICAL SYSTEM

Analyzing the stability of equilibria in a nonlinear dynamical system is an essential part of the training of a scientist interested in modeling biological systems. However, it is not always part of the curriculum of traditional science programs. This appendix represents a brief introduction to this topic to ensure the text can be read in a self-contained manner. The material here is not new and can be skipped by a reader familiar with the analysis of dynamical systems, such as at the level of the introductory texts *Nonlinear Dynamics and Chaos* (Strogatz 1994) or *Population Biology: Models and Concepts* (Hastings 1997).

B.2.1   *The one-dimensional example*

Microbial population dynamics with implicit resources and a fixed washout rate can be written as

$$\frac{dN}{dt} = rN \left(1 - \frac{N}{K}\right) - \omega N, \tag{B.1}$$

where N is the density, r is the maximal growth rate, ω is the density-independent death rate due to washout, and K is the density at which net growth goes to zero in the limit of vanishing washout. Formally, this system can be analyzed for any nonnegative value of N. As was shown previously, the dynamics of this system lead to either the death of the population or its persistence at an equilibrium. Here, the system is used as an example to show how the stability of these equilibria can be studied (in one-dimensional systems), in part as a prelude to analysis of stability of higher-dimensional dynamical systems. In this example, the description utilizes the convention to suppress the time dependency of time-varying populations, for example, writing x instead of $x(t)$ but noting the (constant) equilibrium values of such populations with an asterisk, for example, x^*.

Recall that the term *fixed-point equilibrium* refers to those population densities (in this case the density N) such that there is *no time rate of change*. Formally, this means that for a dynamical system that tracks changes in some population $x(t)$ of the form

$$\frac{dx}{dt} = f(x) \tag{B.2}$$

the equilibria correspond to values of $x = x^*$ such that $f(x^*) = 0$. There are two equilibria in the case of the preceding model:

$$\text{Death}: \quad N^* = 0, \tag{B.3}$$

$$\text{Persistence}: N^* = K\left(1 - \frac{\omega}{r}\right). \tag{B.4}$$

The local stability of a given equilibrium can be considered in terms of the response of the system to small perturbations. The system is termed *unstable* if the size of the perturbation increases with time, whereas it is termed *stable* if the size of the perturbation decreases with time.

In the biological example, consider the stability of the equilibrium $N^* = 0$. This equilibrium has no individuals of the microbial population—one can think of it as a test tube with sterile media. Then, the "perturbation" can be thought of as the addition of a small inoculum of a microbial population, $N(t = 0) = n \ll K$. What happens in such a case? If the media is conducive to growth, then the microbial population will grow in number. If the media is not conducive to growth, then the microbial population will decrease in number. In the former case, density increases from a small perturbation, and so the fixed point is deemed unstable, whereas in the latter case, density decreases from a small perturbation, and so the fixed point is deemed stable. This intuition can be formalized, at least for nonlinear dynamical systems of a single variable, using a graphical, a perturbative, and a linearization approach.

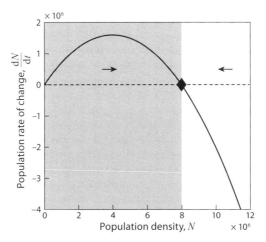

FIGURE B.1. Population dynamics, dN/dt, as a function of population density for the chemostat growth model where $r = 1$, $K = 10^7$, and $\omega = 0.2$. The dashed line denotes $dN/dt = 0$, the diamond denotes equilibrium $N^* = K(1 - \omega/r)$, and the shaded region denotes values of N where population growth rate is positive.

A useful way to assess the stability of a one-dimensional nonlinear dynamical system is to plot the rate of change of the population as a function of the population. In this case, this corresponds to visualizing dN/dt as a function of N (see the solid line in Figure B.1). On the same plot, it is useful to draw a horizontal line corresponding to the value of $dN/dt = 0$ (see the dashed line in Figure B.1). In this way, one can rapidly identify those regions of N for which the population rate of change is positive—the solid line should lie above the dashed line. Similarly, regions for which the population rate of change is negative correspond to those where the solid line lies below the dashed line. Finally, the values for which the population rate of change is zero are identified by finding the intersections of the solid and dashed lines.

With this graphical scheme in place, the following procedure can then be used to determine the dynamics of $N(t)$ given some initial value of $N(t = 0) = N_0$. First, determine whether the population is in an increasing or decreasing region (as just described). If increasing, then the population will continue to increase until it approaches a point where there is no net rate of change. If decreasing, then the population will continue to decrease until it approaches a point where there is no net rate of change. The two possible equilibrium points are at $N^* = 0$ and at $N^* = K(1 - \omega/r)$. In the example, all values between $0 < N < K(1 - \omega/r)$ correspond to increases in the population, and all values that satisfy $N > K(1 - \omega/r)$ correspond to decreases in the population. It is evident that the equilibrium point $N^* = 0$ is unstable, as small increases in the population away from this point lead to even further increases. Similarly, it is evident that the equilibrium point $N^* = K(1 - \omega/r)$ is stable, as small decreases or increases in the population away from this point lead to subsequent increases or decreases back toward the equilibrium. One-dimensional nonlinear dynamical system can be classified in this way. However, this graphical approach is difficult to extend to higher-dimensional dynamical systems. Moreover, it does not specify the quantitative rate of change of the population away from or toward an equilibrium.

Tracking the dynamics of perturbations represents an alternative approach to characterizing the equilibria in a population model. As noted previously, there are two equilibria in this model. Consider first a small perturbation $N(t = 0) = n \ll K$, equivalent to adding a small inoculum of a bacterial population to an otherwise sterile flask. The dynamics of the population follow:

$$\frac{dn}{dt} = rn\left(1 - \frac{n}{K}\right) - \omega n \tag{B.5}$$

$$= rn - rn\frac{n}{K} - \omega n. \tag{B.6}$$

Because $n \ll K$, any term of the form n/K is much smaller than terms with only a density of n. The dynamics of the perturbation $n(t)$ in this limit can be approximated as

$$\frac{dn}{dt} \approx rn - \omega n = (r - \omega)n. \tag{B.7}$$

In other words, for very small perturbations, the population should grow exponentially so long as $r > \omega$ and decrease exponentially so long as $r < \omega$. The stability of the $N^* = 0$ equilibria is determined by the ratio r/ω,

$$\text{Unstable:} \qquad \frac{r}{\omega} > 1. \tag{B.8}$$

$$\text{Stable:} \qquad \frac{r}{\omega} < 1, \tag{B.9}$$

When $r = \omega$, this borderline case is termed *neutrally stable*, in that small perturbations are not expected either to grow or to decay over time.

Alternatively, consider the situation in which a system is perturbed away from the equilibrium $N^* = K(1 - \omega/r)$. What are the dynamics of n in the case when $n = N^* - N$? Replacement of $N(t) = N^* - n(t)$ yields

$$\frac{d(N^* - n)}{dt} = r\left(N^* - n\right)\left(1 - \frac{N^* - n}{K}\right) - \omega(N^* - n), \tag{B.10}$$

which can be rewritten as

$$-\frac{dn}{dt} = rN^*\left(1 - \frac{N^*}{K}\right) - \omega N^* + \frac{rnN^*}{K} - rn\left(1 - \frac{N^* - n}{K}\right) + \omega n. \tag{B.11}$$

However, the definition of the equilibrium N^* is that $rN^*\left(1 - \frac{N^*}{K}\right) - \omega N^* = 0$, so that

$$\frac{dn}{dt} = -\frac{rnN^*}{K} + rn\left(1 - \frac{N^* - n}{K}\right) - \omega n \tag{B.12}$$

$$= -\frac{rnN^*}{K} + rn\left(1 - \frac{N^*}{K}\right) - \omega n + \frac{rn^2}{K} \tag{B.13}$$

$$\approx (r - \omega)n - 2\frac{rnN^*}{K}, \tag{B.14}$$

where the final line is derived, as before, by retaining terms to order n. Substitution of the value of N^* yields

$$\frac{dn}{dt} = -(r - \omega)n. \tag{B.15}$$

In other words, a small perturbation will decrease whenever $r > \omega$ and increase when $r < \omega$. Because the perturbation is a reduction in population density below that of the steady state, biologically this can be interpreted as implying that the persistence equilibrium is stable whenever the density-independent growth rate r is greater than the density-independent death rate ω. Notice that the two conditions coincide. Moreover, the equilibrium $N^* = K\left(1 - \frac{\omega}{r}\right)$ exists only (i.e., is positive) when $r > \omega$. The overall conclusion from this analysis is that a population in which density-independent birth exceeds death has two fixed points. Of the two fixed points, one is unstable, corresponding to the absence of the population, and one is stable, corresponding to the persistence of the population.

The preceding analysis is somewhat cumbersome, as it depends on expanding the nonlinear dynamics and setting all higher-order terms to zero. Formally, this is equivalent to *linearizing* the nonlinear dynamics near the equilibria points—a method that scales up to more complex systems with more than one variable. For a system whose dynamics are $dx/dt = f(x)$, the behavior of the system near $x = x^*$ can be described in terms of some perturbation $x(t) = x^* + u(t)$. The dynamics near the fixed point $x = x^*$ can be approximated by a Taylor approximation of the function $f(x)$ in terms of the perturbation $u = x - x^*$ such that

$$\frac{dx^*}{dt} + \frac{du(t)}{dt} = f(x) \tag{B.16}$$

$$\frac{dx^{\cancel{*}}}{dt} + \frac{du}{dt} = \cancel{f(x^*)} + \frac{df}{dx}\bigg|_{x=x^*} u + \mathcal{O}\left(u^2\right) \tag{B.17}$$

$$\frac{du}{dt} \approx \frac{df}{dx}\bigg|_{x=x^*} u, \tag{B.18}$$

because $f(x^*) = 0$, and higher-order terms are ignored when u is small. The perturbation will increase or decrease depending on the change of the system dynamics with respect to changes in the population near the fixed point: $df/dx|_{x=x^*}$. When local increases in the population correspond to increases in the rate of change, the system is unstable and vice versa. This is the algebraic equivalent of the graphical approach illustrated earlier in this section. In the case of the microbial population growth model,

$$\frac{du}{dt} = (r - \omega)u - 2\frac{r u N^*}{K}, \tag{B.19}$$

exactly as in Eq. B.15 (albeit with the variable n standing in for the perturbation rather than u). Substituting in N^* yields

$$\frac{du}{dt} = -(r - \omega)u \tag{B.20}$$

for dynamics near $N = K\left(1 - \frac{\omega}{r}\right)$, and

$$\frac{du}{dt} = (r - \omega)u \tag{B.21}$$

for dynamics near $N = 0$. As before, the linearization method predicts that persistence is stable when $r > \omega$, and population extinction is stable when $r < \omega$. The linearization approach can be extended to analysis of multivariable systems.

B.2.2   *Two variables and beyond*

Recall the simplified model of virus-host dynamics introduced in Chapter 3 as Eq. 3.16:

$$\frac{\mathrm{d}N}{\mathrm{d}t} = rN\left(1 - N/K\right) - \phi NV - \omega N, \tag{B.22}$$

$$\frac{\mathrm{d}V}{\mathrm{d}t} = \tilde{\beta}\phi NV - \omega V, \tag{B.23}$$

where $\tilde{\beta} = \beta - 1$. There are three equilibria of the form (N^*, V^*): $(0,0)$, $\left(K(1 - \frac{\omega}{r}), 0\right)$, and $\left(\frac{\omega}{\tilde{\beta}\phi}, \frac{r}{\phi}\left(1 - \frac{\omega}{\tilde{\beta}\phi K}\right) - \frac{\omega}{\phi}\right)$. The local stability of each of these equilibria can be evaluated using a linearization approach, analogous to that used in the analysis of the one-dimentional microbial population growth model.

The key insight in conducting a linear stability analysis is to recall that the rate of change of populations can be expressed, near an equilibrium, by approximating the nonlinear response as a linear response in terms of two (or more) variables. Formally, consider the generic dynamical system

$$\frac{\mathrm{d}x}{\mathrm{d}t} = f(x, y), \tag{B.24}$$

$$\frac{\mathrm{d}y}{\mathrm{d}t} = g(x, y). \tag{B.25}$$

The population state (x^*, y^*) is an equilibrium if $f(x^*, y^*) = 0 = g(x^*, y^*)$. The local stability of an equilibrium can be evaluated by considering the perturbations u and v such that $x(t) = x^* + u(t)$ and $y(t) = y^* + v(t)$. The dynamics of these perturbations can be written as

$$\frac{\mathrm{d}u}{\mathrm{d}t} = f(x^*, y^*) + \left.\frac{\partial f}{\partial x}\right|_{x^*, y^*} u + \left.\frac{\partial f}{\partial y}\right|_{x^*, y^*} v + \mathcal{O}(u^2, uv, v^2), \tag{B.26}$$

$$\frac{\mathrm{d}v}{\mathrm{d}t} = g(x^*, y^*) + \left.\frac{\partial g}{\partial x}\right|_{x^*, y^*} u + \left.\frac{\partial g}{\partial y}\right|_{x^*, y^*} v + \mathcal{O}(u^2, uv, v^2). \tag{B.27}$$

Because the local stability analysis is near the equilibrium, these two equations can be approximated as

$$\frac{du}{dt} \approx \frac{\partial f}{\partial x}\bigg|_{x^*,y^*} u + \frac{\partial f}{\partial y}\bigg|_{x^*,y^*} v, \tag{B.28}$$

$$\frac{dv}{dt} \approx \frac{\partial g}{\partial x}\bigg|_{x^*,y^*} u + \frac{\partial g}{\partial y}\bigg|_{x^*,y^*} v. \tag{B.29}$$

The local dynamics of the perturbation (u, v) can be written as

$$\frac{d}{dt}\begin{bmatrix} u \\ v \end{bmatrix} = J \begin{bmatrix} u \\ v \end{bmatrix}, \tag{B.30}$$

where

$$J = \begin{bmatrix} \frac{\partial f}{\partial x} & \frac{\partial f}{\partial y} \\ \frac{\partial g}{\partial x} & \frac{\partial g}{\partial y} \end{bmatrix} \tag{B.31}$$

is termed the *Jacobian*. The Jacobian is evaluated at a specific point, corresponding to the equilibrium under investigation: (x^*, y^*).

The Jacobian encodes information on the rate of change of population dynamics with respect to changes in the populations. The Jacobian can be evaluated at a given equilibrium yielding, in this case, four numbers—each corresponding to a given element of a matrix. In this way, a nonlinear dynamical system in two dimensions can be linearized, yielding a linear matrix equation with coefficients that depend on the equilibria (which themselves depend on parameter values). The solution to a linear system of the form $dx/dt = Ax$ for a single variable is one of exponential growth or decay; that is, $x(t) = x_0 e^{\lambda t}$. The same holds for two- and higher-dimensional systems. For example, consider the candidate solution $(u(t), v(t)) = (u_0, v_0)e^{\lambda t}$, in which case $du/dt = \lambda u$ and $dv/dt = \lambda v$, such that Eq. B.30 can be rewritten as

$$\lambda \begin{bmatrix} u \\ v \end{bmatrix} = J \begin{bmatrix} u \\ v \end{bmatrix}. \tag{B.32}$$

For $u_0 \neq 0$ and $v_0 \neq 0$, the only feasible values of λ are those that satisfy:

$$\det(J - \lambda I) = 0, \tag{B.33}$$

where I is the 2×2 identity matrix, and det refers to the determinant of the matrix.

Consider the motivating example: that of virus-host dynamics. The Jacobian can be expressed in terms of the variables N and V as

$$J = \begin{bmatrix} r - 2rN/K - \phi V - \omega & -\phi N \\ \tilde{\beta}\phi V & \tilde{\beta}\phi N - \omega \end{bmatrix}. \tag{B.34}$$

For the $(0, 0)$ equilibrium, the Jacobian is

$$J = \begin{bmatrix} r - \omega & 0 \\ 0 & -\omega \end{bmatrix}, \tag{B.35}$$

such that the eigenvalues are $\lambda_1 = r - \omega$ and $\lambda_2 = -\omega$. Observe that $\lambda_1 > 0$ when $r > \omega$, such that the trivial steady state is unstable to perturbations so long as the density-independent growth rate of hosts exceeds the density-independent death rate of hosts. The fact that $\lambda_2 < 0$ can be interpreted to mean that if only viruses are added, then the viral population will decrease exponentially with a decay rate ω. The $(0, 0)$ equilibrium is a saddle point on the boundary of the phase space.

For the equilibrium in which hosts are present but viruses are absent (i.e., $(K(1 - \omega/r), 0)$), the Jacobian is

$$J = \begin{bmatrix} -(r - \omega) & -\phi K(1 - \omega/r) \\ 0 & \tilde{\beta}\phi K(1 - \omega/r) - \omega \end{bmatrix}. \tag{B.36}$$

Because this is an upper-triangular matrix, the eigenvalues are equivalent to the diagonal elements of J: $\lambda_1 = -(r - \omega)$ and $\lambda_2 = \tilde{\beta}\phi K(1 - \omega/r) - \omega$. As before, $\lambda_1 < 0$ when $r > \omega$, so that the host-only equilibrium is stable so long as as the density-independent growth rate of hosts exceeds the density-independent death rate of hosts. The sign of λ_2 depends on whether the coexistence steady state is feasible. When feasible, $\lambda_2 > 0$, suggesting that addition of viruses destabilizes the host-only equilibrium. When infeasible, $\lambda_2 < 0$, so that a perturbation of a small population of viruses will decay to extinction. Stability analysis of the coexistence equilibrium confirms this result.

In summary, local analysis of stability predicts that the washout equilibrium and the host-only equilibrium are unstable, while the coexistence equilibrium is stable, whenever it exists. The global dynamics of this particular system agree with this local stability analysis.

More generally, the stability of a fixed point can be classified based on the sign of the real components. If one of the eigenvalues has a positive real component, then the fixed point is unstable. If none of the eigenvalues has a nonnegative real component, then the fixed point is stable. The type of stability can be further classified based on whether the eigenvalues have a nonzero imaginary component. Complex eigenvalues suggest that dynamics will oscillate. For example, dynamics may oscillate away from or toward a fixed point, given instability or stability, respectively.

B.2.3   *Stability analysis of NIV dynamics*

The Jacobian of the nonlinear ODE of NIV dynamics introduced in Eq. 3.27 is

$$
J = \begin{bmatrix} r - 2rN/K - rI/K - \phi V - \omega & -rN/K & -\phi N \\ \phi V & -\eta - \omega & \phi N \\ -\phi V & \beta \eta & -\phi N - \omega \end{bmatrix}, \qquad \text{(B.37)}
$$

where the * are suppressed. This Jacobian needs to be evaluated at the fixed point, as specified in Eq. 3.34.

In the case of the virus-free equilibrium, this Jacobian can be written as

$$
J = \begin{bmatrix} -r - \omega & -r & -\phi \tilde{K} \\ 0 & -\eta - \omega & \phi \tilde{K} \\ 0 & \beta \eta & -\phi \tilde{K} - \omega \end{bmatrix}_{(N=K(1-\omega/r), I=0, V=0)}, \qquad \text{(B.38)}
$$

where $\tilde{K} = K(1 - \omega/r)$. The stability of the virus-free equilibrium is determined by the sign of the eigenvalues. One eigenvalue is $\lambda_1 = -r - \omega < 0$, implying that perturbations in uninfected hosts are stable so long as $r > \omega$, which is an essential presumption of the present analysis. The other two eigenvalues, $\lambda_{2,3}$, are solutions to the following equation:

$$
\text{Det} \begin{bmatrix} -\eta - \omega - \lambda & \phi \tilde{K} \\ \beta \eta & -\phi \tilde{K} - \omega - \lambda \end{bmatrix} = 0. \qquad \text{(B.39)}
$$

Notice that the trace of the Jacobian of this two-component submatrix is negative. The stability of the system is determined based on the sign of the determinant:

$$
(-\eta - \omega)(-\phi \tilde{K} - w) - \phi \tilde{K} \beta \eta. \qquad \text{(B.40)}
$$

If the determinant is positive, then the system is stable; if the determinant is negative, then the system is unstable. Given this criterion, the virus-free equilibrium is predicted to be unstable when

$$
\phi \tilde{K} \beta \eta > (\eta + \omega)(\phi \tilde{K} + \omega). \qquad \text{(B.41)}
$$

The text provides an intuitive alternative to this algebraic condition.

B.3 IMPLICIT RESOURCE DYNAMICS AS A LIMIT OF EXPLICIT RESOURCE DYNAMICS

The use of implicit resource dynamics in modeling cellular growth is common—whether without or with viruses, as introduced in Eq. 3.16. Yet, the particular form of logistic growth warrants examination as a limit of a mechanistic model of nutrient uptake and conversion into cell biomass. Recall the chemostat model in Eq. 3.1:

$$\frac{dR}{dt} = \omega J_0 - f(R)N - \omega R,$$
$$\frac{dN}{dt} = \epsilon f(R)N - \phi NV - \omega N, \tag{B.42}$$
$$\frac{dV}{dt} = \beta \phi NV - \phi NV - \omega V,$$

where $f(R)$ is the functional response. Consider a Type I functional response, where $f(R) = cR$. Assume that the resource dynamics in this system are much faster than that of either the viral or the host population. In other words, given a current host population N and viral population V, the resources should rapidly converge to

$$R^q(N, V) = \frac{\omega J_0}{\omega + cN}, \tag{B.43}$$

where the subscript q denotes that this equation represents a quasi-equilibrium of the system. The interpretation is that resource concentration declines with increasing host populations. With resource levels implicitly defined in terms of host abundance, it is possible to rewrite the population dynamics strictly in terms of N and V:

$$\frac{dN}{dt} = \frac{\omega \epsilon c J_0 N}{\omega + cN} - \phi NV - \omega N, \tag{B.44}$$

$$\frac{dV}{dt} = \beta \phi NV - \phi NV - \omega V. \tag{B.45}$$

This assumption holds strictly in the limit that resource dynamics are much faster than both population and virus dynamics; otherwise, they should be recognized as an approximation. In the limit that $N(t) \ll \omega/c$, the dynamics can be further approximated as

$$\frac{dN}{dt} = \overbrace{rN(1 - N/K)}^{\text{logistic growth}} - \phi NV - \omega N,$$
$$\frac{dV}{dt} = \beta \phi NV - \phi NV - \omega V, \tag{B.46}$$

where $K = \omega/c$ and $r = \epsilon c J_0$. The implicit form of the dynamics is presented as Eq. 3.16. A similar approach may be taken with a Type II response, such that $f(R) = \frac{\gamma R}{Q+R}$. In the limit that $Q \gg R$, then $f(R) = \gamma R/Q = cR$. In that case and in that limit, $K = \omega Q/\gamma$ and $= \epsilon \gamma J_0/Q$. It is important to keep in mind that the link between the mechanistic and phenomenological parameters holds strictly only in certain limits. An extended discussion of the challenges in interpreting model dynamics given implicit dynamics can be found in Mallet (2012).

B.4   ON POISSON PROCESSES AND MEAN FIELD MODELS

This monograph focuses on nonlinear ODEs to describe the interactions of biological populations and abiotic resources. These interactions have an underlying stochastic structure. By way of example, consider a decay process in which a virus particle loses its ability to infect host cells at a probability rate per unit time m. This is a first-order decay process, such that if there are V_0 viruses at time $t = 0$, then each decays independently. The probability that a given virus decays in some infinitesimal unit of time Δt is $m \Delta t$. The probability that a virus has not yet decayed after k such intervals is $(1 - m \Delta t)^k$. Therefore, the expected number of viruses after time $t = k \Delta t$ is

$$\bar{V}(t) = V_0 \lim_{\Delta t \to 0} (1 - m \Delta t)^k \tag{B.47}$$

$$= V_0 e^{-mt}. \tag{B.48}$$

The distribution of the time t at which a Poisson process takes place given an underlying rate m per unit time is exponentially distributed with rate constant m.

Another way of thinking about the same problem is to consider the change in the number of viruses in the population. The probability that there are exactly v viruses at some future time $t + \Delta t$, $P_v(t + \Delta t)$ depends on whether there are either v or $v + 1$ viruses at the present time, t, or formally,

$$P_v(t + \Delta t) = P_v(t) + \overbrace{P_{v+1}(t)(v+1)m \Delta t}^{\text{decay of one of } v+1 \text{ viruses}} - \overbrace{P_v(t) V v \Delta t}^{\text{decay of one of } v \text{ viruses}} , \tag{B.49}$$

which holds so long as $v \geq 1$. The change in probability arises because there is a slightly greater chance that the system with $v + 1$ viruses will transition into one with v viruses than that the system with v viruses will transition into one with $v - 1$ viruses. This equation can be converted into a dynamical system by taking a Taylor expansion:

$$P_v(t + \Delta t) \approx P_v(t) + \frac{dP_v(t)}{dt} \Delta t. \tag{B.50}$$

Then, the master equation for this process can be written as

$$\frac{dP_v}{dt} = m(v+1)P_{v+1} - mvP_v, \tag{B.51}$$

where time is now taken implicitly. The average number of viruses at any given point is a sum over all configurations:

$$\bar{V} = \sum_{v=0}^{V_0} vP_v, \tag{B.52}$$

given the condition $\sum_{v=0}^{V_0} P_v = 1$. The time rate of change of the average number of viruses is

$$\frac{d\bar{V}}{dt} = \sum_{v=0}^{V_0} v \frac{dP_v}{dt} \tag{B.53}$$

$$= \sum_{v=0}^{V_0} m \left(v(v+1)P_{v+1} - v^2 P_v \right) \tag{B.54}$$

$$= \sum_{v=0}^{V_0} m \left((v-1)v P_v - v^2 P_v \right) \tag{B.55}$$

$$= -m \sum_{v=0}^{V_0} v P_v, \tag{B.56}$$

such that

$$\frac{d\bar{V}}{dt} = -m\bar{V}. \tag{B.57}$$

The solution is $\bar{V}(t) = V_0 e^{-mt}$, just as before. In practice, the master-equation approach can be scaled up to more complex systems.

Of note, a recurring technique in this monograph is to consider the relative probability that one stochastic event will occur before another, for example, given a Poisson process with rate k_1 and another independent Poisson process with rate k_2. The probability that event 1 will occur before event 2 can be expressed as

$$P_{1<2} = \int_0^\infty \overbrace{k_1 e^{-k_1 t_1} \, dt_1}^{\text{event 1 occurs at } t_1} \quad \overbrace{e^{-k_2 t_1}}^{\text{event 2 has not yet occurred by } t_1} \quad , \tag{B.58}$$

which is equivalent to $k_1/(k_1 + k_2)$. The intuition behind this result is that the total rate at which independent events happen is the sum of their individual rates. Yet, the probability that a particular event will occur is equal to its relative rate. This logic extends to the case when there are more than two independent Poisson processes.

B.5   A NOTE ON SIMULATING DYNAMICAL SYSTEMS

The analysis of nonlinear dynamical systems in biology requires numerical simulation. This may be a relief to some and a disappointment to others. To keep the focus on the biology, rather than on the numerical methods, this book contains no explicit code. However, code was written! In fact, all the numerical simulations in this book were done using MATLAB. The two advantages of using the MATLAB toolbox for the analysis of dynamical systems are that (i) the numerical integration uses a "higher-order" scheme, and (ii) the toolbox is highly optimized. The same advantages could have been achieved by other software libraries.[1] The use of a higher-order scheme is the essential advantage. The term *higher* implies a relative value. What is in fact "higher" about the numerical integration scheme, and what are the "lower" alternatives? Consider a model of a microbial population growing to its carrying capacity:

$$\frac{dN}{dt} = rN \left(1 - \frac{N}{K} \right), \tag{B.59}$$

where N is the density, r is the density-independent growth rate, and K is the carrying capacity. Although an exact solution to this equation can be derived, such exact solutions are rarely available. Instead, for most systems in ecology, including those considered in this book, even the experienced modeler will turn to numerical simulation. The use of MATLAB and other packages for simulating dynamical systems is not part of the core training in the biosciences. Some may be tempted to simulate a dynamical system of the form $dx/dt = f(x)$ by performing the following steps as part of a script or in

---

[1] The open-source movement has developed many fine alternatives for numerical integration of dynamical systems including libraries in Octave, python, and so on. Developing one's own numerical integration scheme is rarely necessary and, frankly, should not be encouraged except for pedagogical or specialized purposes.

a spreadsheet:

1. Initialize $x(0) = x_0$ and choose a step size Δt.
2. Calculate the rate of change of x, that is, $f(x)$.
3. Increment x, as follows:

$$x(t + \Delta t) = x(t) + \Delta t \frac{dx}{dt}, \tag{B.60}$$

$$x(t + \Delta t) = x(t) + \Delta t f(x(t)). \tag{B.61}$$

4. Repeat steps 2–3 until a maximum time is reached.

This procedure is termed the *Euler* method of numerical integration after the famous eighteenth-century mathematician who discovered it, Leonhard Euler. The intuition behind the method is to approximate the population at the next time step by assuming the population changes at a constant rate over the period Δt. These steps seem logical; indeed, many students have the very pleasant experience of rediscovering them. Nonetheless, the approach is no longer recommended. Let's see why.

First, once a Δt is chosen, the first few values of the predicted growth curve can be calculated:

$$x(0) = x_0, \tag{B.62}$$

$$x(\Delta t) = x_0 + \Delta t f(x_0), \tag{B.63}$$

$$x(2\Delta t) = x(\Delta t) + \Delta t f(x(\Delta t)). \tag{B.64}$$

In the case of logistic growth, the first few values of this numerical "integration" of the dynamics via Euler's method are

$$N(0) = N_0, \tag{B.65}$$

$$N(\Delta t) = N_0 + r \Delta t N_0 (1 - N_0/K), \tag{B.66}$$

$$N(2\Delta t) = N_0 + r \Delta t N(\Delta t)(1 - N(\Delta t)/K) \tag{B.67}$$

$$= N_0 + r N_0(1 - N_0/K)(\Delta t) +$$

$$r (1 - N(\Delta t)/K). \tag{B.68}$$

How does this result compare with the exact solution? Thankfully, in the case of logistic growth, an exact solution is known. It can be derived by moving the terms with N to the left-hand side of Eq. B.59 and those with t to the

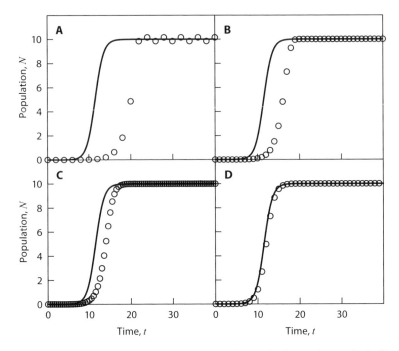

FIGURE B.2. Effect of step size on accuracy of the Euler integration method of the logistic equation given $r = 1$ and $K = 10^{10}$. The step size, Δt, is 2, 1, 0.5, and 0.1 for A–D respectively. (D) The results of the Euler integration method are subsampled 10:1 to facilitate comparison with theory.

right-hand side, normalizing densities in terms of N/K, and then integrating both sides. The solution is

$$\hat{N}(t) = \frac{KCe^{rt}}{1 + Ce^{rt}}, \tag{B.69}$$

where $C = x_0/(1 - x_0)$, where $x_0 = N_0/K$. Here, $\hat{N}(0) = N_0$ is the initial density, and $\hat{N}(t \to \infty) = K$ is the expected final density. In Figure B.2, the values of $N(t)$ calculated via the Euler numerical integration method are compared with exact solutions. There are a few take-away points here. First, when Δt is large, not only is the integration inaccurate, but it can lead to spurious qualitative dynamics (e.g., oscillations versus convergence to a steady state). Second, when Δt is reduced, the results of the Euler scheme appear to approach the true logistic growth curve. However, there is a more insidious issue with the Euler integration scheme, even when it appears to be accurate.

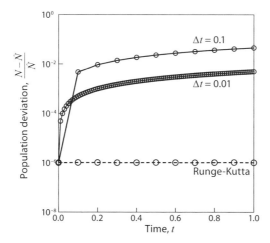

FIGURE B.3. Deviation between $N(t)$ and $\hat{N}(t)$. $N(t)$ is calculated via the Euler integration scheme with $\Delta t = 0.1$ (top) and $\Delta t = 0.01$ (middle), as well as via a higher-order Runge-Kutta scheme with relative tolerance of 10^{-8} (bottom). All time series of $N(t)$ were simulated using $r = 1$, $K = 10^{10}$, and an initial value of $10^5 \left(1 + 10^{-6}\right)$, that is, one in a million deviation from the true initial value.

Consider instead the *stability* of the Euler integration scheme. That is to say, how does the deviation between $N(t)$ as calculated via the Euler integration scheme change over time compared with the theoretical expectation? Figure B.3 compares the Euler scheme against the logistic curve, quantifying the deviation in terms of the fractional difference between the two curves for the cases $\Delta t = 0.01$ and $\Delta t = 0.1$. Although the curves are initially identical, the deviation is already apparent after one time step, because $N(\Delta t) \neq \hat{N}(\Delta t)$. Moreover, although the Euler scheme would appear to be accurate for each of these cases, it also unstable irrespective of the value of Δt. For comparison, Figure B.3 shows the deviation resulting from a modern method of numerical integration, the so-called Runge-Kutta higher-order scheme (formally, ode45 in MATLAB). The Runge-Kutta methods have errors, but their errors are proportional to the step size raised to a higher power. When the step size becomes sufficiently small, these errors become far smaller than in the Euler method.

Although the nature of the logistic equation leads to accurate solutions over long times, this is a special case. Generally speaking, the instability of the Euler integration scheme can lead to discordant time dependence of the state variables compared with the exact solutions. This instability makes it fundamentally unsuitable for numerical analysis of nonlinear population dynamics—the kind of systems that are characteristic of virus-host interactions and of many other systems in ecology. The take-home message is, do not use a first-order Euler scheme to numerically integrate a nonlinear dynamical system of the kind analyzed in this book. Instead, use a standard higher-order integration scheme, ideally from a vetted source.

B.6   ANALYSIS OF A POPULATION DYNAMICS MODEL WITH REINFECTION OF INFECTED HOSTS

Consider the model of susceptible hosts, infected hosts, and viruses:

$$
\frac{dN}{dt} = rN\left(1 - \frac{N+I}{K}\right) - \phi NV - \omega N,
$$

$$
\frac{dI}{dt} = \phi NV - \eta I - \omega I, \tag{B.70}
$$

$$
\frac{dV}{dt} = \beta \eta I - \phi NV - \overbrace{\rho \phi I V}^{\text{infection of infected cells}} - \omega V,
$$

as introduced in the text as Eq. 3.42. The steady-state densities are

$$
N^* = \frac{-b + \sqrt{b^2 - 4ac}}{2a}, \tag{B.71}
$$

where

$$
a = r\rho\phi^2 r\phi^2 \tilde{\beta}, \tag{B.72}
$$

$$
b = -r\rho\phi^2 K + r\phi\omega + \phi K \tilde{\beta}\phi(\eta + \omega) + \omega K \rho\phi^2, \tag{B.73}
$$

$$
c = -\omega\phi K(\eta + \omega), \tag{B.74}
$$

$\tilde{\beta} = \beta\eta/(\eta + \omega) - 1$, and

$$
V^* = \frac{\tilde{\beta}\phi N - \omega}{\rho\phi N \frac{\phi}{\eta+\omega}}, \tag{B.75}
$$

$$
I^* = \frac{\phi N^* V^*}{\eta + \omega}. \tag{B.76}
$$

Despite the opaqueness of this full solution, the results of the perturbation theory do coincide with the exact solutions for small ρ, and even intermediate ρ given sufficiently low ecological likelihood of multiple infections (see Figure B.4).

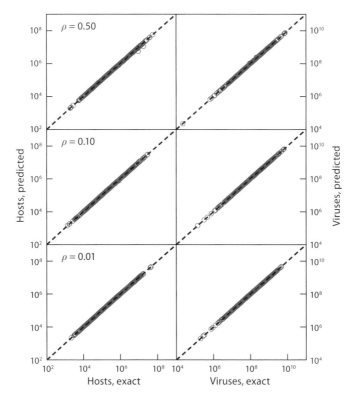

FIGURE B.4. Comparison of perturbation theory prediction of steady-state host (N) (left) and viral (V) (right) densities versus exact solutions. Dashed lines denote the 1:1 line. Open circles denote 1000 random samples based on the same parameter sets as in Figure 3.8 with $\eta = 1$. Perturbation theory is discussed in the text.

Evolutionary Dynamics of Viruses or Microbes, but Not Both

Consider two alternative models of the emergence of mutations in a population: (i) acquired and (ii) independent. The independent mode, proposed by Luria and Delbrück, leads to a number of interesting conceptual problems. As noted in Chapter 4, simulations of this mode depend on the extent to which cell divisions in the population are synchronized. The two limit cases are that of complete synchronization and complete desynchronization. In the former case, the population doubles in size from a single or small number of individuals via a series of $\approx \log_2 N_f$ divisions to a final size of N_f. For each cell division, one of the offspring is mutated with a probability μ. All daughters of mutant cells are themselves mutants. This model is equivalent to one in which the time between divisions is fixed at age a. In the latter case, there are $N_f - 1$ events to consider, and at each event, one of the individuals in the population is selected to divide. The offspring is mutated with a probability μ. This model is equivalent to one in which the time between divisions is exponentially distributed with average age of division a. The objective of simulations is to quantify the probability that there are $0 \le N' < N_f$ mutant cells in a population growing from a single susceptible cell to a total population of N_f cells. The stochastic process can be simulated and then averaged to calculate $p(N'|N_f)$. To accelerate the processes of simulating and averaging models, it is preferable to define a master equation to follow the change in $p(m|n)$, the probability that m of n cells are resistant, where $n \le N_f$. The benefit of this approach is that one can explicitly calculate the entire probability distribution. Optimization would be required to scale up the following method, which scales in complexity as $\mathcal{O}^2$.

In the case of synchronous cell division, if m of n cells after g generations are resistant, then there will be between $2m$ and $2m + (n - m)$ resistant cells in the $g + 1$ generation. The former occurs when no new mutants arise, and the latter when all the offspring of the $(n - m)$ susceptible cells are new mutants. This definition also helps make evident the principle that the contribution of early mutants to the total number of mutants will be exponentially larger. Accordingly, the probability that there will be m resistant cells in the $g + 1$ generation is therefore

$$p(m|g + 1) = \sum_{m'=0}^{[m/2]} p(m'|g)B(m - 2m'|2^g - m', \mu), \qquad (C.1)$$

where $[\ldots]$ represents the largest positive integer that is less than or equal to its argument, and $B(m, n, \mu)$ is the binomial distribution for the occurrecnce of m events out of n trials, where each event (mutation) occurs with probability μ. The derivation of this master equation is as follows. Consider the population after g generations. It has $n = 2^g$ cells. There can be anywhere from $m' = 0$ to $m' = 2^g - 1$ resistant cells. The sum evaluates transitions given all these possible states, $p(m'|g)$. If there are m' resistant cells, then there are $2^g - m'$ susceptible cells. The m' cells will divide, leading to $2m'$ resistant cells at generation $g + 1$. Therefore, to transition from m' resistant cells at generation g to m resistant cells at generation $g + 1$ requires that exactly $m - 2m'$ new mutants be generated during the division of the $2^g - m'$ susceptible cells. The probability of this occurrence is given by the earlier binomial distribution. Simulation of the master equation commences with the condition that $p(0|0) = 1$, where the first individual is susceptible, and finishes when $g = \log_2 N_f$.

In the case of asynchronous cell division, if m of n cells are resistant, then there will be either m or $m + 1$ resistant cells of the $n + 1$ cells in the population. The probability that m of $n + 1$ cells are resistant obeys the master equation

$$p(m|n + 1) = p(m|n) \left(\frac{n - m}{n}(1 - \mu) \right) + p(m - 1|n) \left(\frac{m - 1}{n} + \frac{n - m}{n} \mu \right).$$
$$(C.2)$$

The first term denotes the probability that a susceptible cell was selected to divide and did not mutate. The second term is the addition of two events: either a resistant cell divided, or a susceptible cell divided and mutated. Both events lead to an increase in the number of resistant cells. Simulation of the master equation commences with the condition $p(0|1) = 1$; that is, the first individual is susceptible, and finishes when $n = N_f$, the final size of the population.

C.2   INVASION CRITERION FOR MUTANT VIRUSES WITH DISTINCT LIFE HISTORY TRAITS

Consider the following model of a susceptible host population with density N that can be infected by viruses from two subpopulations, the resident V and the mutant V'. The full dynamics, as introduced in Eq. 4.31 are

$$\frac{dN}{dt} = rN(1 - N/K) - \phi NV - \phi' NV' - \omega N,$$

$$\frac{dV}{dt} = \beta \phi NV - \phi NV - \omega V, \qquad (C.3)$$

$$\frac{dV'}{dt} = \beta' \phi' NV' - \phi NV' - \omega V'.$$

The invasion dynamics of a rare mutant virus depend on the stability of the fixed point $(N^*, V^*, 0)$, where

$$N^* = \frac{\omega}{\tilde{\beta}\phi},$$
$$V^* = \frac{r(1 - N^*/K) - \omega}{\phi}, \qquad (C.4)$$

where $\tilde{\beta} = \beta - 1$. The Jacobian of Eq. 4.31 is

$$J = \begin{bmatrix} r - 2rN/K - \phi V - \phi' V' - \omega & -\phi N & -\phi' N \\ \tilde{\beta}\phi V & \tilde{\beta}\phi N - \omega & 0 \\ \tilde{\beta}'\phi' V' & 0 & \tilde{\beta}'\phi' N - \omega \end{bmatrix}. \qquad (C.5)$$

The Jacobian as evaluated at the fixed point is

$$J = \begin{bmatrix} r - 2rN^*/K - \phi V^* - \omega & -\phi N^* & -\phi' N^* \\ \tilde{\beta}\phi V^* & \tilde{\beta}\phi N^* - \omega & 0 \\ 0 & 0 & \tilde{\beta}'\phi' N^* - \omega \end{bmatrix} \qquad (C.6)$$

The eigenvalues, λ, are determined by solving the equation

$$det(J - \lambda I) = 0. \qquad (C.7)$$

Recall that the dynamics of dV'/dt can be written in the form $V' f(N, V)$. The corresponding structure of J implies that two of the eigenvalues will correspond to dynamics in the N–V boundary plane. It was shown in Chapter 2 that this resource-explicit model is stable whenever it exists. The third eigenvalue can be read from the bottom-right element and is $\lambda = \tilde{\beta}'\phi'N^* - \omega$. In words, the fixed point without mutant viruses will be unstable to virus invasion when $\lambda > 0$, or $\frac{\tilde{\beta}'\phi'}{\tilde{\beta}\phi}\omega > \omega$, or, compactly:

$$\tilde{\beta}'\phi' > \tilde{\beta}\phi. \tag{C.8}$$

This condition coincides with Eq. 4.29 in the text.

C.3   DERIVING THE CANONICAL EQUATION OF ADAPTIVE DYNAMICS

This appendix derives the canonical equation of adaptive dynamics following the notation and approach of Boettiger et al. (2010). The original derivation can be found in Dieckmann and Law (1996). Within the framework of adaptive dynamics the probability that individuals of the resident population have a particular trait value x at time t follows the master equation (van Kampen 2001)

$$\frac{d}{dt}P(x,t) = \int dx' \left[w(x|x')P(x',t) - w(x'|x)P(x,t) \right], \tag{C.9}$$

where $w(x'|x)$ is the probability per unit time of making the transition $x \to x'$. The transition probability rate, $w(x'|x) = \mathcal{M}(x',x)\mathcal{D}(x',x)$, is the product of the mutation generation rate, $\mathcal{M}(x',x)$, and the probability, $\mathcal{D}(x',x)$, that mutants survive to outcompete the resident type.

The dynamical equation for the change in the trait of the resident population, $d\hat{x}/dt$, can be derived using a master-equation formalism. Using the master equation to replace $d/dt\, P(x,t)$ in Eq. 4.42 and performing a change of variables we obtain

$$\frac{d\hat{x}(t)}{dt} = \int dx \int dx' \cdot [x' - x]w(x'|x)P(x,t). \tag{C.10}$$

A formal means of treating fluctuations in this model has recently been developed using the linear noise approximation (Boettiger et al. 2010). Ignoring fluctuations, it is possible to rewrite the differential equation for the mean trait as

$$\frac{d\hat{x}(t)}{dt} = \int dx' \cdot [x' - \hat{x}]w(x'|\hat{x}), \tag{C.11}$$

such that deriving the evolution of the mean phenotypic trait requires specification of the transition probability rate, $w(x'|\hat{x}) = \mathcal{M}(x',\hat{x})\mathcal{D}(x',\hat{x})$.

In adaptive dynamics, small mutations in phenotypes are characterized by a normal distribution, such that the rate at which a mutant with phenotype x' is generated by an individual with phenotype $\hat{x}$ is

$$M(x', \hat{x}) = \frac{\mu b(\hat{x}) N^*(\hat{x}) e^{-(x'-\hat{x})^2/(2\sigma_\mu^2)}}{\sqrt{2\pi\sigma_\mu^2}}, \tag{C.12}$$

where μ is the mutation probability per birth, $b(\hat{x})$ is the per capita birth rate, $N^*(\hat{x})$ is the equilibrium population density, and σ_μ is the standard deviation of the mutational distribution. The use of a normal distribution is valid only insofar as the system remains in the small-fluctuation regime; for exceptions and consequences see Boettiger et al. (2010). The ecology of the model enters into two terms: $b(\hat{x})$ and $N^*(\hat{x})$, which set the rate at which mutants are generated, whereas the phenotypic values of those mutants are set by the standard deviation of the mutational distribution, σ_μ. This takes care of $\mathcal{M}$, but what about $\mathcal{D}$—the likelihood that a small mutant population escapes stochastic extinction? Here, we apply the standard result from branching processes to derive the survival probability (Feller 1968):

$$\mathcal{D}(x', \hat{x}) = \begin{cases} 1 - \frac{d(x',\hat{x})}{b(x',\hat{x})} & \text{if } b > d \\ 0 & \text{otherwise} \end{cases}. \tag{C.13}$$

As introduced in the text, define the invasion fitness when rare, $s_{x'}(x)$, to be the difference between the birth- and death rates of a rare mutant with trait x' given an environment set by a resident with trait x. With this definition, and given positive survival probability, then

$$\mathcal{D}(x', \hat{x}) = 1 - \frac{d(x', \hat{x})}{b(x', \hat{x})}$$

$$= \frac{s_{x'}(\hat{x})}{b(x', \hat{x})}$$

$$\approx \frac{x' - \hat{x}}{b(\hat{x})} \frac{\partial s_{x'}(\hat{x})}{\partial x'}\bigg|_{x'=\hat{x}}, \tag{C.14}$$

where the approximation uses a Taylor expansion around the point $x' = \hat{x}$ for which $s_{\hat{x}}(\hat{x}) = 0$.

The transition probability per unit time is the product of these two factors:

$$w(x'|\hat{x}) = \mu N^*(\hat{x}) \left.\frac{\partial s_{x'}(\hat{x})}{\partial x'}\right|_{x'=\hat{x}} \frac{\left[x' - \hat{x}\right] e^{-(x'-\hat{x})^2/(2\sigma_\mu^2)}}{\sqrt{2\pi\sigma_\mu^2}}. \tag{C.15}$$

Incorporating this into Eq. C.11, we obtain

$$\frac{\mathrm{d}\hat{x}(t)}{\mathrm{d}t} = \mu N^*(\hat{x}) \left.\frac{\partial s_{x'}(\hat{x})}{\partial x'}\right|_{x'=\hat{x}} \int \mathrm{d}x' \frac{\left[x' - \hat{x}\right]^2 e^{-(x'-\hat{x})^2/(2\sigma_\mu^2)}}{\sqrt{2\pi\sigma_\mu^2}}. \tag{C.16}$$

This integral can be evaluated by recognizing that mutants invade with either greater or smaller trait values, but not both, except at evolutionarily singular points. Therefore, the integration is performed over only half its range. The integral is equal to half the the variance of the mutational distribution. Thus, integration yields

$$\frac{\mathrm{d}\hat{x}(t)}{\mathrm{d}t} = \frac{1}{2}\mu\sigma_\mu^2 N^*(\hat{x})\frac{\partial s_{x'}(\hat{x})}{\partial x'}|_{x'=\hat{x}}. \tag{C.17}$$

This equation describes the evolutionary dynamics of a monomorphic phenotypic trait in a population. It is known as the canonical equation of adaptive dynamics (Dieckmann and Law 1996) and describes dynamics in which fluctuations are small.

C.4   SIMULATING EVOLUTIONARY DYNAMICS

In many circumstances, the analysis of models of evolutionary dynamics must include numerical simulations. Many of the same challenges that applied to the simulation of ecological dynamics also apply to the simulation of evolutionary dynamics. For example, evolutionary dynamics are often nonlinear, requiring the use of robust stable numerical packages for integration. There are also distinct approaches to the simulation of evolutionary dynamics, two of which are introduced here in the context of virus-host interactions.

Recall the model of virus-host dynamics in a chemostat with implicit resources introduced in Chapter 3 as Eq. 3.16:

$$\frac{\mathrm{d}N}{\mathrm{d}t} = rN(1 - N/K) - \phi NV - \omega N, \tag{C.18}$$

$$\frac{\mathrm{d}V}{\mathrm{d}t} = \beta(\tau)\phi NV - \phi NV - \omega V. \tag{C.19}$$

In the text, the following trade-off relationship between burst size and latent period was used:

$$\beta = \rho\,(\tau - \tau_e)\,e^{-\omega\tau}, \tag{C.20}$$

where τ_e is the eclipse period, and ρ is the virion production rate per unit time after the eclipse period. The equilibrium of this model (derived in Chapter 3) is

$$N^* = \frac{\omega}{(\beta - 1)\phi}, \tag{C.21}$$

$$V^* = \frac{r}{\phi}\left(1 - \frac{\omega}{(\beta - 1)\phi K}\right). \tag{C.22}$$

Consider a virus with trait τ', whose dynamics are

$$\frac{dV'}{dt} = \beta(\tau')\phi N V' - \phi N V' - \omega V'. \tag{C.23}$$

To adopt the language of adaptive dynamics, let $x = \tau$ and $x' = \tau'$, such that the per capita growth rate of the mutant virus is

$$s_{x'}(x) = \big(\beta(x') - 1\big)\,\phi N^*(x) - \omega, \tag{C.24}$$

where $N^*(x)$ indicates that the mutant is experiencing an environment set by the resident type. The sign of the invasion fitness, $s_{x'}(x)$, determines the outcome. Substitution for $N^*(x)$ yields

$$s_{x'}(x) = \omega\left(\frac{\beta(x') - 1}{\beta(x) - 1}\right). \tag{C.25}$$

Therefore, $s_{x'}(x)$ is positive whenever $\beta(x') > \beta(x)$ or, alternatively, when $N^*(x') < N^*(x)$. The mutant type has the potential to invade if it draws down host resources to a lower level; that is, the more efficient virus will win. In more complicated models, the invasion fitness can be density-, frequency-, or time-dependent. It is a unifying concept for thinking about the survival of a rare type. How can one simulate evolutionary dynamics of the evolvable phenotype x—that is to say, the latent period, τ, and corresponding life history trait, β?

The canonical equation of adaptive dynamics describes the long-term change in phenotypes of a monomorphic population. In the current example, the selective derivative can be calculated explicitly:

$$\frac{\partial s_{x'}(\hat{x})}{\partial x'}\bigg|_{x'=\hat{x}} = \rho\omega e^{-\omega\tau}\left(\frac{1-\omega(\tau-\tau_e)}{\rho(\tau-\tau_e)e^{-\omega\tau}-1}\right). \tag{C.26}$$

In the text, Figure 4.10 shows a comparison between the canonical equation and stochastic simulations of the model. The canonical equation was simulated using ode45 in MATLAB. The stochastic simulations require a bit more explanation.

To simulate evolutionary dynamics in the "slow" limit, to which adaptive dynamics is intended to apply, requires the separation of ecological dynamics from evolutionary dynamics. An event-driven simulation can be utilized, as follows. First, an initial resident type is introduced into the system. In the present context, that denotes a virus with latent period τ. Given this resident type, the ecological equilibrium is calculated: $(N^*(\tau), V^*(\tau))$. The separation of scales implies that mutants emerge from reproduction events that occur at this equilibrium. The rate of viral mutant introduction is $\mu\beta\phi N^*V^*$. The time of the next mutant appearance is then sampled from an exponential distribution with this rate parameter. At this time, the latent period of the mutant is determined and set equal to the latent period of the resident plus a random deviate. For example, in this problem, a suitable choice is a normal distribution with a small standard deviation on the order of 1 minute difference in latent period. The per capita birth- and death rates of the mutant are then calculated. If the death rate, $d(x'|x)$, exceeds the birth rate, $b(x'|x)$, the mutant is removed, and the process repeats. If the birth rate exceeds the death rate, there is a chance that the mutant can invade. The probability of invasion is calculated based on the standard branching process theory, $p = 1 - d/b$. In this model context, the probability of invasion of a mutant virus with trait τ' given a resident virus with trait τ is

$$p = 1 - \frac{\phi N^*(\tau)+\omega}{\beta(\tau')\phi N^*(\tau)}. \tag{C.27}$$

If the mutant invades, then the "fast" ecological dynamics takes place (but is not simulated explicitly), such that a new ecological equilibrium is determined in which the mutant replaces and becomes the new resident. At this point the process repeats, albeit the rate of viral mutant introduction is set by the updated equilibrium densities of N^* and V^*. The result of this process is a series of events, (t, x^*), corresponding to the trait value of the resident at each event time. This stochastic trajectory can be compared directly with the solution of the canonical equation of adaptive dynamics.

Ocean Viruses: On Their Abundance, Diversity, and Target Hosts

Consider a three-channel color image acquired from an epifluorescence microscope. Given a suitably diluted sample, individual particles will appear mostly separated, for example, as bright green spots against a dark background. A standard pipeline for counting virus particles automatically includes the following steps:

- The color image is transformed into an intensity image $I(x, y)$, where the intensity at each pixel ranges from 0, the black background, presumably not part of a virus particle, to 255, the bright foreground, potentially part of a virus particle.
- The image is segmented, that is, divided into foreground and background. There are many approaches to segmentation, the simplest of which is to identify a suitable global threshold, that is, a critical intensity value, such that the foreground is defined as $F = I > I_c$. Each pixel at position (x, y) whose intensity $I(x, y)$ exceeds I_c is set to $F(x, y) = 1$; otherwise, it is set to $F(x, y) = 0$. Thresholds can be set in a user-assisted, semiautomated, or fully automated fashion. If there is not significant variation in the intensity of potential particles, then a unique I_c value can be determined by *Otsu's method* (Otsu 1979). Prepackaged code for evaluating all possible candidate thresholds $0 \leq I_c \leq 255$ is readily available in most standard mathematical libraries, including MATLAB, R, and SciPython.
- Connected clusters of foreground pixels are identified in the segmented image. These clusters are candidates to be labeled as virus-like particles. One approach is to cluster pixels that share at least one of their eight

neighbors. An alternative approach is to use a watershed or similar algorithm that can divide connected clusters, particularly in cases where the geometry of the cluster resembles a dumbbell, implying two overlapping particles. Watershed algorithms are standard components of most image-processing packages.

- The quantitative features of each distinct cluster are calculated, such as, the diameter equivalent and the eccentricity.
- Particles are then classified as virus-like particles if their features meet or exceed an explicit criterion. One such criterion would be that viruses must be less than or equal to 350 nm in diameter and possess a circular shape as measured by the eccentricity of the best-fitting ellipse to the cluster.
- The total estimated number of virus-like particles will vary based on the criteria used to label clusters.

Virus-Host Dynamics in a Complex Milieu

E.1 A MULTITROPHIC MODEL TO QUANTIFY THE EFFECTS OF MARINE VIRUSES ON MICROBIAL FOOD WEBS AND ECOSYSTEM PROCESSES

The contents of this technical appendix are quoted verbatim from Weitz et al. (2015). For completeness, the key model equations are presented here along with two sets of baseline parameters that lead to coexistence with oscillatory and fixed-point dynamics.

We propose the following systems of equations to represent the dynamic changes of biotic populations, including heterotrophs (H), cyanobacteria (C), eukaryotic autotrophs (E), zooplankton (Z), and viruses (V_i, where $i = H, C$, and E), along with organic and inorganic nutrients (x_{on} and x_{in}, respectively). The systems of equations (E1–E9) are nonlinear, coupled ODEs. Eq. E10 makes explicit the export from the system to higher trophic levels. We use units of particles/L for all populations and μmol/L for nutrient concentrations. Hence, conversion factors, q, denote the equivalent nitrogen content of cells. Definitions of parameters are in Table E.1. Parameters include those associated with interactions and with the nutrient content of each biotic population on a per-cell basis.

Heterotrophs:

$$\dot{H} = \overbrace{\frac{\mu_H H x_{on}}{x_{on} + K_{on}}}^{\text{H growth}} - \overbrace{\phi_{VH} H V_H}^{\text{viral lysis}} - \overbrace{\psi_{ZH} H Z}^{\text{grazing}} - \overbrace{m_{on,H} H}^{\text{organic loss}} - \overbrace{m_{in,H} H}^{\text{respiration}} \quad \text{(E.1)}$$

Cyanobacteria:

$$\dot{C} = \overbrace{\frac{\mu_C C x_{in}}{x_{in} + K_{in,C}}}^{\text{C growth}} - \overbrace{\phi_{VC} C V_C}^{\text{viral lysis}} - \overbrace{\psi_{ZC} C Z}^{\text{grazing}} - \overbrace{m_{on,C} C}^{\text{organic loss}} - \overbrace{m_{in,C} C}^{\text{respiration}} \quad \text{(E.2)}$$

Euk. autos:

$$\dot{E} = \overbrace{\frac{\mu_E E x_{in}}{x_{in} + K_{in,E}}}^{\text{E growth}} - \overbrace{\phi_{VE} E V_E}^{\text{viral lysis}} - \overbrace{\psi_{ZE} E Z}^{\text{grazing}} - \overbrace{m_{on,E} E}^{\text{organic loss}} - \overbrace{m_{in,E} E}^{\text{respiration}} \tag{E.3}$$

Zooplankton:

$$\dot{Z} = p_g \overbrace{\left(\frac{q_H}{q_Z} \psi_{ZH} H Z + \frac{q_C}{q_Z} \psi_{ZC} C Z + \frac{q_E}{q_Z} \psi_{ZE} E Z \right)}^{\text{grazing}} - \overbrace{m_Z Z}^{\text{respiration}} - \overbrace{m_{ZP} Z^2}^{\text{consumption}} \tag{E.4}$$

Viruses of H:

$$\dot{V}_H = \overbrace{\beta_H \phi_{VH} H V_H}^{\text{lysis}} - \overbrace{m_{VH} V_H}^{\text{decay}} \tag{E.5}$$

Viruses of C:

$$\dot{V}_C = \overbrace{\beta_C \phi_{VC} C V_C}^{\text{lysis}} - \overbrace{m_{VC} V_C}^{\text{decay}} \tag{E.6}$$

Viruses of E:

$$\dot{V}_E = \overbrace{\beta_E \phi_{VE} E V_E}^{\text{lysis}} - \overbrace{m_{VE} V_E}^{\text{decay}} \tag{E.7}$$

Organic N:

$$\dot{x}_{on} = - \overbrace{\frac{q_H}{\epsilon_H} \frac{\mu_H H x_{on}}{x_{on} + K_{on}}}^{\text{H growth}} + \overbrace{q_V m_{VH} V_H + q_V m_{VC} V_C + q_V m_{VE} V_E}^{\text{viral decay}}$$

$$+ \overbrace{(q_H - q_V \beta_H) \phi_{VH} H V_H}^{\text{H lysis by viruses}} + \overbrace{(q_C - q_V \beta_C) \phi_{VC} C V_C}^{\text{C lysis by viruses}} + \overbrace{(q_E - q_V \beta_E) \phi_{VE} E V_E}^{\text{E lysis by viruses}}$$

$$+ \overbrace{q_H m_{on,H} H}^{\text{loss of H}} + \overbrace{q_C m_{on,C} C}^{\text{loss of E}} + \overbrace{q_E m_{on,E} E}^{\text{loss of C}}$$

$$+ \overbrace{p_{on} q_H \psi_{ZH} H Z}^{\text{H grazing by Z}} + \overbrace{p_{on} q_C \psi_{ZC} C Z}^{\text{C grazing by Z}} + \overbrace{p_{on} q_E \psi_{ZE} E Z}^{\text{E grazing by Z}} \tag{E.8}$$

Inorganic N:

$$
\dot{x}_{in} = \overbrace{-\omega(x_{in} - x_{sub})}^{\text{import}} + \overbrace{\frac{q_H(1-\epsilon_H)}{\epsilon_H}\frac{\mu_H H x_{on}}{x_{on} + K_{on}}}^{\text{H growth}} - \overbrace{\frac{q_C \mu_C C x_{in}}{x_{in} + K_{in,C}}}^{\text{C growth}} - \overbrace{\frac{q_E \mu_E E x_{in}}{x_{in} + K_{in,E}}}^{\text{E growth}}
$$

$$
+ \overbrace{q_Z m_z Z}^{\text{respiration}} + \overbrace{q_H m_{in,H} H}^{\text{respiration of H}} + \overbrace{q_C m_{in,C} C}^{\text{respiration of E}} + \overbrace{q_E m_{in,E} E}^{\text{respiration of C}}
$$

$$
+ \overbrace{p_{in} q_H \psi_{ZH} H Z}^{\text{H grazing by Z}} + \overbrace{p_{in} q_C \psi_{ZC} C Z}^{\text{C grazing by Z}} + \overbrace{p_{in} q_E \psi_{ZE} E Z}^{\text{E grazing by Z}} \tag{E.9}
$$

Export :

$$
J_{out} = \overbrace{p_{ex} q_Z m_{ZP} Z^2}^{\text{consumption}} \tag{E.10}
$$

TABLE E.1. Parameters used in simulations of attracting and limit-cycle ecosystem dynamics. Reprinted from Weitz et al. (2015) with permission.

Event	Variable	Meaning	Units	Attracting	Limit cycle
H growth	μ_H	Max H growth rate	day^{-1}	0.51	0.61
	K_{on}	Half-saturation constant	μmol/L	0.94	0.68
	ϵ_H	Efficiency	N/A	0.20	0.19
C growth	μ_C	Max C growth rate	day^{-1}	1.97	0.68
	$K_{in,C}$	Half-saturation constant	μmol/L	0.053	0.059
E growth	μ_E	Max E growth rate	day^{-1}	7.5	5.4
	$K_{in,E}$	Half-saturation constant	μmol/L	5.4	5.8
Viral lysis	ϕ_{VH}	Lysis rate	L/(virus·day)	8.0×10^{-12}	4.3×10^{-11}
	ϕ_{VC}	Lysis rate	ibid.	9.9×10^{-11}	1.9×10^{-11}
	ϕ_{VE}	Lysis rate	ibid.	5.8×10^{-11}	6.7×10^{-11}
	β_H	Burst size	N/A	25	15
	β_C	Burst size	N/A	16	82
	β_E	Burst size	N/A	440	370
Viral decay	m_{VH}	Decay rate	day^{-1}	0.17	0.59
	m_{VC}	Decay rate	day^{-1}	0.63	0.76
	m_{VE}	Decay rate	day^{-1}	0.058	0.56
Zooplankton grazing	ψ_{ZH}	Grazing rate	L/(zoopl·day)	1.9×10^{-6}	2.2×10^{-5}
	ψ_{ZC}	Grazing rate	ibid.	2.1×10^{-5}	4.2×10^{-5}
	ψ_{ZE}	Grazing rate	ibid.	1.5×10^{-6}	2.8×10^{-5}
	p_g	Fraction for growth	N/A	0.4	0.4
	p_{on}	Fraction egested	N/A	0.3	0.3
	p_{in}	Fraction respired	N/A	0.3	0.3
Grazer respiration	m_Z	Basal respiration	day^{-1}	0.048	0.051
Consumption by higher predators	m_{ZP}	Mortality rate	L/(cells·day)	3.3×10^{-7}	2.8×10^{-5}
	p_{ex}	Fraction exported	N/A	0.49	0.37
Import	ω	Surface-deep mixing rate	1/day	0.016	0.077
	x_{sub}	Deep inorganic N conc.	μmol N/L	7.7	2.67
Nutrient levels	q_H	Nitrogen content of H	μmol N/cell	8.8×10^{-10}	1.4×10^{-9}
	q_C	Nitrogen content of C	ibid.	3.9×10^{-9}	3.0×10^{-9}
	q_E	Nitrogen content of E	ibid.	2.1×10^{-7}	6.7×10^{-8}
	q_Z	Nitrogen content of Z	ibid.	2.3×10^{-4}	3.9×10^{-4}
	q_V	Nitrogen content of V	ibid.	2.0×10^{-12}	2.2×10^{-12}
Cellular loss	$m_{in,H}$	H respiration	day^{-1}	0.0069	0.0067
	$m_{in,C}$	C respiration	day^{-1}	0.0050	0.016
	$m_{in,E}$	E respiration	day^{-1}	0.0043	0.0023
	$m_{on,H}$	H organic loss	day^{-1}	0.014	0.034
	$m_{on,C}$	C organic loss	day^{-1}	0.022	0.0063
	$m_{on,E}$	E organic loss	day^{-1}	0.011	0.022

Bibliography

Abedon, S. T., ed. (2008). *Bacteriophage Ecology: Population Growth, Evolution, and Impact of Bacterial Viruses*. Cambridge University Press, Cambridge.

———(2000). The murky origin of Snow White and her T-even dwarfs. *Genetics*, 155:481–486.

Abedon, S. T., and Culler, R. R. (2007). Bacteriophage evolution given spatial constraint. *Journal of Theoretical Biology*, 48:111–119.

Abedon, S. T., Herschler, T. D., and Stopar, D. (2001). Bacteriophage latent-period evolution as a response to resource availability. *Applied and Environmental Microbiology*, 67:4233–4241.

Abedon, S. T., Kuhl, S. J., Blasdel, B. G., and Kutter, E. M. (2011). Phage treatment of human infections. *Bacteriophage*, 1(2):66–85.

Abedon, S. T., and Yin, J. (2006). Bacteriophage plaques: Theory and analysis. In Kropinski, M.C.A., ed., *Bacteriophages: Methods and Protocols*, Humana Press, Totowa, NJ.

Abrams, P. A. (2000). The evolution of predator–prey interactions. *Annual Review of Ecology, Evolution, and Systematics*, 31:79–105.

Ackermann, H.-W. (2011). The first phage electron micrographs. *Bacteriophage*, 1:225–227.

Ackermann, H.-W. and Heldal, M. (2010). Basic electron microscopy of aquatic viruses. *Manual of Aquatic Viral Ecology*. Waco, American Society of Limnology and Oceanography, 182–192.

Adams, M. H. (1959). *Bacteriophages*. Interscience, New York.

Albert, R., and Barabasi, A. L. (2002). Statistical mechanics of complex networks. *Reviews of Modern Physics*, 74:47–97.

Allen, M. J., Martinez-Martinez, J., Schroeder, D. C., Somerfield, P. J., and Wilson, W. H. (2007). Use of microarrays to assess viral diversity: From genotype to phenotype. *Environmental Microbiology*, 9(4):971–982.

Allesina, S., Alonso, D., and Pascual, M. (2008). A general model for food web structure. *Science*, 320:658–661.

Almeida-Neto, M., Guimarães, P., Guimarães, P. R., Loyola, R. D., and Ulrich, W. (2008). A consistent metric for nestedness analysis in ecological systems: Reconciling concept and measurement. *Oikos*, 117:1227–1239.

Anderson, P. W. (1972). More is different. *Science*, 177(4047):393–396.

Angly, F., Rodriguez-Brito, B., Bangor, D., McNairnie, P., Breitbart, M., Salamon, P., Felts, B., Nulton, J., Mahaffy, J., and Rohwer, F. (2005). PHACCS, an online tool for estimating the structure and diversity of uncultured viral communities using metagenomic information. *BMC Bioinformatics*, 6:41.

Ankrah, N.Y.D., May, A. L., Middleton, J. L., Jones, D. R., Hadden, M. K., Gooding, J. R., LeCleir, G. R., Wilhelm, S. W., Campagna, S. R., and Buchan, A. (2014).

Phage infection of an environmentally relevant marine bacterium alters host metabolism and lysate composition. *ISME Journal*, 8:1089–1100.

Arkin, A., Ross, J., and McAdams, H. H. (1998). Stochastic kinetic analysis of developmental pathway bifurcation in phage λ–infected *Escherichia coli* cells. *Genetics*, 149:1633–1648.

Armstrong, R. A., and McGehee, R. (1980). Competitive exclusion. *American Naturalist*, 115:151–170.

Atmar, W., and Patterson, B. D. (1993). The measure of order and disorder in the distribution of species in fragmented habitat. *Oecologia*, 96:373–382.

Aurell, E., Brown, S., Johanson, J., and Sneppen, K. (2002). Stability puzzles in phage λ. *Physical Review E*, 65:051914.

Aurell, E., and Sneppen, K. (2002). Epigenetics as a first exit problem. *Physical Review Letters*, 88:048101.

Avrani, S., Schwartz, D. A., and Lindell, D. (2012). Virus–host swinging party in the oceans: Incorporating biological complexity into paradigms of antagonistic coexistence. *Mobile Genetic Elements*, 2:88–95.

Avrani, S., Wurtzel, O., Sharon, I., Sorek, R., and Lindell, D. (2011). Genomic island variability facilitates *Prochlorococcus*-virus coexistence. *Nature*, 474:604–608.

Azam, F., and Long, R. A. (2001). Oceanography: Sea snow microcosms. *Nature*, 414(6863):495–498.

Bamford, D. H., Ravantti, J. J., Rönnholm, G., Laurinavičius, S., Kukkaro, P., Dyall-Smith, M., Somerharju, P., Kalkkinen, N., and Bamford, J. K. (2005). Constituents of SH1, a novel lipid-containing virus infecting the halophilic euryarchaeon *Haloarcula hispanica*. *Journal of Virology*, 79:9097–9107.

Barber, M. J. (2007). Modularity and community detection in bipartite networks. *Physical Review E*, 76:066102.

Barenblatt, G. I. (2003). *Scaling*. Cambridge University Press, Cambridge.

Barr, J. J., Auro, R., Furlan, M., Whiteson, K. L., Erb, M. L., Pogliano, J., Stotland, A., Wolkowicz, R., Cutting, A. S., Doran, K. S., Salamon, P., Youle, M., and Rohwer, F. (2013). Bacteriophage adhering to mucus provide a non-host-derived immunity. *Proceedings of the National Academy of Sciences of the United States of America*, 110:10771–10776.

Barrangou, R., Fremaux, C., Deveau, H., Richards, M., Boyaval, P., Moineau, S., Romero, D. A., and Horvath, P. (2007). CRISPR provides acquired resistance against viruses in prokaryotes. *Science*, 315:1709–1712.

Bascompte, J., and Jordano, P. (2007). Plant-animal mutualistic networks: The architecture of biodiversity. *Annual Review of Ecology, Evolution, and Systematics*, 38:567–593.

———(2013). *Mutualistic Networks*. Princeton University Press, Princeton, NJ.

Bascompte, J., Jordano, P., Melián, C. J., and Olesen, J. M. (2003). The nested assembly of plant–animal mutualistic networks. *Proceedings of the National Academy of Sciences of the United States of America*, 100:9383–9387.

Baudoux, A.-C., Hendrix, R. W., Lander, G. C., Bailly, X., Podell, S., Paillard, C., Johnson, J. E., Potter, C. S., Carragher, B., and Azam, F. (2012). Genomic and functional analysis of Vibrio phage SIO-2 reveals novel insights into ecology and evolution of marine siphoviruses. *Environmental Microbiology*, 14:2071–2086.

Beckett, S. J., Boulton, C. A., and Williams, H.T.P. (2014). FALCON: Nestedness statistics for bipartite networks. *F1000 Research*, 3:185.

Beckett, S. J., and Williams, H. T. (2013). Coevolutionary diversification creates nested-modular sturcture in phage-bacteria interaction networks. *Interface Focus*, 3:20130033.

Bednarz, M., Halliday, J. A., Herman, C., and Golding, I. (2014). Revisiting bistability in the lysis/lysogeny circuit of bacteriophage lambda. *PLoS One*, 9:e100876.

Benner, I., Diner, R. E., Lefebvre, S. C., Li, D., Komada, T., Carpenter, E. J., and Stillman, J. H. (2013). *Emiliania huxleyi* increases calcification but not expression of calcification-related genes in long-term exposure to elevated temperature and pCO_2. *Philosophical Transactions of the Royal Society of London B: Biological Sciences*, 368(1627).

Beretta, E., and Kuang, Y. (1998). Modeling and analysis of a marine bacteriophage infection. *Mathematical Biosciences*, 149:57–76.

Beretta, E., Solimano, F., and Tang, Y. (2002). Analysis of a chemostat model for bacteria and virulent bacteriophage. *Discrete and Continuous Dynamical Systems*, 2:495–520.

Berg, H. C., and Purcell, E. M. (1977). Physics of chemoreception. *Biophysical Journal*, 20:193–219.

Bergh, O., Borhseim, K. Y., Bratbak, G., and Heldal, M. (1989). High abundance of viruses found in aquatic environments. *Nature*, 340:467–468.

Bersier, L., Dixon, P., and Sugihara, G. (1999). Scale-invariant or scale-dependent behavior of the link density property in food webs: A matter of sampling effort? *American Naturalist*, 153:676–682.

Bertani, G. (2004). Lysogeny at mid-twentieth century: P1, P2 and other experimental systems. *Journal of Bacteriology*, 186:595–600.

Bertilsson, S., Berglund, O., Karl, D. M., and Chisholm, S. W. (2003). Elemental composition of marine *Prochlorococcus* and *Synechococcus*: Implications for the ecological stoichiometry of the sea. *Limnology and Oceanography*, 48:1721–1731.

Bidle, K. D., and Vardi, A. (2011). A chemical arms race at sea mediates algal host-virus interactions. *Current Opinion in Microbiology*, 14:449–457.

Bielke, L., Higgins, S., Donoghue, A., Donoghue, D., and Hargis, B. M. (2007). Salmonella host range of bacteriophages that infect multiple genera. *Poultry Science*, 86:2536–2540.

Biller, S. J., Schubotz, F., Roggensack, S. E., Thompson, A., Summons, R. E., and Chisholm, S. W. (2014). Bacterial vesicles in marine ecosystems. *Science*, 343:183–186.

Boettiger, C., Dushoff, J., and Weitz, J. S. (2010). Fluctuation domains in adaptive evolution. *Theoretical Population Biology*, 77:6–13.

Bohannan, B. J., and Lenski, R. E. (1999). Effect of prey heterogeneity on the response of a model food chain to resource enrichment. *American Naturalist*, 153:73–82.

Bohannan, B.J.M., Kerr, B., Jessup, C. M., Hughes, J. B., and Sandvik, G. (2002). Trade-offs and coexistence in microbial microcosms. *Antonie van Leeuwenhoek*, 81:107–115.

Bohannan, B.J.M., and Lenski, R. E. (1997). Effect of resource enrichment on a chemostat community of bacteria and bacteriophage. *Ecology*, 78:2303–2315.

———(2000). Linking genetic change to community evolution: Insights from studies of bacteria and bacteriophage. *Ecology Letters*, 3:362–377.

Bolotin, A., Quinquis, B., Sorokin, A., and Ehrlich, S. D. (2005). Clustered regularly interspaced short palindrome repeats (CRISPRs) have spacers of extrachromosomal origin. *Microbiology*, 151:2551–2561.

Bolton, J. R., and Cotton, C. A. (2008). *The Ultraviolet Disinfection Handbook*. American Water Works Association, Denver, CO.

Bonachela, J. A., and Levin, S. A. (2014). Evolutionary comparison between viral lysis rate and latent period. *Journal of Theoretical Biology*, 345:32–42.

Breitbart, M. (2012). Marine viruses: Truth or dare. *Annual Review of Marine Science*, 4:425–448.

Breitbart, M., Salamon, P., Andresen, B., Mahaffy, J. M., Segall, A. M., Mead, D., Azam, F., and Rohwer, F. (2002). Genomic analysis of uncultured marine viral communities. *Proceedings of the National Academy of Sciences of the United States of America*, 99:14250–14255.

Brown, C. M., Lawrence, J. E., and Campbell, D. A. (2006). Are phytoplankton population density maxima predictable through analysis of host and viral genomic DNA content? *Journal of the Marine Biological Association of the United Kingdom*, 86:491–498.

Brown, J. H. (1995). *Macroecology*. University of Chicago Press, Chicago.

Brown, J. H., and West, G. B. (2000). *Scaling in Biology*. Oxford University Press, New York.

Brum, J. (2005). Concentration, production and turnover of viruses and dissolved DNA pools at Stn ALOHA, North Pacific Subtropical Gyre. *Aquatic Microbial Ecology*, 41:103–113.

Brum, J. R., Schenck, R. O., and Sullivan, M. B. (2013). Global morphological analysis of marine viruses shows minimal regional variation and dominance of non-tailed viruses. *ISME Journal*, 7:1738–1751.

Brussaard, C.P.D., Payet, J. P., Winter, C., and Weinbauer, M. G. (2010). Quantification of aquatic viruses by flow cytometry. In Wilhelm, S. W., and Weinbauer, M. G., eds., *Manual of Aquatic Viral Ecology*, 102–109. American Society of Limnology and Oceanography, Waco, TX.

Brussaard, C.P.D., Wilhelm, S. W., Thingstad, F., Weinbauer, M. G., Bratbak, G., Heldal, M., Kimmance, S. A., Middelboe, M., Nagasaki, K., Paul, J. H., Schroeder, D. C., Suttle, C. A., Vaqué, D., and Wommack, K. E. (2008). Global-scale processes with a nanoscale drive: The role of marine viruses. *ISME Journal*, 2:575–578.

Bruttin, A., Desiere, F., Lucchini, S., Foley, S., and Brüssow, H. (1997). Characterization of the lysogeny DNA module from the temperate *Streptococcus thermophilus* bacteriophage phi Sfi21. *Virology*, 233:136–148.

Buckling, A., and Brockhurst, M. (2012). Bacteria–virus coevolution. *Advances in Experimental Medicine and Biology*, 751:347–370.

Buckling, A., and Rainey, P. B. (2002a). Antagonistic coevolution between a bacterium and a bacteriophage. *Proceedings of the Royal Society of London Series B–Biologial Sciences*, 269:931–936.

———(2002b). The role of parasites in sympatric and allopatric host diversification. *Nature*, 420:496–499.

Bull, J. J., Millstein, J., Orcutt, J., and Wichman, H. A. (2006). Evolutionary feedback mediated through population density, illustrated with viruses in chemostats. *American Naturalist*, 167:E39–E51.

Bull, J. J., and Molineux, I. J. (2007). Predicting evolution from genomics: Experimental evolution of bacteriophage T7. *Heredity*, 100:453–463.

Burch, C. L., and Chao, L. (2000). Evolvability of an RNA virus is determined by its mutational neighborhood. *Nature*, 406:625–628.

Burnet, F. M. (1930). Bacteriophage activity and the antigenic structure of bacteria. *Journal of Pathology and Bacteriology*, 33:647–664.

Cairns, J., Stent, G. S., and Watson, J. D., eds. (2007). *Phage and the Origins of Molecular Biology*. Cold Spring Harbor Laboratory Press, Cold Spring Harbor, NY.

Calendar, R. L., ed. (2005). *The Bacteriophages*. Oxford University Press, New York.

Campbell, A. (1961). Conditions for the existence of bacteriophage. *Evolution*, 15:153–165.

———(1976). Significance of constitutive integrase synthesis. *Proceedings of the National Academy of Sciences of the United States of America*, 73:887–890.

Carpenter, S. R., Kitchell, J. F., and Hodgson, J. R. (1985). Cascading trophic interactions and lake productivity. *BioScience*, 35:634–639.

Case, T. J. (2000). *An Illustrated Guide to Theoretical Ecology*. Oxford University Press, Oxford.

Ceballos, R. M., Marceau, C. D., Marceau, J. O., Morris, S., Clore, A. J., and Stedman, K. M. (2012). Differential virus host-ranges of the Fuselloviridae of hyperthermophilic Archaea: Implications for evolution in extreme environments. *Frontiers in Microbiology*, 3(295).

Chang, C. Y., Nam, K., and Young, R. (1995). S gene expression and the timing of lysis by bacteriophage lambda. *Journal of Bacteriology*, 177:3283–3294.

Chao, A. (1984). Non-parametric estimation of the number of classes in a population. *Scandinavian Journal of Statistics*, 11:265–270.

Chao, A., Gotelli, N. J., Hsieh, T. C., Sander, E. L., Ma, K. H., Colwell, R. K., and Ellison, A. M. (2014). Rarefaction and extrapolation with Hill numbers: A framework for sampling and estimation in species diversity studies. *Ecological Monographs*, 84(1):45–67.

Chao, L., Levin, B. R., and Stewart, F. M. (1977). A complex community in a simple habitat: An experimental study with bacteria and phage. *Ecology*, 58:369–378.

Charnov, E. L. (1993). *Life History Invariants*. Oxford University Press, Oxford.

Childs, L. M., England, W. E., Young, M. J., Weitz, J. S., and Whitaker, R. J. (2014). CRISPR-induced distributed immunity in microbial populations. *PLoS One*, 9:e101710.

Childs, L. M., Held, N. L., Young, M. J., Whitaker, R. J., and Weitz, J. S. (2012). Multiscale model of CRISPR-induced coevolutionary dynamics: Diversification at the interface of Lamarck and Darwin. *Evolution*, 66(7):2015–2029.

Chisholm, S. W. (1992). Phytoplankton size. In Falkowski, P. G., and Woodhead, A. D., eds., *Primary Productivity and Biogeochemical Cycles in the Sea*, 213–237. Plenum Press, New York.

Clokie, M.R.J., and Kropinski, A. M., eds. (2009a). *Bacteriophages: Methods and Protocols*. Volume 1: *Isolation, Characterization, and Interactions*. Humana Press, Totowa, NJ.

———(2009b). *Bacteriophages: Methods and Protocols*, Volume 2: *Molecular and Applied Aspects*. Humana Press.

Clokie, M.R.J., and Mann, N. H. (2006). Marine cyanophages and light. *Environmental Microbiology*, 8:2074–2082.

Cohen, J. E. (1978). *Food Webs and Niche Space*. Princeton University Press, Princeton, NJ.

Cohen, J. E., Briand, F., and Newman, C. M. (1990). *Community Food Webs: Data and Theory*. Springer-Verlag, Berlin.

Cohen, S. S. (1951). The synthesis of nucleic acid by virus-infected bacteria. *Bacteriological Reviews*, 15:131–146.

Cole, L. C. (1954). The population consequences of life history phenomena. *Quarterly Review of Biology*, 29:103–137.

Colwell, R., Mao, C., and Chang, J. (2004). Interpolating, extrapolating, and comparing incidence-based species accumulation curves. *Ecology*, 85:2717–2727.

Colwell, R. R., and Huq, A. (1994). In Wachsmuth, I. K., Blake, P. A., and Olsvik, O., eds., Vibrio cholerae *and Cholera: Molecular to Global Perspectives*, 117–133. American Society of Microbiology, Washington, DC.

Connell, J. H. (1970). On the role of natural enemies in preventing competitive exclusion in some marine animals and in rain forest trees. In den Boer, P. J., and Gradwell, G. R., eds., *Dynamics of Populations: Proceedings of the Advanced Study Institute on Dynamics of Numbers in Populations, Oosterbeek, the Netherlands*. Pudoc.

Contois, D. E. (1959). Kinetics of bacterial growth: Relationship between population density and specific growth rate of continuous cultures. *Journal of General Microbiology*, 21:40–50.

Cortez, M. H., and Ellner, S. P. (2010). Understanding rapid evolution in predator-prey interactions using the theory of fast-slow dynamical systems. *American Naturalist*, 176:E109–27.

Cortez, M. H., and Weitz, J. S. (2014). Coevolution can reverse predator-prey cycles. *Proceedings of the National Academy of Sciences*, 111: 7486–7491

Court, D. L., Oppenheim, A. B., and Adhya, S. L. (2007). A new look at bacteriophage λ genetic networks. *Journal of Bacteriology*, 189:298–304.

Culley, A. I., and Welschmeyer, N. A. (2002). The abundance, distribution, and correlation of viruses, phytoplankton, and prokaryotes along a Pacific Ocean transect. *Limnology and Oceanography*, 47:1508–1513.

Daegelen, P., Studier, F. W., Lenski, R. E., Cure, S., and Kim, J. F. (2009). Tracing ancestors and relatives of *Escherichia coli* B, and the derivation of B strains REL606 and BL21(DE3). *Journal of Molecular Biology*, 394:634–643.

Danovaro, R., Corinaldesi, C., Dell'anno, A., Fuhrman, J. A., Middelburg, J. J., Noble, R. T., and Suttle, C. A. (2011). Marine viruses and global climate change. *FEMS Microbiology Reviews*, 35:993–1034.

de Kruif, P. (2002). *Microbe Hunters*. Harcourt, San Diego, CA.

De Paepe, M., and Taddei, F. (2006). Viruses' life history: Towards a mechanistic basis of a trade-off between survival and reproduction among phages. *PLoS Biology*, 4:e193.

Delbrück, M. (1946). Bacterial viruses or bacteriophages. *Biological Reviews*, 21:30–40.

Deltcheva, E., Chylinski, K., Sharma, C. M., Gonzales, K., Chao, Y., Pirzada, Z. A., Eckert, M. R., Vogel, J., and Charpentier, E. (2011). CRISPR RNA maturation by trans-encoded small RNA and host factor RNase III. *Nature*, 471(7340):602–607.

Deng, L., Gregory, A., Yilmaz, S., Poulos, B. T., Hugenholtz, P., and Sullivan, M. B. (2012). Contrasting life strategies of viruses that infect photo– and heterotrophic bacteria, as revealed by viral tagging. *mBio*, 3:e00373–12.

Deng, L., Ignacio-Espinoza, J. C., Gregory, A., Poulos, B. T., Weitz, J. S., Hugenholtz, P., and Sullivan, M. B. (2014). Viral tagging reveals discrete populations in *Synechococcus* viral genome sequence Space. *Nature*, 513:242–245.

Deng, X., Phillippy, A., Li, Z., Salzberg, S., and Zhang, W. (2010). Probing the pan-genome of *Listeria monocytogenes*: New insights into intraspecific niche expansion and genomic diversification. *BMC Genomics*, 11:500.

Dennehy, J. J. (2012). What can phages tell us about host-pathogen coevolution? *International Journal of Evolutionary Biology*, 2012:396165.

Denno, R. F., and Lewis, D. (2009). Predator-prey interactions. In Levin, S. A., ed., *The Princeton Guide to Ecology*. Princeton University Press, Princeton, NJ.

Dercole, F., and Rinaldi, S. (2008). *Analysis of Evolutionary Processes: The Adaptive Dynamics Approach and Its Applications*. Princeton University Press, Princeton, NJ.

d'Herelle, F. (1917). Sur un microbe invisible antagoniste des bacilles dysentèriques. *Comptes rendus de l'Académie des Sciences*, 165.

Dieckmann, U., and Doebeli, M. (1999). On the origin of species by sympatric speciation. *Nature*, 400:354–357.

Dieckmann, U., and Law, R. (1996). The dynamical theory of coevolution: A derivation from stochastic ecological processes. *Journal of Mathematical Biology*, 34:579–612.

Dimmock, N. J., Easton, A. J., and Leppard, K. N. (2007). *Introduction to Modern Virology*, 5th ed. Blackwell, Malden, MA.

Dodd, I. B., Perkins, A. J., Tsemitsidis, D., and Egan, J. B. (2001). Octamerization of CI repressor is needed for effective repression of PRM and efficient switching from lysogeny. *Genes & Development*, 15:3013–3022.

Doebeli, M. (2011). *Adaptive Diversification*. Princeton University Press, Princeton, NJ.

Doebeli, M., and Dieckmann, U. (2000). Evolutionary branching and sympatric speciation caused by different types of ecological interactions. *American Naturalist*, 156:S77–101.

Dormann, C. F., Gruber, B., and Fruend, J. (2008). Introducing the bipartite package: Analysing ecological networks. *R News*, 8:8–11.

Duffy, M. A., and Sivars-Becker, L. (2007). Rapid evolution and ecological host-parasite dynamics. *Ecology Letters*, 10:44–53.

Duffy, S., Shackelton, L. A., and Holmes, E. C. (2008). Rates of evolutionary change in viruses: Patterns and determinants. *Nature Reviews Genetics*, 9:267–76.

Dulbecco, R. (1949). Reactivation of ultraviolet-inactivated bacteriophage by visible light. *Nature*, 163:949–950.

Dulbecco, R. (1950). Experiments on photoreactivation of bacteriophages inactivated with ultraviolet radiation. *Journal of Bacteriology*, 59:329–347.

Dutilh, B. E., Cassman, N., McNair, K., Sanchez, S. E., Silva, G. G. Z., Boling, L., Barr, J. J., Speth, D. R., Seguritan, V., Aziz, R. K., Felts, B., Dinsdale, E. A., Mokili, J. L., and Edwards, R. A. (2014). A highly abundant bacteriophage discovered in the unknown sequences of human faecal metagenomes. *Nature Communications*, 5:4498.

Echols, H. (1975). Constitutive integrative recombination by bacteriophage lambda. *Virology*, 64:557–559.

Edwards, R. A., and Rohwer, F. (2005). Viral metagenomics. *Nature Reviews Microbiology*, 3:504–510.

Eigen, M. (1971). Selforganization of matter and the evolution of biological macromolecules. *Naturwissenschaften*, 58:465–523.

Eigen, M., McCaskill, J., and Schuster, P. (1989). The molecular quasi-species. *Advances in Chemical Physics*, 75:149–263.

Einstein, A. (1956). *Investigations on the Theory of the Brownian Movement*. Dover Press. Mineola, NY.

Ellis, E. L., and Delbrück, M. (1939). The growth of bacteriophage. *Journal of General Physiology*, 22:365–384.

Elton, C. S., and Nicholson, M. (1942). Fluctuations in numbers of the muskrat (*Ondatra zibethica*) in Canada. *Journal of Animal Ecology*, 11:96–126.

Erdős, P., and Rényi, A. (1960). On the evolution of random graphs. *Publications of the Mathematical Institute of the Hungarian Academy of Sciences*, 5:17–61.

Fagerbakke, K. M., Heldal, M., and Norland, S. (1996). Content of carbon, nitrogen, oxygen, sulfur and phosphorus in native aquatic and cultured bacteria. *Aquatic Microbial Ecology*, 10(1):15–27.

Faruque, S. M., Islam, M. J., Ahmad, Q. S., Faruque, A. S., Sack, D. A., Nair, G. B., and Mekalanos, J. J. (2005a). Self-limiting nature of seasonal cholera epidemics: Role of host-mediated amplification of phage. *Proceedings of the National Academy of Sciences of the United States of America*, 102:6119–6124.

Faruque, S. M., Naser, I. B., Islam, M. J., Faruque, A. S. G., Ghosh, A. N., Nair, G. B., Sack, D. A., and Mekalanos, J. J. (2005b). Seasonal epidemics of cholera inversely correlate with the prevalence of environmental cholera phages. *Proceedings of the National Academy of Sciences of the United States of America*, 102:1702–1707.

Fauquet, C. M., Mayo, M. A., Maniloff, J., Desselberger, U., and Ball, L. A., eds. (2005). *Virus Taxonomy: Classification and Nomenclature of Viruses*. Elsevier, Amsterdam.

Feller, W. (1968). *An Introduction to Probability Theory and Its Applications*, Vol. 1. Wiley, New York.

Fischer, M. G., Allen, M. J., Wilson, W. H., and Suttle, C. A. (2010). Giant virus with a remarkable complement of genes infects marine zooplankton. *Proceedings of the National Academy of Sciences of the United States of America*, 107:19508–19513.

Fischetti, V. A., Nelson, D., and Schuch, R. (2006). Reinventing phage therapy: Are the parts greater than the sum? *Nature Biotechnology*, 24:1508–1511.

Flores, C. O., Meyer, J. R., Valverde, S., Farr, L., and Weitz, J. S. (2011). Statistical structure of host-phage interactions. *Proceedings of the National Academy of Sciences of the United States of America*, 108:E288–E297.

Flores, C. O., Poisot, T., Valverde, S., and Weitz, J. S. (2014). BiMAT: A MATLAB ®package to facilitate the analysis and visualization of bipartite networks. *arXiv preprint arXiv:1406.6732.*

Flores, C. O., Valverde, S., and Weitz, J. S. (2013). Multi-scale structure and geographic drivers of cross-infection within marine bacteria and phages. *ISME Journal*, 7:520–532.

Follows, M. J., and Dutkiewicz, S. (2011). Modeling diverse communities of marine microbes. *Annual Review of Marine Science*, 3:427–451.

Follows, M. J., Dutkiewicz, S., Grant, S., and Chisholm, S. W. (2007). Emergent biogeography of microbial communities in a model ocean. *Science*, 315:1843–6.

Forde, S. E., Beardmore, R. E., Gudelj, I., Arkin, S. S., Thompson, J. N., and Hurst, L. D. (2008). Understanding the limits to generalizability of experimental evolutionary models. *Nature*, 455:220–223.

Forde, S. E., Thompson, J. N., and Bohannan, B.J.M. (2004). Adaptation changes in space and time in a coevolving host-parasitoid interaction. *Nature*, 431:841–844.

Fort, J., and Méndez, V. (2002). Time-delayed spread of viruses in growing plaques. *Physical Review Letters*, 89:178101.

Forterre, P. (2013). The virocell concept and environmental microbiology. *ISME Journal*, 7(2):233–236.

Forterre, P., Soler, N., Krupovic, M., Marguet, E., and Ackermann, H.-W. (2013). Fake virus particles generated by fluorescence microscopy. *Trends in Microbiology*, 21:1–5.

Fortuna, M. A., Stouffer, D. B., Olesen, J. M., Jordano, P., Mouillot, D., Krasnov, B. R., Poulin, R., and Bascompte, J. (2010). Nestedness versus modularity in ecological networks: Two sides of the same coin? *Journal of Animal Ecology*, 79:811–817.

Fortunato, S., and Barthélemy, M. (2007). Resolution limit in community detection. *Proceedings of the National Academy of Sciences of the United States of America*, 104:36–41.

Franks, P. S. (2002). NPZ models of plankton dynamics: Their construction, coupling to physics, and application. *Journal of Oceanography*, 58:379–387.

Fraser, C., Alm, E. J., Polz, M. F., Spratt, B. G., and Hanage, W. P. (2009). The bacterial species challenge: Making sense of genetic and ecological diversity. *Science*, 323:741–746.

Fuhrman, J. A. (1999). Marine viruses and their biogeochemical and ecological effects. *Nature*, 399:541–8.

Fujieki, L. A. (2014). Hawaii ocean time series data organization and graphical system (HOT-DOGS). http://hahana.soest.hawaii.edu/hot/hot-dogs/index.html.

Fulton, J. M., Fredricks, H. F., Bidle, K. D., Vardi, A., Kendrick, B. J., DiTullio, G. R., and Van Mooy, B.A.S. (2014). Novel molecular determinants of viral susceptibility and resistance in the lipidome of *Emiliania huxleyi*. *Environmental Microbiology*, 16:1137–1149.

Gale, W. A., and Sampson, G. (1995). Good-Turing smoothing without tears. *Journal of Quantitative Linguistics*, 2:217–237.

Gallet, R., Kannoly, S., and Wang, I.-N. (2011). Effects of bacteriophage traits on plaque formation. *BMC Microbiology*, 11:181.

Gandon, S., and Vale, P. F. (2014). The evolution of resistance against good and bad infections. *Journal of Evolutionary Biology*, 27:303–312.

Geritz, S. A., Gyllenberg, M., Jacobs, F. J., and Parvinen, K. (2002). Invasion dynamics and attractor inheritance. *Journal of Mathematical Biology*, 44:548–560.

Geritz, S.A.H., and Gyllenberg, M. (2005). Seven answers from adaptive dynamics. *Journal of Evolutionary Biology*, 18:1174–1177.

Geritz, S. A. H., Kisdi, E., Meszéna, G., and Metz, J. A. J. (1998). Evolutionarily singular strategies and the adaptive growth and branching of the evolutionary tree. *Evolutionary Ecology*, 12:35–57.

Geritz, S.A.H., Metz, J. A. J., Kisdi, E., and Meszéna, G. (1997). The dynamics of adaptation and evolutionary branching. *Physical Review Letters*, 78:2024–2027.

Ghai, R., Mizuno, C. M., Picazo, A., Camacho, A., and Rodriguez-Valera, F. (2013). Metagenomics uncovers a new group of low GC and ultra-small marine Actinobacteria. *Scientific Reports*, 3:2471.

Gobler, C. J., Hutchins, D. A., Fisher, N. S., Cosper, E. M., and Sañudo Wilhelmy, S. (1997). Release and bioavailability of C, N, P, Se, and Fe following viral lysis of a marine chrysophyte. *Limnology and Oceanography*, 42:1492–1504.

Golding, I. (2011). Decision making in living cells: Lessons from a simple system. *Annual Review of Biophysics*, 40:63–80.

Gómez, P., and Buckling, A. (2011). Bacteria-phage antagonistic coevolution in soil. *Science*, 332:106–109.

Gonzalez, J. M., and Suttle, C. A. (1993). Grazing by marine nanoflagellates on viruses and virus-sized particles: Ingestion and digestion. *Marine Ecology Progress Series*, 94:1–10.

Good, I. J. (1953). The population frequences of species and the estimation of population parameters. *Biometrika*, 40:237–264.

Goodsell, D. S. (2010). *The Machinery of Life*. Copernicus Books, New York.

Grayson, P., Han, L., Winther, T., and Phillips, R. (2007). Real-time observations of single bacteriophage DNA ejections in vitro. *Proceedings of the National Academy of Sciences of the United States of America*, 104:14652–14657.

Gudelj, I., Weitz, J. S., Ferenci, T., Claire Horner-Devine, M., Marx, C. J., Meyer, J. R., and Forde, S. E. (2010). An integrative approach to understanding microbial diversity: From intracellular mechanisms to community structure. *Ecology Letters*, 13:1073–1084.

Gutenkunst, R. N., Waterfall, J. J., Casey, F. P., Brown, K. S., Myers, C. R., and Sethna, J. P. (2007). Universally sloppy parameter sensitivities in systems biology models. *PLoS Computational Biology*, 3:e189.

Hadas, H., Einav, M., Fishov, I., and Zaritsky, A. (1997). Bacteriophage T4 development depends on the physiology of its host *Escherichia coli*. *Microbiology*, 143:179–85.

Haegeman, B., Hamelin, J., Moriarty, J., Neal, P., Dushoff, J., and Weitz, J. S. (2013). Robust estimation of microbial diversity in theory and in practice. *ISME Journal*, 7(6):1092–1101.

Haerter, J. O., Mitarai, N., and Sneppen, K. (2014). Phage and bacteria support mutual diversity in a narrowing staircase of coexistence. *ISME Journal*, 8:2317–2326.

Haerter, J. O., and Sneppen, K. (2012). Spatial structure and Lamarckian adaptation explain extreme genetic diversity at CRISPR locus. *mBio*, 3:e00126-12.

Haerter, J. O., Trusina, A., and Sneppen, K. (2011). Targeted bacterial immunity buffers phage diversity. *Journal of Virology*, 85:10554–10560.

Hagan, M. F. (2014). Modeling viral capsid assembly. In *Advances in Chemical Physics*. Vol. 155, 1–68. Wiley, New York.

Hale, C. R., Zhao, P., Olson, S., Duff, M. O., Graveley, B. R., Wells, L., Terns, R. M., and Terns, M. P. (2009). RNA-guided RNA cleavage by a CRISPR RNA-Cas protein complex. *Cell*, 139:945–956.

Hansen, P. J., Bjornsen, P. K., and Hansen, B. W. (1997). Zooplankton grazing and growth: Scaling within the 2-2,000-μm body size range. *Limnology and Oceanography*, 42:687–704.

Hastings, A. (1997). *Population Biology: Concepts and Models*. Springer, New York.

Hatfull, G. F. (2012). The secret lives of mycobacteriophages, In Labocka, M., and Szybalski, Bacteriophages, Part A, vol. 82 of Advances in Virus Research, 179–278. Elsevier, Wattham, MA.

Hatfull, G. F., Cresawn, S. G., and Hendrix, R. W. (2008). Comparative genomics of the mycobacteriophages: Insights into bacteriophage evolution. *Research in Microbiology*, 159:332–339.

Haurwitz, R. E., Jinek, M., Wiedenheft, B., Zhou, K., and Doudna, J. A. (2010). sequence- and structure-specific RNA processing by a CRISPR endonuclease. *Science*, 329:1355–1358.

Heilmann, S., Sneppen, K., and Krishna, S. (2010). Sustainability of virulence in a phage-bacterial ecosystem. *Journal of Virology*, 84:3016–3022.

Heilmann, S., Sneppen, K., and Krishna, S. (2012). Coexistence of phage and bacteria on the boundary of self-organized refuges. *Proceedings of the National Academy of Sciences of the United States of America*, 109:12828–12833.

Held, N. L., and Whitaker, R. J. (2009). Viral biogeography revealed by signatures in *Sulfolobus islandicus* genomes. *Environmental Microbiology*, 11:457–466.

Heldal, M., and Bratbak, G. (1991). Production and decay of viruses in aquatic environments. *Marine Ecology Progress Series*, 72:205–212.

Hill, M. O. (1973). Diversity and evenness: A unifying notation and its consequences. *Ecology*, 54:427–432.

Hiltunen, T., Hairston, N. G., Hooker, G., Jones, L. E., and Ellner, S. P. (2014). A newly discovered role of evolution in previously published consumer-resource dynamics. *Ecology Letters*, 17:915–23.

Holmengen, N., and Seip, K. L. (2009). Cycle lengths and phase portrait characteristics as probes for predator-prey interactions: Comparing simulations and observed data. *Canadian Journal of Zoology*, 87:20–30.

Holmengen, N., Seip, K. L., Boyce, M., and Stenseth, N. C. (2009). Predator-prey coupling: Interaction between mink *Mustela vison* and muskrat *Ondatra zibethicus* across Canada. *Oikos*, 118:440–448.

Holmfeldt, K., Howard-Varona, C., Solonenko, N., and Sullivan, M. B. (2014). Contrasting genomic patterns and infection strategies of two co-existing *Bacteroidetes* podovirus genera. *Environmental Microbiology*, 16:2501–2513.

Holmfeldt, K., Middelboe, M., Nybroe, O., and Riemann, L. (2007). Large variabilities in host strain susceptibility and phage host range govern interactions between lytic marine phages and their *Flavobacterium* hosts. *Applied and Environmental Microbiology*, 73:6730–6739.

Holmfeldt, K., Odić, D., Sullivan, M. B., Middelboe, M., and Riemann, L. (2012). Cultivated single stranded DNA phages that infect marine *Bacteroidetes* prove difficult to detect with DNA binding stains. *Applied and Environmental Microbiology*, 78:892–894.

Horvath, P., and Barrangou, R. (2010). CRISPR/Cas, the immune system of Bacteria and Archaea. *Science*, 327:167–170.

Høyland-Kroghsbo, N. M., Mærkedahl, R. B., and Svenningsen, S. L. (2013). A quorum-sensing-induced bacteriophage defense mechanism. *mBio*, 4:e00362–12

Hu, B., Margolin, W., Molineux, I. J., and Liu, J. (2013). The bacteriophage T7 virion undergoes extensive structural remodeling during infection. *Science*, 339:576–9.

Hunter, M. D., and Price, P. W. (1992). Playing chutes and ladders: Heterogeneity and the relative roles of bottom-up and top-down forces in natural communities. *Ecology*, 73:724–732.

Hurwitz, B. L., and Sullivan, M. B. (2013). The Pacific Ocean Virome (POV): A marine viral metagenomic dataset and associated protein clusters for quantitative viral ecology. *PLoS One*, 8:e57355.

Hurwitz, B. L., Westveld, A. H., Brum, J. R., and Sullivan, M. B. (2014). Modeling ecological drivers in marine viral communities using comparative metagenomics and network analyses. *Proceedings of the National Academy of Sciences of the United States of America*, 111:10714–10719.

Hutchinson, G. E. (1957). Concluding remarks. In *Population Studies: Animal Ecology and Demography*, vol. 22 of *Cold Spring Harbor Symposia on Quantitative Biology*, 415–427.

Hyman, P., and Abedon, S. T. (2009). Practical methods for determining phage growth parameters. In Clokie, M.R.J. and Kropinski, A. M., eds., *Bacteriophages*, vol. 501 of *Methods in Molecular Biology*, 175–202, Humana Press, Totowa, NJ.

Hyman, P., and Abedon, S. T. (2010). Bacteriophage host range and bacterial resistance. *Advances in Applied Microbiology*, vol. 70, 217–248.

Iglesias-Rodríguez, M. D., Brown, C. W., Doney, S. C., Kleypas, J., Kolber, D., Kolber, Z., Hayes, P. K., and Falkowski, P. G. (2002). Representing key phytoplankton functional groups in ocean carbon cycle models: Coccolithophorids. *Global Biogeochemical Cycles*, 16(4):47-1–47-20.

Ignacio-Espinoza, J. C., Solonenko, S. A., and Sullivan, M. B. (2013). The global virome: Not as big as we thought? *Current Opinion in Virology*, 3:566–571.

Inamdar, M. M., Gelbart, W. M., and Phillips, R. (2006). Dynamics of DNA ejection from bacteriophage. *Biophysical Journal*, 91:411–420.

Ivanovska, I., Wuite, G., Jönsson, B., and Evilevitch, A. (2007). Internal DNA pressure modifies stability of WT phage. *Proceedings of the National Academy of Sciences of the United States of America*, 104:9603–9608.

Jackson, M. P., Newland, J. W., Holmes, R. K., and O'Brien, A. D. (1987). Nucleotide sequence analysis of the structural genes for Shiga-like toxin I encoded by bacteriophage 933J from *Escherichia coli*. *Microbial Pathogenesis*, 2:147–153.

Jackson, R. N., Golden, S. M., van Erp, P.B.G., Carter, J., Westra, E. R., Brouns, S.J.J., van der Oost, J., Terwilliger, T. C., Read, R. J., and Wiedenheft, B. (2014). Crystal structure of the CRISPR RNA-guided surveillance complex from *Escherichia coli*. *Science*, 345(6203):1473–1479.

Jacobsen, A., Bratbak, G., and Heldal, M. (1996). Isolation and characterization of a virus infecting *Phaeocystis pouchetii* (prymnesiophyceae). *Journal of Phycology*, 32:923–927.

Janzen, D. H. (1970). Herbivores and the number of tree species in tropical forests. *American Naturalist*, 104:501–528.

Jensen, E. C., Schrader, H. S., Rieland, B., Thompson, T. L., Lee, K. W., Nickerson, K. W., and Kokjohn, T. A. (1998). Prevalence of broad-host-range lytic bacteriophages of *Sphaerotilus natans*, *Escherichia coli*, and *Pseudomonas aeruginosa*. *Applied and Environmental Microbiology*, 64:575–580.

Jetz, W., Wilcove, D., and Dobson, A. (2007). Projected impacts of climate and land-use change on the global diversity of birds. *PLoS Biology*, 5:e157.

Jiang, W., Li, Z., Zhang, Z., Baker, M. L., Prevelige, P. E., and Chiu, W. (2003). Coat protein fold and maturation transition of bacteriophage p22 seen at subnanometer resolutions. *Nature Structural & Molecular Biology*, 10(2):131–135.

Joh, R. I., and Weitz, J. S. (2011). To lyse or not to lyse: Transient-mediated stochastic fate determination in cells infected by bacteriophages. *PLoS Computational Biology*, 7:e1002006.

John, S. G., Mendez, C. B., Deng, L., Poulos, B., Kauffman, A.K.M., Kern, S., Brum, J., Polz, M. F., Boyle, E. A., and Sullivan, M. B. (2011). A simple and efficient method for concentration of ocean viruses by chemical flocculation. *Environmental Microbiology Reports*, 3:195–202.

Jones, L. E., and Ellner, S. P. (2007). Effects of rapid prey evolution on predator-prey cycles. *Journal of Mathematical Biology*, 55:541–73.

Jover, L. F., Cortez, M. H., and Weitz, J. S. (2013). Mechanisms of multi-strain coexistence in host-phage systems with nested infection networks. *Journal of Theoretical Biology*, 332:65 – 77.

Jover, L. F., Effler, T. C., Buchan, A., Wilhelm, S. W., and Weitz, J. S. (2014). The elemental composition of virus particles: implications for marine biogeochemical cycles. *Nature Reviews Microbiology*, 12:519–528.

Kang, I., Oh, H.-M., Kang, D., and Cho, J.-C. (2013). Genome of a SAR116 bacteriophage shows the prevalence of this phage type in the oceans. *Proceedings of the National Academy of Sciences of the United States of America*, 110:12343–12348.

Katsura, I., and Hendrix, R. (1984). Length determination in bacteriophage lambda tails. *Cell*, 39:691–698.

Kausche, G. A., Pfankuch, E., and Ruska, H. (1939). Die sichtbarmachung von pflanzlichem virus im Übermik-roskop. *Naturwissenschaften*, 27:292–299.

Keeling, C. D., Whorf, T. P., Wahlen, M., and van der Plicht, J. (1995). Interannual extremes in the rate of rise of atmospheric carbon dioxide since 1980. *Nature*, 375:667–670.

Kerr, B., Neuhauser, C., Bohannan, B.J.M., and Dean, A. M. (2006). Local migration promotes competitive restraint in a host–pathogen "tragedy of the commons." *Nature*, 442:75–78.

Kerr, B., Riley, M. A., Feldman, M. W., and Bohannan, B.J.M. (2002). Local dispersal promotes biodiversity in a real-life game of rock-paper-scissors. *Nature*, 418:171–174.

Kessler, D. A., and Levine, H. (2013). Large population solution of the stochastic Luria-Delbrück evolution model. *Proceedings of the National Academy of Sciences of the United States of America*, 110:11682–11687.

Klausmeier, C. A. (2008). Floquet theory: A useful tool for understanding nonequilibrium dynamics. *Theoretical Ecology*, 1:153–161.

Klausmeier, C. A., Litchman, E., and Levin, S. A. (2004). Phytoplankton growth and stoichiometry under multiple nutrient limitation. *Limnology and Oceanography*, 49:1463–1470.

Kleiber, M. (1947). Body size and metabolic rate. *Physiological Reviews*, 27:511–541.

——(1961). *The Fire of Life: An Introduction to Animal Energetics*. Wiley, New York.

Koelle, K., Cobey, S., Grenfell, B. T., and Pascual, M. (2006). Epochal evolution shapes the phylodynamics of interpandemic influenza A (H3N2) in humans. *Science*, 314:1898–1903.

Koelle, K., Khatri, P., Kamradt, M., and Kepler, T. B. (2010). A two-tiered model for simulating the ecological and evolutionary dynamics of rapidly evolving viruses, with an application to influenza. *Journal of the Royal Society Interface*, 7:1257–1274.

Koonin, E. V., and Wolf, Y. I. (2009). Is evolution Darwinian or/and Lamarckian? *Biology Direct*, 4:42.

Koonin, E. V., Wolf, Y. I., and Karev, G. P. (2002). The structure of the protein universe and genome evolution. *Nature*, 420:218–223.

Korytowski, D. A., and Smith, H. L. (2014). How nested infection networks in host-phage communities come to be. *arXiv*, page 1406.5461.

Koskella, B., and Meaden, S. (2013). Understanding bacteriophage specificity in natural microbial communities. *Viruses*, 5:806–23.

Koskella, B., Thompson, J. N., Preston, G. M., and Buckling, A. (2011). Local biotic environment shapes the spatial scale of bacteriophage adaptation to bacteria. *American Naturalist*, 177:440–451.

Kostyuchenko, V. A., Chipman, P. R., Leiman, P. G., Arisaka, F., Mesyanzhinov, V. V., and Rossmann, M. G. (2005). The tail structure of bacteriophage T4 and its mechanism of contraction. *Nature Structural & Molecular Biology*, 12(9):810–813.

Kourilsky, P. (1973). Lysogenization by bacteriophage lambda: 1. Multiple infection and lysogenic response. *Molecular and General Genetics*, 122:183–195.

Kourilsky, P. (1975). Lysogenization by bacteriophage lambda: II. Identification of genes involved in the multiplicity dependent processes. *Biochimie*, 56:1511–1516.

Kourilsky, P., and Knapp, A. (1975). Lysogenization by bacteriophage lambda: III. Multiplicity dependent phenomena occuring upon infection by lambda. *Biochimie*, 56:1517–1523.

Kraiss, J. P., Gelbart, S. M., and Juhasz, S. E. (1973). A comparison of three mycobacteriophages. *Journal of General Virology*, 20:75–87.

Kristensen, D. M., Cai, X., and Mushegian, A. (2011). Evolutionarily conserved orthologous families in phages are relatively rare in their prokaryotic hosts. *Journal of Bacteriology*, 193:1806–1814.

Kristensen, D. M., Waller, A. S., Yamada, T., Bork, P., Mushegian, A. R., and Koonin, E. V. (2013). Orthologous gene clusters and taxon signature genes for viruses of prokaryotes. *Journal of Bacteriology*, 195:941–950.

Krüger, D. H., and Schroeder, C. (1981). Bacteriophage T3 and bacteriophage T7 virus–host cell interactions. *Microbiological Reviews*, 45:9–51.

Labrie, S. J., Samson, J. E., and Moineau, S. (2010). Bacteriophage resistance mechanisms. *Nature Reviews Microbiology*, 8:317–327.

Lack, D. (1947). The significance of clutch-size. *Ibis*, 89(2):302–352.

Ladero, V., García, P., Bascarán, V., Herrero, M., Alvarez, M. A., and Suárez, J. E. (1998). Identification of the repressor-encoding gene of the *Lactobacillus* bacteriophage A2. *Journal of Bacteriology*, 180:3474–3476.

Langlet, J., Gaboriaud, F., and Gantzer, C. (2007). Effects of pH on plaque forming unit counts and aggregation of MS2 bacteriophage. *Journal of Applied Microbiology*, 103(5):1632–1638.

La Scola, B., Audic, S., Robert, C., Jungang, L., de Lamballerie, X., Drancourt, M., Birtles, R., Claverie, J.-M., and Raoult, D. (2003). A giant virus in amoebae. *Science*, 299:2033.

La Scola, B., Desnues, C., Pagnier, I., Robert, C., Barrassi, L., Fournous, G., Merchat, M., Suzan-Monti, M., Forterre, P., Koonin, E., and Raoult, D. (2008). The virophage as a unique parasite of the giant mimivirus. *Nature*, 455:100–104.

Laybourn-Parry, J., Marshall, W. A., and Madan, N. J. (2007). Viral dynamics and patterns of lysogeny in saline Antarctic lakes. *Polar Biology*, 30(3):351–358.

Leadbetter, J. R. (2003). Cultivation of recalcitrant microbes: Cells are alive, well and revealing their secrets in the 21st century laboratory. *Current Opinion in Microbiology*, 6:274–281.

Legendre, M., Bartoli, J., Shmakova, L., Jeudy, S., Labadie, K., Adrait, A., Lescot, M., Poirot, O., Bertaux, L., Bruley, C., Couté, Y., Rivkina, E., Abergel, C., and Claverie, J.-M. (2014). Thirty-thousand-year-old distant relative of giant icosahedral DNA viruses with a pandoravirus morphology. *Proceedings of the National Academy of Sciences of the United States of America*, 111:4274–4279.

Legović, T., and Cruzado, A. (1997). A model of phytoplankton growth on multiple nutrients based on the Michaelis-Menten-Monod uptake, Droop's growth and Liebig's law. *Ecological Modelling*, 99:19–31.

Leibold, M. A. (1989). Resource edibility and the effects of predators and productivity on the outcome of trophic interactions. *American Naturalist*, 134(6): 922–949.

Lennon, J. T., Khatana, S. A. M., Marston, M. F., and Martiny, J.B.H. (2007). Is there a cost of virus resistance in marine cyanobacteria? *ISME Journal*, 1:300–312.

Lennon, J. T., and Martiny, J.B.H. (2008). Rapid evolution buffers ecosystem impacts of viruses in a microbial food web. *Ecology Letters*, 11:1178–1188.

Lenski, R. E. (1984). Coevolution of bacteria and phage: Are there endless cycles of bacterial defenses and phage counterdefenses? *Journal of Theoretical Biology*, 108:319–25.

———(1988a). Dynamics of interactions between bacteria and virulent bacteriophage. *Advances in Microbial Ecology*, 10:1–44.

———(1988b). Experimental studies of pleiotropy and epistasis in *Escherichia coli*. I. Variation in competitive fitness among mutants resistant to virus T4. *Evolution*, 42:425–432.

Lenski, R. E., and Levin, B. R. (1985). Constraints on the coevolution of bacteria and virulent phage: A model, some experiments, and predictions for natural communities. *American Naturalist*, 125:585–602.

Levin, B. R. (2010). Nasty viruses, costly plasmids, population dynamics, and the conditions for establishing and maintaining CRISPR-mediated adaptive immunity in bacteria. *PLoS Genetics*, 6:e1001171.

Levin, B. R., and Bull, J. J. (2004). Population and evolutionary dynamics of phage therapy. *Nature Reviews Microbiology*, 2:166–173.

Levin, B. R., Stewart, F. M., and Chao, L. (1977). Resource-limited growth, competition, and predation: A model and experimental studies with bacteria and bacteriophage. *American Naturalist*, 111:3–24.

Levin, S. A. (1970). Community equilibria and stability, and an extension of the competitive exclusion principle. *American Naturalist*, 104(939):413–423.

———(1992). The problem of pattern and scale in ecology. *Ecology*, 73:1943–67.

Lindberg, A. A. (1973). Bacteriophage receptors. *Annual Reviews Microbiology*, 27:205–241.

Lindell, D., Jaffe, J., Johnson, Z., Church, G., and Chisholm, S. (2005). Photosynthesis genes in marine viruses yield proteins during host infection. *Nature*, 438:86–89.

Lindell, D., Jaffe, J. D., Coleman, M. L., Futschik, M. E., Axmann, I. M., Rector, T., Kettler, G., Sullivan, M. B., Steen, R., Hess, W. R., Church, G. M., and Chisholm, S. W. (2007). Genome-wide expression dynamics of a marine virus and host reveal features of co-evolution. *Nature*, 449:83–86.

Lindell, D., Sullivan, M., Johnson, Z., Tolonen, A., Rohwer, F., and Chisholm, S. (2004). Transfer of photosynthesis genes to and from *Prochlorococcus viruses*. *Proceedings of the National Academy of Sciences of the United States of America*, 101:11013–11018.

Lomas, M. W., Burke, A. L., Lomas, D. A., Bell, D. W., Shen, C., Dyhrman, S. T., and Ammerman, J. W. (2010). Sargasso Sea phosphorus biogeochemistry: An important role for dissolved organic phosphorus (DOP). *Biogeosciences*, 7:695–710.

Lotka, A. (1925). *Elements of Physical Biology*. Dover, New York. Reprinted 1956 edition.

Luria, S. E. (1945). Mutations of bacterial viruses affecting their host range. *Genetics*, 30:84–99.

Luria, S. E., and Delbrück, M. (1943). Mutations of bacteria from virus sensitivity to virus resistance. *Genetics*, 28:491–511.

Lwoff, A. (1953). Lysogeny. *Bacteriology Reviews*, 17:269–337.

Mallet, J. (2012). The struggle for existence: How the notion of carrying capacity, K, obscures the links between demography, Darwinian evolution, and speciation. *Evolutionary Ecology Research*, 14:627–665.

Marraffini, L. A., and Sontheimer, E. J. (2010). CRISPR interference: RNA-directed adaptive immunity in bacteria and archaea. *Nature Reviews Genetics*, 11:181–190.

Marston, M. F., Pierciey, F. J., Shepard, A., Gearin, G., Qi, J., Yandava, C., Schuster, S. C., Henn, M. R., and Martiny, J.B.H. (2012). Rapid diversification of coevolving marine *Synechococcus* and a virus. *Proceedings of the National Academy of Sciences of the United States of America*, 109:4544–4549.

Martinez, N. D., Hawkins, B. A., Dawah, H. A., and Feifarek, B. P. (1999). Effects of sampling effort on characterization of food-web structure. *Ecology*, 80:1044–1055.

Martiny, A. C., Pham, C.T.A., Primeau, F. W., Vrugt, J. A., Moore, J. K., Levin, S. A., and Lomas, M. W. (2013). Strong latitudinal patterns in the elemental ratios of marine plankton and organic matter. *Nature Geoscience*, 6(4):279–283.

Maslov, S., and Sneppen, K. (2014). Well-temperate phage. http://arxiv.org/pdf/1308.1646v1.pdf.

Matteson, A. R., Loar, S. N., Pickmere, S., DeBruyn, J. M., Ellwood, M. J., Boyd, P. W., Hutchins, D. A., and Wilhelm, S. W. (2012). Production of viruses during a spring phytoplankton bloom in the South Pacific Ocean near of New Zealand. *FEMS Microbiology Ecology*, 79:709–719.

May, R. M. (1988). How many species are there on Earth? *Science*, 241:1441–1449.

May, R. M., and Nowak, M. A. (1995). Coinfection and the evolution of parasite virulence. *Proceedings of the Royal Society of London B: Biological Sciences*, 261(1361):209–215.

Maynard, N. D., Birch, E. W., Sanghvi, J. C., Chen, L., Gutschow, M. V., and Covert, M. W. (2010). A forward-genetic screen and dynamic analysis of lambda phage host-dependencies reveals an extensive interaction network and a new anti-viral strategy. *PLoS Genetics*, 6:e1001017.

McDaniel, L., Breitbart, M., Mobberley, J., Long, A., Haynes, M., Rohwer, F., and Paul, J. H. (2008). Metagenomic analysis of lysogeny in Tampa Bay: Implications for prophage gene expression. *PLoS One*, 3.

McMahon, T. A. (1973). Size and shape in biology. *Science*, 179:1201–1204.

Meaden, S., and Koskella, B. (2013). Exploring the risks of phage application in the environment. *Frontiers in Microbiology*, 4:358.

Memmott, J. (1999). The structure of a plant-pollinator food web. *Ecology Letters*, 2:276–280.

Menge, D. N. L., and Weitz, J. S. (2009). Dangerous nutrients: Evolution of phytoplankton resource uptake subject to virus attack. *Journal of Theoretical Biology*, 257:104–115.

Merril, C., Scholl, D., and Adhya, S. L. (2003). The prospect for bacteriophage therapy in western medicine. *Nature Reviews Drug Discovery*, 6:489–497.

Meyer, J. R., Dobias, D. T., Weitz, J. S., Barrick, J. E., Quick, R. T., and Lenski, R. E. (2012). Repeatability and contingency in the evolution of a key innovation in phage lambda. *Science*, 335:428–432.

Middelboe, M. (2000). Bacterial growth rate and marine virus-host dynamics. *Microbial Ecology*, 40:114–124.

Middelboe, M., Riemann, L., Steward, G., Hansen, V., and Nybroe, O. (2003). Virus-induced transfer of organic carbon between marine bacteria in a model community. *Aquatic Microbial Ecology*, 33:1–10.

Miki, T., Nakazawa, T., Yokokawa, T., and Nagata, T. (2008). Functional consequences of viral impacts on bacterial communities: A food-web model analysis. *Freshwater Biology*, 53:1142–1153.

Milo, R., Jorgensen, P., Moran, U., Weber, G., and Springer, M. (2010). BioNumbers: The database of key numbers in molecular and cell biology. *Nucleic Acids Research*, 38:D750–D753.

Minot, S., Bryson, A., Chehoud, C., Wu, G. D., Lewis, J. D., and Bushman, F. D. (2013). Rapid evolution of the human gut virome. *Proceedings of the National Academy of Sciences of the United States of America*, 110:12450–12455.

Mitchell, G. J., Nelson, D. C., and Weitz, J. S. (2010). Quantifying enzymatic lysis: Estimating the combined effects of chemistry, physiology and physics. *Physical Biology*, 7:046002.

Mizoguchi, K., Morita, M., Fischer, C. R., Yoichi, M., Tanji, Y., and Unno, H. (2003). Coevolution of bacteriophage PP01 and *Escherichia coli* O157:H7 in continuous culture. *Applied and Environmental Microbiology*, 69:170–176.

Moebus, K. (1980). A method for the detection of bacteriophages from ocean water. *Helgoländer Meeresuntersuchungen*, 34:1–14.

Moebus, K., and Nattkemper, H. (1981). Bacteriophage sensitivity patterns among bacteria isolated from marine waters. *Helgoländer Meeresuntersuchungen*, 34:375–385.

Molineux, I. J., and Debabrata, P. (2013). Popping the cork: Mechanisms of phage genome ejection. *Nature Reviews Microbiology*, 11:194–204.

Monod, J. (1949). The growth of bacterial cultures. *Annual Review of Microbiology*, 3:371–394.

Mora, C., Tittensor, D. P., Adl, S., Simpson, A. G. B., and Worm, B. (2011). How many species are there on earth and in the ocean? *PLoS Biology*, 9:e1001127.

Morelli, M. J., ten Wolde, P. R., and Allen, R. J. (2009). DNA looping provides stability and robustness to the bacteriophage λ switch. *Proceedings of the National Academy of Sciences of the United States of America*, 106:8101–8106.

Morris, W. F. (2009). Life history. In Levin, S. A., ed., *The Princeton Guide to Ecology*, Princeton University Press, Princeton, NJ.

Murdoch, W. W., Briggs, C. J., and Nisbet, R. M. (2003). *Consumer-Resource Dynamics*. Princeton University Press, Princeton, NJ.

Murray, A. G., and Jackson, G. A. (1992). Viral dynamics: A model of the effects of size, shape, motion and abundance of single-celled planktonic organisms and other particles. *Marine Ecology Progress Series*, 89:103–116.

Needham, D. M., Chow, C.-E. T., Cram, J. A., Sachdeva, R., Parada, A., and Fuhrman, J. A. (2013). Short-term observations of marine bacterial and viral communities : Patterns, connections and resilience. *ISME Journal*, 7:1274–1285.

Newman, M. E. J. (2003). The structure and function of complex networks. *SIAM Review*, 45:167–256.

Noble, R., and Fuhrman, J. (1998). Use of SYBR Green I for rapid epifluorescence counts of marine viruses and bacteria. *Aquatic Microbial Ecology*, 14:113–118.

Nowak, M. A. (2006). *Evolutionary Dynamics: Exploring the Equations of Life*. Belknap Press of the Harvard University Press, Cambridge, MA.

Nowak, M. A., and May, R. M. (1994). Superinfection and the evolution of parasite virulence. *Proceedings of the Royal Society of London B: Biological Sciences*, 255(1342):81–89.

———(2000). *Virus Dynamics: Mathematical Principles of Immunology and Virology*. Oxford University Press, Oxford.

Oppenheim, A. B., Kobiler, O., Stavans, J., Court, D. L., and Adhya, S. (2005). Switches in bacteriophage lambda development. *Annual Review of Genetics*, 39:409–429.

Otsu, N. (1979). A threshold selection method from gray-level histograms. *IEEE Transactions on Systems, Man and Cybernetics*, SMC-9:62–66.

Pace, M., Cole, J., Carpenter, S., and Kitchell, J. (1999). Trophic cascades revealed in diverse ecosystems. *Trends in Ecology and Evolution*, 14:483–488.

Pace, N. R. (1997). A molecular view of microbial diversity and the biosphere. *Science*, 276:734–740.

Parada, P., Herndl, G. J., and Weinbauer, M. G. (2006). Viral burst size of heterotrophic prokaryotes in aquatic systems. *Journal of the Marine Biological Association of the United Kingdom*, 86:613–621.

Parry, B. R., Surovtsev, I. V., Cabeen, M. T., Hern, C.S.O., and Dufresne, E. R. (2014). The bacterial cytoplasm has glass-like properties and is fluidized by metabolic activity. *Cell*, 156:183–194.

Paterson, S., Vogwill, T., Buckling, A., Benmayor, R., Spiers, A. J., Thomson, N. R., Quail, M., Smith, F., Walker, D., Libberton, B., Fenton, A., Hall, N., and Brockhurst, M. A. (2010). Antagonistic coevolution accelerates molecular evolution. *Nature*, 464:275–278.

Paul, J. H., and Weinbauer, M. (2010). Detection of lysogeny in marine environments. In Wilhelm, S. W., Weinbauer, M. G., and Suttle, C. A., eds., *Manual of Aquatic Viral Ecology*, 30–33. ASLO Press, Waco, TX.

Payet, J. P., and Suttle, C. A. (2013). To kill or not to kill: The balance between lytic and lysogenic viral infection is driven by trophic status. *Limnology and Oceanography*, 58:465–474.

Pearl, S., Gabay, C., Kishony, R., Oppenheim, A., and Balaban, N. Q. (2008). Nongenetic individuality in the host-phage interaction. *PLoS Biology*, 6:957–964.

Peters, R. H. (1983). *The Ecological Implications of Body Size*. Cambridge University Press, Cambridge.

Petrov, A. S., and Harvey, S. C. (2007). Structural and thermodynamic principles of viral packaging. *Structure*, 15:21–27.

Piarroux, R., Barrais, R., Faucher, B., Haus, R., Piarroux, M., Magloire, R., and Raoult, D. (2011). Understanding the cholera epidemic, Haiti. *Emerging Infectious Diseases*, 17:1161–1168.

Poisot, T., Canard, E., Mouquet, N., and Hochberg, M. E. (2012). A comparative study of ecological specialization estimators. *Methods in Ecology and Evolution*, 3:537–544.

Poisot, T., Lepennetier, G., Martinez, E., Ramsayer, J., and Hochberg, M. E. (2011). Resource availability affects the structure of a natural bacteria-bacteriophage community. *Biology Letters*, 7:201–204.

Polis, G. A., Myers, C. A., and Holt, R. D. (1989). The ecology and evolution of intraguild predation: Potential competitors that eat each other. *Annual Review of Ecology and Systematics*, 20(1):297–330.

Poorvin, L., Rinta-Kanto, J. M., Hutchins, D. A., and Wilhelm, S. W. (2004). Viral release of iron and its bioavailability to marine plankton. *Limnology and Oceanography*, 49:1734–1741.

Poullain, V., Gandon, S., Brockhurst, M. A., Buckling, A., and Hochberg, M. E. (2008). The evolution of specificity in evolving and coevolving antagonistic interactions between a bacteria and its phage. *Evolution*, 62:1–11.

Prangishvili, D. (2013). The wonderful world of archaeal viruses. *Annual Review of Microbiology*, 67:565–585.

Prangishvili, D., Forterre, P., and Garrett, R. A. (2006). Viruses of the Archaea: A unifying view. *Nature Reviews Microbiology*, 4(11):837–848.

Prosser, J. I., et al., (2007). The role of ecological theory in microbial ecology. *Nature Reviews Microbiology*, 5:384–392.

Ptashne, M. (2004). *A Genetic Switch: Phage Lambda Revisited*, 3rd ed. Cold Spring Harbor Laboratory Press, Cold Spring Harbor, NY.

Purohit, P. K., Inamdar, M. M., Grayson, P. D., Squires, T. M., Kondev, J., and Phillips, R. (2005). Forces during bacteriophage DNA packaging and ejection. *Biophysical Journal*, 88(2):851–866.

Rabinovitch, A., Hadas, H., Einav, M., Melamed, Z., and Zaritsky, A. (1999). Model for bacteriophage T4 development in *Escherichia coli*. *Journal of Bacteriology*, 181:1677–1683.

Raoult, D., Audic, S., Robert, C., Abergel, C., Renesto, P., Ogata, H., Scola, B. L., Suzan, M., and Claverie, J.-M. (2004). The 1.2-megabase genome sequence of mimivirus. *Science*, 306:1344–1350.

Rappé, M. S., and Giovannoni, S. J. (2003). The uncultured microbial majority. *Annual Review of Microbiology*, 57:369–394.

Raytcheva, D. A., Haase-Pettingell, C., Piret, J. M., and King, J. (2011). Intracellular assembly of cyanophage syn5 proceeds through a scaffold-containing procapsid. *Journal of Virology*, 85:2406–2415.

Reader, R. W., and Siminovitch, L. (1971). Lysis defective mutants of bacteriophage lambda: Genetics and physiology of S cistron mutants. *Journal of Virology*, 43:607–622.

Redfield, A. C., Ketchum, B. H., and Richards, F. A. (1963). The influence of organisms on the composition of sea-water. In Hill, M. N., ed., *The Composition of Sea-water: Comparative and Descriptive Oceanography*. Vol. 2 of *The Sea*, 26–77. Harvard University Press, Cambridge, MA.

Révet, B., von Wilcken-Bergmann, B., Bessert, H., Barker, A., and Muller-Hill, B. (1999). Four dimers of λ repressor bound to two suitably spaced pairs of lambda operators form octamers and DNA loops over large distances. *Current Biology*, 9(3):151–154.

Riedel, S. (2005). Edward Jenner and the history of smallpox and vaccination. *BUMC Proceedings*, 18:21–25.

Rodin, S. N., and Ratner, V. M. (1983a). Some theoretical aspects of protein coevolution in the ecosystem "phage-bacteria." I. The problem. *Journal of Theoretical Biology*, 100:185–195.

——(1983b). Some theoretical aspects of protein coevolution in the ecosystem "phage-bacteria." II. The deterministic model of microevolution. *Journal of Theoretical Biology*, 100:197–210.

Rodriguez-Valera, F., Martin-Cuadrado, A.-B., Rodriguez-Brito, B., Pašić, L., Thingstad, T. F., Rohwer, F. and Mira, A. (2009). Explaining microbial population genomics through phage predation. *Nature Reviews Microbiology*, 7:828–836.

Roff, D. A. (2002). *Life History Evolution*. Sinauer, Sunderland, MA.

Rohwer, F. (2003). Global phage diversity. *Cell*, 113(2):141.

Rohwer, F., and Edwards, R. (2002). The phage proteomic tree: A genome-based taxonomy for phage. *Journal of Bacteriology*, 184:4529–4535.

Rokney, A., Kobiler, O., Amir, A., Court, D. L., Stavans, J., Adhya, S., and Oppenheim, A. B. (2008). Host responses influence on the induction of lambda prophage. *Molecular Microbiology*, 68:29–36.

Rosenblueth, A., and Wiener, N. (1945). The role of models in science. *Philosophy of Science*, 12:316–321.

Rosenzweig, M. L., and MacArthur, R. H. (1963). Graphical representation and stability conditions of predator-prey interactions. *American Naturalist*, 97:209–223.

Rosner, J. L. (1972). Formation, induction, and curing of bacteriophage p1 lysogens. *Virology*, 48:679–689.

Rothenberg, E., Sepulveda, L. A., Skinner, S. O., Zeng, L., Selvin, P. R., and Golding, I. (2011). Single-virus tracking reveals a spatial receptor-dependent search mechanism. *Biophysical Journal*, 100:2875 – 2882.

Ruardij, P., Veldhuis, M.J.W., and Brussaard, C.P.D. (2005). Modeling the bloom dynamics of the polymorphic phytoplankter *Phaeocystis globosa*: Impact of grazers and viruses. *Harmful Algae*, 4:941–963.

Rusch, D. B., Halpern, A. L., Sutton, G., Heidelberg, K. B., Williamson, S., Yooseph, S., Wu, D., Eisen, J. A., Hoffman, J. M., Remington, K., Beeson, K., Tran, B., Smith, H., Baden-Tillson, H., Stewart, C., Thorpe, J., Freeman, J., Andrews-Pfannkoch, C., Eguiarte, L. E., Karl, D. M., SathyenDranath, S., Platt, T., Bermingham, E., Gallardo, V., Tamayo-Castillo, G., Ferrari, M. R., and Strausberg, R. L. (2007). The Sorcerer II Global Ocean Sampling expedition: Northwest Atlantic through eastern tropical Pacific. *PLoS Biology*, 5:e77.

Sadarjoen, I. A., and Post, F. H. (1999). Geometric methods for vortex extraction. In *Data Visualization'99*, 53–62. Springer, Vienna.

Sanjuán, R., Nebot, M. R., Chirico, N., Mansky, L. M., and Belshaw, R. (2010). Viral mutation rates. *Journal of Virology*, 84:9733–9748.

Santillán, M., and Mackey, M. C. (2004). Why the lysogenic state of phage λ is so stable: a mathematical modeling approach. *Biophysical Journal*, 86:75–84.

Sauer, R. T., Ross, M. J., and Ptashne, M. (1982). Cleavage of the λ and P22 repressors by recA protein. *Journal of Biological Chemistry*, 257:4458–4462.

Schleifer, K. H., and Kandler, O. (1972). Peptidoglycan types of bacterial cell walls and their taxonomic implications. *Bacteriological Reviews*, 36:407–477.

Schloss, P. D., and Handelsman, J. (2005). Introducing DOTUR, a computer program for defining operational taxonomic units and estimating species richness. *Applied and Environmental Microbiology*, 71:1501–1506.

Schmidt-Nielsen, K. (1984). *Scaling: Why Is Animal Size So Important?* Cambridge University Press, Cambridge.

Schrag, S. J., and Mittler, J. E. (1996). Host-parasite coexistence: The role of spatial refuges in stabilizing bacteria-phage interactions. *American Naturalist*, 148:348–377.

Shao, Y., and Wang, I.-N. (2008). Bacteriophage adsorption rate and optimal lysis time. *Genetics*, 482:471–482.

Shelford, E. J., Middelboe, M., Møller, E. F., and Suttle, C. A. (2012). Virus-driven nitrogen cycling enhances phytoplankton growth. *Aquatic Microbial Ecology*, 66:41–46.

Sieber, M., and Gudelj, I. (2014). Do-or-die life cycles and diverse post-infection resistance mechanisms limit the evolution of parasite host ranges. *Ecology Letters*, 17:491–498.

Sieber, M., Robb, M., Forde, S. E., and Gudelj, I. (2014). Dispersal network structure and infection mechanism shape diversity in a coevolutionary bacteria-phage system. *ISME Journal*, 8(3):504–514.

Singh, A., and Dennehy, J. J. (2014). Stochastic holin expression can account for lysis time variation in the bacteriophage λ. *Journal of the Royal Society Interface*, 11(95):2014140.

Smith, H. (2011). *An Introduction to Delay Differential Equations with Applications to the Life Sciences*. Springer, New York.

Smith, H. L., and Thieme, H. R. (2012). Persistence of bacteria and phages in a chemostat. *Journal of Mathematical Biology*, 64:951–979.

Smith, H. L., and Waltman, P. (1995). *The Theory of the Chemostat: Dynamics of Microbial Competition*. Cambridge University Press, Cambridge.

Smith, J. M. (1973). The stability of predator-prey systems. *Ecology*, 54:384–391.

Sneppen, K. (2014). *Models of Life: Dynamics and Regulation in Biological Systems*. Cambridge University Press, Cambridge.

Snyder, J. C., and Young, M. J. (2011). Advances in understanding archaea-virus interactions in controlled and natural environments. *Current Opinion in Microbiology*, 14:497–503.

Sorek, R., Kunin, V., and Hugenholtz, P. (2008). CRISPR: A widespread system that provides acquired resistance against phages in bacteria and archaea. *Nature Reviews Microbiology*, 6:181–186.

St. Pierre, F., and Endy, D. (2008). Determination of cell fate selection during phage lambda infection. *Proceedings of the National Academy of Sciences of the United States of America*, 105(52):20705–20710.

Staley, J. T., and Konopka, A. (1985). Measurement of in situ activities of nonphotosynthetic microorganisms in aquatic and terrestrial habitats. *Annual Review of Microbiology*, 39:321–346.

Stearns, S. C. (1976). Life-history tactics: A review of the ideas. *Quarterly Review of Biology*, 51(1): 3–47.

Stenholm, A. R., Dalsgaard, I., and Middelboe, M. (2008). Isolation and characterization of bacteriophages infecting the fish pathogen *Flavobacterium psychrophilum*. *Applied and Environmental Microbiology*, 74:4070–4078.

Sterner, R. W., and Elser, J. J. (2002). *Ecological Stoichiometry*. Princeton University Press, Princeton, NJ.

Steward, G. F., Culley, A. I., Mueller, J. A., Wood-Charlson, E. M., Belcaid, M., and Poisson, G. (2012). Are we missing half of the viruses in the ocean? *ISME Journal*, 7:672–679.

Stewart, F. M., and Levin, B. R. (1984). The population biology of bacterial viruses: Why be temperate. *Theoretical Population Biology*, 26:93–117.

Stock, C. A., Dunne, J. P., and John, J. G. (2014). Global-scale carbon and energy flows through the marine planktonic food web: An analysis with a coupled physical-biological model. *Progress in Oceanography*, 120:1–28.

Strogatz, S. (1994). *Nonlinear Dynamics and Chaos*. Addison Wesley, Reading, MA.

Strogatz, S. H. (2001). Exploring complex networks. *Nature*, 410:268–276.

Strzepek, R. F., Maldonado, M. T., Higgins, J. L., Hall, J., Safi, K., Wilhelm, S. W., and Boyd, P. W. (2005). Spinning the Ferrous Wheel: The importance of the microbial community in an iron budget during the FeCycle experiment. *Global Biogeochemical Cycles*, 19:GB4S26.

Sulakvelidze, A., Alavidze, Z., and Morris, J. G., Jr. (2001). Bacteriophage therapy. *Antimicrobial Agents and Chemotherapy*, 45:649–659.

Sullivan, M. B., Coleman, M., Weigele, P., Rohwer, F., and Chisholm, S. W. (2005). Three *Prochlorococcus* cyanophage genomes: Signature features and ecological interpretations. *PLoS Biology*, 3:e144.

Sullivan, M. B., Lindell, D., Lee, J. A., Thompson, L. R., Bielawski, J. P., and Chisholm, S. W. (2006). Prevalence and evolution of core photosystem II genes in marine cyanobacterial viruses and their hosts. *PLoS Biology*, 4:e234.

Sullivan, M. B., Waterbury, J. B., and Chisholm, S. W. (2003). Cyanophage infecting the oceanic cyanobacterium *Prochlorococcus*. *Nature*, 424:1047–1051.

Summers, W. C. (1999). *Félix d'Herelle and the Origins of Molecular Biology*. Yale University Press, New Haven, CT.

Suttle, C. A. (2005). Viruses in the sea. *Nature*, 437:356–361.

Suttle, C. A. (2007). Marine viruses: Major players in the global ecosystem. *Nature Reviews Microbiology*, 5:801–812.

Suttle, C. A., and Chan, A. M. (1993). Marine cyanophages infecting oceanic and coastal strains of *Synechococcus*: Abundance, morphology, cross-infectivity and growth characteristics. *Marine Ecology Progress Series*, 92:99–109.

Suttle, C. A., and Chen, F. (1992). Mechanisms and rates of decay of marine viruses in seawater. *Applied and Environmental Microbiology*, 58:3721–3729.

Suttle, C. A., and Fuhrman, J. A. (2010). Enumeration of virus particles in aquatic or sediment samples by epifluorescence microscopy. In Wilhelm, S. W., Weinbauer, M. G., and Suttle, C. A., eds., *Manual of Aquatic Viral Ecology*, 145–153, American Society of Limnology and Oceanography, Waco, TX.

Suzan-Monti, M., La Scola, B., and Raoult, D. (2006). Genomic and evolutionary aspects of mimivirus. *Virus Research*, 117:145–155.

Svenningsen, S. L., and Semsey, S. (2014). Commitment to lysogeny is preceded by a prolonged period of sensitivity to the late lytic regulator Q in bacteriophage λ. *Journal of Bacteriology*.

Tatusov, R. L., Galperin, M. Y., Natale, D. A., and Koonin, E. V. (2000). The COG database: A tool for genome-scale analysis of protein functions and evolution. *Nucleic Acids Research*, 28:33–36.

Taylor, B. P., Cortez, M. H., and Weitz, J. S. (2014). The virus of my virus is my friend: Ecological effects of virophage with alternative modes of coinfection. *Journal of Theoretical Biology*, 354:124–136.

Terborgh, J., and Estes, J. A., eds. (2010). *Trophic Cascades: Predators, Prey, and the Changing Dynamics of Nature*. Island Press, Washington, DC.

Thingstad, T., and Lignell, R. (1997). Theoretical models for the control of bacterial growth rate, abundance, diversity and carbon demand. *Aquatic Microbial Ecology*, 13:19–27.

Thingstad, T. F. (2000). Elements of a theory for the mechanisms controlling abundance, diversity, and biogeochemical role of lytic bacterial viruses in aquatic systems. *Limnology and Oceanography*, 45:1320–1328.

Thingstad, T. F., Våge, S., Støresund, J. E., Sandaa, R.-A., and Giske, J. (2014). A theoretical analysis of how strain-specific viruses can control microbial species diversity. *Proceedings of the National Academy of Sciences of the United States of America*, 111:7813–7818.

Tilman, D. (1985). The resource-ratio hypothesis of plant succession. *American Naturalist*, 125:827–852.

———(1994). Competition and biodiversity in spatially structured habits. *Ecology*, 75:2–16.

Trun, N., and Trempy, J. (2009). *Fundamental Bacterial Genetics*. Wiley, New York.

Turner, P. E., and Chao, L. (1999). Prisoner's dilemma in an RNA virus. *Nature*, 398:441–443.

Twort, T. W. (1915). An investigation on the nature of ultramicroscopic viruses. *Lancet*, 186:1241–1243.

Ulrich, W., Almeida-Neto, M., and Gotelli, N. J. (2009). A consumer's guide to nestedness analysis. *Oikos*, 118:3–17.

van der Oost, J., Westra, E. R., Jackson, R. N., and Wiedenheft, B. (2014). Unravelling the structural and mechanistic basis of CRISPR-Cas systems. *Nature Reviews Microbiology*, 12(7):479–492.

Van Etten, J. L., Lane, L. C., and Dunigan, D. D. (2010). DNA viruses: The really big ones (giruses). *Annual Review of Microbiology*, 64:83–99.

Van Etten, J. L., Lane, L. C., and Meints, R. H. (1991). Viruses and viruslike particles of eukaryotic algae. *Microbiological Reviews*, 55:586–620.

van Kampen, N. G. (2001). *Stochastic Processes in Physics and Chemistry*. Elsevier Science, Amsterdam.

van Valen, L. (1973). A new evolutionary law. *Evolutionary Theory*, 1:1–30.

Vardi, A., Haramaty, L., Mooy, B. A. S. V., Fredricks, H. F., Kimmance, S. A., and Larsen, A. (2012). Host-virus dynamics and subcellular controls of cell fate in a natural coccolithophore population. *Proceedings of the National Academy of Sciences of the United States of America*, 109:19327–19332.

Violle, C., Navas, M.-L., Vile, D., Kazakou, E., Fortunel, C., Hummel, I., and Garnier, E. (2007). Let the concept of trait be functional! *Oikos*, 116:882–892.

Volterra, V. (1926). Fluctuations in the abundance of a species considered mathematically. *Nature*, 118:558–60.

Von Borries, B., Ruska, E., and Ruska, H. (1938). Bakterien und virus in übermikroskopischer aufnahme. *Klin Wochenschr*, 17:921–925.

Waldor, M. K., and Mekalanos, J. J. (1996). Lysogenic conversion by a filamentous phage encoding cholera toxin. *Science*, 272:1910–1914.

Wang, I.-N. (2006). Lysis timing and bacteriophage fitness. *Genetics*, 172:17–26.

Wang, I.-N., Dykhuizen, D. E., and Slobodkin, L. B. (1996). The evolution of phage lysis timing. *Evolutionary Ecology*, 10:545–558.

Wang, Z., and Goldenfeld, N. (2010). Fixed points and limit cycles in the population dynamics of lysogenic viruses and their hosts. *Physical Review E*, 82:011918.

Ward, B. B. (2002). How many species of prokaryotes are there? *Proceedings of the National Academy of Sciences of the United States of America*, 99:10234–10236.

Waterbury, J. B., and Valois, F. W. (1993). Resistance to co-occurring phages enables marine *Synechococcus* communities to coexist with cyanophages abundant in seawater. *Applied and Environmental Microbiology*, 59:3393–3399.

Watts, D. (2012). *Everything Is Obvious: How Common Sense Fails Us*. Random House, New York.

Waxman, D., and Gavrilets, S. (2005). 20 questions on adaptive dynamics. *Journal of Evolutionary Biology*, 18:1139–1154.

Weed, L. L., and Cohen, S. S. (1951). The utilization of host pyrimidines in the synthesis of bacterial viruses. *Journal of Biological Chemistry*, 192:693–700.

Wei, Y., Kirby, A., and Levin, B. R. (2011). The population and evolutionary dynamics of *Vibrio cholerae* and its bacteriophage: Conditions for maintaining phage-limited communities. *American Naturalist*, 178:715–725.

Weinbauer, M. (2004). Ecology of prokaryotic viruses. *FEMS Microbiology Reviews*, 28:127–181.

Weinbauer, M. G., Bonilla-Findji, O., Chan, A. M., Dolan, J. R., Short, S. M., Simek, K., Wilhelm, S. W., and Suttle, C. A. (2011). *Synechococcus* growth in the ocean may depend on the lysis of heterotrophic bacteria. *Journal of Plankton Research*, 33:1465–1476.

Weinbauer, M. G., and Peduzzi, P. (1994). Frequency, size and distribution of bacteriophages in different marine bacterial morphotypes. *Marine Ecology Progress Series*, 108:11–20.

Weinbauer, M. G., Wilhelm, S. W., Suttle, C. A., and Garza, D. R. (1997). Photoreactivation compensates for UV damage and restores infectivity to natural marine virus communities. *Applied and Environmental Microbiology*, 63:2200–2205.

Weinberger, A. D., Sun, C. L., Pluciński, M. M., Denef, V. J., Thomas, B. C., Horvath, P., Barrangou, R., Gilmore, M. S., Getz, W. M., and Banfield, J. F. (2012). Persisting viral sequences shape microbial CRISPR-based immunity. *PLoS Computational Biology*, 8:e1002475.

Weitz, J. S., and Dushoff, J. (2008). Alternative stable states in host–phage dynamics. *Theoretical Ecology*, 1:13–19.

Weitz, J. S. et al. (2015). A multitrophic model to quantify the effects of marine viruses on microbial food webs and ecosystem processes. *ISME Journal*, doi:10.1038/ismej.2014.220.

Weitz, J. S., Hartman, H., and Levin, S. A. (2005). Coevolutionary arms races between bacteria and bacteriophage. *Proceedings of the National Academy of Sciences of the United States of America*, 102:9535–9540.

Weitz, J. S., Mileyko, Y., Joh, R. I., and Voit, E. O. (2008). Collective decision making in bacterial viruses. *Biophysical Journal*, 95:2673–2680.

Weitz, J. S., Poisot, T., Meyer, J. R., Flores, C. O., Valverde, S., Sullivan, M. B., and Hochberg, M. E. (2013). Phage-bacteria infection networks. *Trends in Microbiology*, 21:82–91.

Wessner, D. R. (2010). Discovery of the giant Mimivirus. *Nature Education*, 3:61.

Wichman, H. A., Badgett, M. R., Scott, L. A., Boulianne, C. M., and Bull, J. J. (1999). Different trajectories of parallel evolution during viral adaptation. *Science*, 285:422–424.

Wikner, J., Vallino, J. J., Smith, D. C., and Azam, F. (1993). Nucleic acids from the host bacterium as a major source of nucleotides for three marine bacteriophages. *FEMS Microbiology Ecology*, 12:237–248.

Wilhelm, S. W., and Suttle, C. A. (1999). Viruses and nutrient cycles in the sea. *BioScience*, 49:781–788.

Wilhelm, S. W., Weinbauer, M. G., and Suttle, C. A., eds. (2010). *Manual of Aquatic Viral Ecology*. ASLO Press, Waco, TX.

Wilke, C. O. (2005). Quasispecies theory in the context of population genetics. *BMC Evolutionary Biology*, 5:44.

Williams, H.T.P. (2013). Phage-induced diversification improves host evolvability. *BMC Evolutionary Biology*, 13:17.

Williams, R. G., and Follows, M. J. (2011). *Ocean Dynamics and the Carbon Cycle: Principles and Mechanisms*. Cambridge University Press, Cambridge.

Williams, R. J., and Martinez, N. D. (2000). Simple rules yield complex food webs. *Nature*, 404:180–183.

Williamson, K. E., Radosevich, M., Smith, D. W., and Wommack, K. E. (2007). Incidence of lysogeny within temperate and extreme soil environments. *Environmental Microbiology*, 9:2563–2574.

Wilson, W. H., Carr, N. G., and Mann, N. H. (1996). The effect of phosphate status on the kinetics of cyanophage infection in the oceanic cyanobacterium *Synechococcus* sp. WH7803. *Journal of Phycology*, 32:506–516.

Winter, C., Bouvier, T., Weinbauer, M. G., and Thingstad, T. F. (2010). Trade-offs between competition and defense specialists among unicellular planktonic organisms: The "killing the winner" hypothesis revisited. *Microbiology and Molecular Biology Reviews*, 74:42–57.

Wirtz, K. W. (2012). Who is eating whom? Morphology and feeding type determine the size relation between planktonic predators and their ideal prey. *Marine Ecology Progress Series*, 445:1–12.

Wommack, K., and Colwell, R. (2000). Virioplankton: Viruses in aquatic ecosystems. *Microbiology and Molecular Biology Reviews*, 64:69–114.

Xiao, C., Chipman, P., Battisti, A., Bowman, V., Renesto, P., Raoult, D., and Rossmann, M. (2005). Cryo-electron microscopy of the giant mimivirus. *Journal of Molecular Biology*, 353:493–496.

Xie, X. S., Choi, P. J., Li, G.-W., Lee, N. K., and Lia, G. (2008). Single-molecule approach to molecular biology in living bacterial cells. *Annual Review of Biophysics*, 37:417–444.

Yooseph, S., Nealson, K. H., Rusch, D. B., Mccrow, J. P., Dupont, C. L., Kim, M., Johnson, J., Montgomery, R., Ferriera, S., Beeson, K., Williamson, S. J., Tovchigrechko, A., Allen, A. E., Zeigler, L. A., Sutton, G., Eisenstadt, E., Rogers, Y.-H., Friedman, R., Frazier, M., and Venter, J. C. (2010). Genomic and functional adaptation in surface ocean planktonic prokaryotes. *Nature*, 468:60–66.

Yoshida, T., Ellner, S. P., Jones, L. E., Bohannan, B.J.M., Lenski, R. E., and N.G.H., Jr. (2007). Cryptic population dynamics: Rapid evolution masks trophic interactions. *PLoS Biology*, 5:e235.

Zeng, L., Skinner, S. O., Zong, C., Sippy, J., Feiss, M., and Golding, I. (2010). Decision making at a subcellular level determines the outcome of bacteriophage infection. *Cell*, 141:682–691.

Zhao, Y., Temperton, B., Thrash, J. C., Schwalbach, M. S., Vergin, K. L., Landry, Z. C., Ellisman, M., Deerinck, T., Sullivan, M. B., and Giovannoni, S. J. (2013). Abundant SAR11 viruses in the ocean. *Nature*, 494:357–360.

Zimmer, C. (2012). *A Planet of Viruses*. University of Chicago Press, Chicago, IL.

Zinser, E. R., Lindell, D., Johnson, Z. I., Futschik, M. E., Steglich, C., Coleman, M. L., Wright, M. A., Rector, T., Steen, R., McNulty, N., Thompson, L. R., and Chisholm, S. W. (2009). Choreography of the transcriptome, photophysiology, and cell cycle of a minimal photoautotroph, *Prochlorococcus*. *PLoS One*, 4(4):e5135.

Index

Abedon, Stephen, xiv, 198

Actinobacter, ubiquitous small type of, 166

adaptive dynamics, 115–18; canonical equation of, 117, 123, 281–83, 284–85; diversification and, 153–56; latent period evolution and, 121, 123, 124

adaptive sweeps, 115

adsorption rate (φ), 49–52; estimation of, 254–55; to infected hosts (ρ), 85–87; matrix of, 148; in multitrophic model, 291; resistance mutations and, 99

age of viral infection, 31

algae. See *Emiliana huxleyi*; *Phaeocysti pouchetii*

algal viruses: of *Chlorella*, 6, 28, 166; EhV, 16, 60, 215; latent periods of, 28

allometry, 7, 9

amoeba: giant viruses infecting, 6–7, 21–22, 166; virus factory in, 22

Anderson, Philip W., 238

antibiotic-resistance cassettes, 72, 256–57

archaea, as hosts of viruses, xvi, 16, 186; CRISPR system in, 156–57; infection cycle in, 25; morphological diversity of viruses in, 8

arms races, coevolutionary, 150, 159

Arrowsmith (Lewis), 5

autotrophic processes: resources derived from, 202; stimulated by viruses, 213. *See also* cyanobacteria; eukaryotic autotrophs

Azam, Farooq, 216

bacteria: beneficial to humans, 4; Gram-positive, carbon requirements of, 172

bacteria, as hosts of viruses, xvi, 16, 186; morphological diversity of viruses in, 8; resistance arising in, 89–97 (*see also* resistant host mutants). *See also* cyanobacteria; *E. coli*; host *entries*; lysogeny; marine bacteria

bacteriophage. *See* phage

Barrangou, Rodolphe, 157

base pair, average molecular formula of, 13

basic reproductive number: of hosts in chemostat, 64; of mutant viruses, 134; of viruses in chemostat, 65; of virus in cross-infection network, 208

batch culture experiments, latent periods in, 120

Bayes's rule, 252

Beckett, Stephen, 156

Beretta, Eduardo, 79, 80

Berg, Howard, 203

Bergh, Øivind, 164, 168

Bertani, Giuseppe, 39

bifurcation points: cryptic oscillations and, 110. *See also* Hopf bifurcations

big-data science, 241, 243

biodiversity. *See* diversification; diversity, viral

Bipartite, Recursively Induced Modules (BRIM) method, 190–91, 194

bipartite infection networks, 188–93, 204–6. *See also* cross-infection networks

blooms: of coccolithophores, 215; of *Emiliana huxleyi*, 60; viruses in, 170

Bohannan, Brendan, 99, 106, 108

Bonachela, Juan, 83, 122

Boolean infection networks, 188–93, 204–6

Børsheim, Knut, 164

bottom-up control, 58–59, 63; alternative coevolution models and, 153; viruses in marine food web and, 234

branching points, 115, 153–54, 156

branching processes, 103, 123, 282, 285

Bratbak, Gunnar, 164, 202

Breitbart, Mya, 183

BRIM method, 190–91, 194

Brown, Chris, 32

Brum, Jennifer, 170, 173, 174

Brussaard, Corina, 239

Buchan, Alison, 172, 239

Buckling, Angus, 127

MONOGRAPHS IN POPULATION BIOLOGY

EDITED BY SIMON A. LEVIN AND HENRY S. HORN